Primate Ecol

Techniques in Ecology and Conservation Series

Series Editor: William J. Sutherland

Bird Ecology and Conservation: A Handbook of Techniques
William J. Sutherland, Ian Newton, and Rhys E. Green

Conservation Education and Outreach Techniques
Susan K. Jacobson, Mallory D. McDuff, and Martha C. Monroe

Forest Ecology and Conservation: A Handbook of Techniques
Adrian C. Newton

Habitat Management for Conservation: A Handbook of Techniques
Malcolm Ausden

Conservation and Sustainable Use: A Handbook of Techniques
E.J. Milner-Gulland and J. Marcus Rowcliffe

Invasive Species Management: A Handbook of Principles and Techniques
Mick N. Clout and Peter A. Williams

Amphibian Ecology and Conservation: A Handbook of Techniques
C. Kenneth Dodd, Jr.

Insect Conservation: A Handbook of Approaches and Methods
Michael J. Samways, Melodie A. McGeoch, and Tim R. New

Remote Sensing for Ecology and Conservation: A Handbook of Techniques
Ned Horning, Julie A. Robinson, Eleanor J. Sterling, Woody Turner, and Sacha Spector

Marine Mammal Ecology and Conservation: A Handbook of Techniques
Ian L. Boyd, W. Don Bowen, and Sara J. Iverson

Carnivore Ecology and Conservation: A Handbook of Techniques
Luigi Boitani and Roger A. Powell

Primate Ecology and Conservation: A Handbook of Techniques
Eleanor J. Sterling, Nora Bynum, and Mary E. Blair

Primate Ecology and Conservation

A Handbook of Techniques

Edited by

Eleanor J. Sterling
Nora Bynum
and
Mary E. Blair

OXFORD
UNIVERSITY PRESS

Great Clarendon Street, Oxford, OX2 6DP,
United Kingdom

Oxford University Press is a department of the University of Oxford.
It furthers the University's objective of excellence in research, scholarship,
and education by publishing worldwide. Oxford is a registered trade mark of
Oxford University Press in the UK and in certain other countries

First Edition published 2013

Impression: 1

British Library Cataloguing in Publication Data

Data available

ISBN   978–0–19–965944–9 (Hbk.)
           978–0–19–965945–6 (Pbk.)

Printed and bound by
CPI Group (UK) Ltd, Croydon, CRO 4YY

Foreword

My first thought as I read this book in manuscript form was how much I wished it had been available when I was embarking on my own Ph.D. research more than 40 years ago. At that time, one simply gleaned what one could from the methods sections of published articles. To be sure, my supervisors offered me good advice and sound practical suggestions, like remembering to take a plentiful supply of pencils and a back-up pencil sharpener. The fact that *Primate Ecology and Conservation: A Handbook of Techniques* exists is helpful in itself, for it brings together so much information in one place. It is also testimony to the vastly increased sophistication of field techniques today.

My second thought as I read was how much I wished this book had been available when I, in turn, became the supervisor of my own students' research. Much more had been written about field techniques by then, and methodology was a far greater focus of attention, but still there was nothing like the *Handbook* to which I could direct them. It is a source of great pride and delight for me that two of those students, Eleanor Sterling and Nora Bynum, are editors of this volume and I can only hope they forgive me for any advice on pencil sharpeners I may have given them.

In selecting a field site and designing a research project, so much depends on context and the question asked. With many decades of experience and expertise under their collective belt, the distinguished contributors to the volume understand this well. Each chapter discusses the merits of a variety of techniques and approaches to the particular challenges of fieldwork, providing a lot of detailed and helpful information. But prescriptive statements are rare and few "correct answers" offered, which is as it should be. Even with the *Handbook* in hand, the researcher heading for the field will have many choices and decisions to make. Happily, those decisions will now be far better informed.

What strikes me above all, however, is the *Handbook*'s scope and ambition. It challenges all of us who study primates in the wild to explore research paths opening up almost daily, as a result of new technologies and conceptual advances in the biological sciences. Thanks to those developments, the distance between the lab and the field is shrinking steadily and research is burgeoning in recent, still "young" fields—from health and physiology to genetics and molecular biology.

The *Handbook*'s title proclaims its greatest and perhaps most important ambition. Decades of effort have largely, but not entirely, laid to rest the false dichotomy between research on the one hand and conservation on the other. The *Handbook* makes another, timely contribution to this endeavor. Discussions of ecology and conservation, and of science and action, are interwoven throughout, and the last three chapters bring the focus directly to conservation in the wild, captive breeding, and the scourge of international trafficking. The *Handbook* is as important in this regard as it is valuable to anyone setting forth to study our closest relatives in the wild.

Alison Richard
Yale University, July 2012

Preface

In this book we provide a comprehensive source that outlines major techniques in the study of primate ecology and conservation. Our target audiences are early career primatologists and graduate students, as well as established researchers and conservation professionals embarking on new research or conservation projects. Our synthesis focuses on new and emerging field methods alongside a comprehensive presentation of laboratory and data analysis techniques, as well as key methods for determining conservation status and conservation management. Importantly, we also discuss data interpretation as well as guiding questions and principles for students and researchers to consider as they plan research projects in primate ecology and conservation.

This volume's particular focus is on innovative ways to study primates in a changing world, in recognition of the fact that in order for primatologists to continue studying primates, we must successfully implement studies that inform our understanding of primate biology as well as primate conservation. Thus, we incorporate consideration of conservation status and threats to primate populations throughout this volume where appropriate. We are donating part of the proceeds from this book to support capacity development for biodiversity conservation.

We are very grateful for the efforts of friends and colleagues who supported the preparation of this volume. The following individuals provided guidance, references, and editorial support: Javier Alvarez, Erin Betley, Rachel Booth, Hannah Burnett, Lindsey Desmul, Kevin Frey, Kevin Frey, Karen Kennea, Ned Horning, and Connie Rogers. Many others (including Susan Alberts, Pat Whitten, Jeanne Altmann, Karen Strier, Carel van Schaik, Diane Brockman, Alison Richard, Agustin Fuentes) helped us to find chapter authors and reviewers. We extend deep thanks to all of the contributors and chapter peer reviewers for their outstanding efforts, and to Alison Richard for writing the Foreword.

A special thank you goes to the series editor, William Sutherland, and Ian Sherman, Lucy Nash, Muhammad Ridwaan, and Helen Eaton of Oxford University Press for their enthusiasm, support, and guidance during the preparation of this volume.

The Editors

Contents

List of contributors

Mary E. Blair Center for Biodiversity and Conservation, American Museum of Natural History, Central Park West at 79th St, New York, NY 10024, USA
mblair1@amnh.org

Scott A. Blumenthal Department of Anthropology, The Graduate Center, City University of New York, New York, NY 10016, USA
SBlumenthal@gc.cuny.edu

Justin S. Brashares Department of Environmental Science, Policy, and Management, University of California, Berkeley, Berkeley, CA 94720, USA
brashares@berkeley.edu

Sarah F. Brosnan Department of Psychology, Neuroscience Institute and Language Research Center, Georgia State University, PO Box 5010, Atlanta, GA 30302, USA
sbrosnan@gsu.edu

Michelle Brown Anthropologisches Institut und Museum, Universität Zürich, Zürich, Switzerland and Department of Anthropology, University of New Mexico, Albuquerque, NM 87131, USA
moowalks@gmail.com

Stephen T. Buckland Centre for Research into Ecological and Environmental Modelling, The Observatory, Buchanan Gardens, University of St Andrews, St Andrews, Fife KY16 9LZ, UK
steve@mcs.st-andrews.ac.uk

Nora Bynum Office of Global Strategy and Programs, Duke University, Box 90036, 101 Allen Bldg, Durham, NC 27708, USA
elb@duke.edu

Margaret Crofoot Smithsonian Tropical Research Institute, Balboa, Ancón, Panamá, República de Panamá, Division of Migration and Immuno-ecology, Max Planck Institute for Ornithology, Radolfzell, Germany and Department of Ecology and Evolutionary Biology, Princeton University, 106a Guyot Hall, Princeton, NJ 08544 USA
crofootm@si.edu

Amanda L. Ellwanger Department of Anthropology, University of Texas at San Antonio, San Antonio, TX 78249, USA
Amanda.ellwanger@utsa.edu

Dean Gibson San Diego Zoo Global, PO Box 120551, San Diego, CA 92112, USA
DGibson@sandiegozoo.org

Kenneth E. Glander Department of Evolutionary Anthropology, Duke University, 130 Science Dr Box 90383, Durham, NC 27708, USA
glander@duke.edu

Charles H. Janson Division of Biological Sciences, University of Montana, Missoula, MT 59812, USA
charles.janson@umontana.edu

Beth Kaplin Department of Environmental Studies, Antioch University New England, 40 Avon Street, Keene, NH 03431, USA
bkaplin@antioch.edu

Cari M. Lewis Department of Anthropology, Indiana University, 701 East Kirkwood Ave., Bloomington, IN 47405, USA
lewiscm@indiana.edu

Joshua Linder Department of Sociology and Anthropology, James Madison University, 800 S. Main St., Harrisonburg VA 22807, USA
linderjm@jmu.edu

Andrew J. Marshall Department of Anthropology, Graduate Group in Ecology, and Animal Behavior Graduate Group, University of California at Davis, One Shields Avenue, Davis, CA 95616, USA
ajmarshall@ucdavis.edu

Colleen McCann Wildlife Conservation Society, Bronx Zoo, 2300 Southern Blvd., Bronx, NY 10460, USA
cmccann@wcs.org

Erin McCreless Department of Ecology and Evolutionary Biology, University of California Santa Cruz, Long Marine Laboratory, 100 Shaffer Road, Santa Cruz, CA 95060, USA
erin.mccreless@gmail.com

Olga L. Montenegro Instituto de Ciencias Naturales, Universidad Nacional de Colombia, Carrera 30 No. 45–03, Edificio 425, Oficina 108, Bogotá, Colombia
olmontenegrod@unal.edu.co

Alba L. Morales-Jimenez Department of Anthropology, Lehman College, Bronx, NY 10468, USA and Fundación Biodiversa Colombia, Bogotá, Colombia
almoralesj@gmail.com

Michael P. Muehlenbein Department of Anthropology, Indiana University, 701 East Kirkwood Ave., Bloomington, IN 47405, USA
mpm1@indiana.edu

K. Anne-Isola Nekaris Nocturnal Primate Research Group, Department of Anthropology and Geography, Oxford Brookes University, Oxford OX3 0BP, UK
anekaris@brookes.ac.uk

Nga Nguyen Department of Anthropology and Environmental Studies Program, California State University Fullerton, 800 N. State College Blvd., Fullerton, CA 92834, USA
nganguyen@fullerton.edu

Andrew J. Plumptre Wildlife Conservation Society, PO Box 7487, Kampala, Uganda
aplumptre@wcs.org

Erin P. Riley Department of Anthropology, San Diego State University, 5500 Campanile Dr., San Diego, CA 92182, USA
epriley@mail.sdsu.edu

E. Johanna Rode Nocturnal Primate Research Group, Department of Anthropology and Geography, Oxford Brookes University, Oxford OX3 0BP, UK
eva.rode-2011@brookes.ac.uk

Jessica M. Rothman Department of Anthropology, Hunter College of the City University of New York, New York, NY 10065, USA
jessica.rothman@hunter.cuny.edu

Sarah C. Sawyer Department of Environmental Science, Policy and Management, University of California, Berkeley, Berkeley, CA 94720, USA
ssawyer@berkeley.edu

Jutta Schmid Institute of Experimental Ecology, University of Ulm, Albert Einstien Alee 11, D 89069 Ulm, Germany
jutta.schmid@uni-ulm.de

Carrie J. Stengel Nocturnal Primate Research Group, Department of Anthropology and Geography, Oxford Brookes University, Oxford OX3 0BP, UK
carriestengel@gmail.com

Eleanor J. Sterling Center for Biodiversity and Conservation, American Museum of Natural History, Central Park West at 79th St, New York, NY 10024, USA
sterling@amnh.org

Erin R. Vogel Department of Anthropology and Center for Human Evolutionary Studies, Rutgers University, 131 George Street, New Brunswick, NJ 08901, USA
erin.vogel@rutgers.edu

Serge Wich Anthropologisches Institut und Museum, Universität Zürich, Zürich, Switzerland and Sumatran Orangutan Conservation Programme (PanEco-YEL), Medan, Sumatra, Indonesia
sergewich1@yahoo.com

Apollinaire William Department of Environmental Studies, Antioch University New England, 40 Avon Street, Keene, NH 03431, USA
awilliam1@antioch.edu

1

Introduction: why a new methods book on primate ecology and conservation?

Eleanor J. Sterling, Nora Bynum, and Mary E. Blair

The field of primatology has a long and rich history, and methods used to study primate species have evolved considerably since Ray Carpenter's pioneering field-work in the 1930s (Carpenter 1935). Gone are the days when primatologists focused solely on the meticulous description of a particular population of primates (e.g., Devore 1965) without analyses grounded in ecological, and in particular socioecological, theory. Primatologists today, in addition to studying the trad-itional behavioral ecology and population biology, use tools and methods such as genome-level genetics, spatial bioinformatics and modeling, endocrinology, and epidemiology. And even within the more traditional areas of inquiry, such as behavioral ecology, methods for studying social networks, for example, are more sophisticated, in initial data collection as well as analysis (see Chapter 5).

In part because primate populations and habitats continue to decline, primat-ologists have inevitably become involved in conservation issues, either through the use of their data for conservation management or via direct action. The foundational text *Primate Conservation Biology* (Cowlishaw and Dunbar 2000) makes a particularly convincing argument for why primates are important conser-vation targets: they play a critical role in many ecosystems, serving as seed dispersers and seed germinators and at times as keystone species that play a critical role in maintaining community structure; they can often represent up to 50% of the total mammalian herbivore/frugivore biomass in forested habitats; primates can also be conservation targets, helping to guide specific conservation action through their role as landscape species, and they can be flagship species, garnering overall conservation attention. At the same time, studies of our closest relatives in the animal kingdom help us reflect on human biology and evolution. At the time of their publication, Cowlishaw and Dunbar (2000) tallied 200–230 species in the Primate order following Groves (1993), and identified 13% as endangered or

Primate Ecology and Conservation: A Handbook of Techniques. First Edition. Edited by Eleanor J. Sterling, Nora Bynum, and Mary E. Blair. © Oxford University Press 2013. Published 2013 by Oxford University Press.

critically endangered and 18% as vulnerable, following the 1996 International Union for the Conservation of Nature (IUCN) Red List of Threatened Species™ (Baillie and Groombridge 1996; see Chapter 16). In 2012, the IUCN recognized around 400 primate taxa (including both species and subspecies) of which 30% are categorized as endangered or critically endangered, and 19% as vulnerable (IUCN 2012). This increase in the percentage of primate populations categorized as threatened with extinction might be real, or it might be related to increased attention and work being done in the field of primate ecology and conservation, or it could also be related to taxonomic inflation. However, it remains clear that if primatologists wish to continue studying primates, we must successfully implement studies that inform both our understanding of primate biology as well as primate conservation.

This volume brings together a group of distinguished primate researchers to synthesize field, laboratory, and conservation management techniques for primate ecology and conservation into a practical empirical reference text with an international scope that is appropriate for graduate students, researchers, and conservation professionals across the globe. Our synthesis focuses on new and emerging field methods alongside a comprehensive presentation of laboratory and data analysis techniques, as well as key methods for determining conservation status and conservation management. This volume's particular focus is on innovative ways to study primates in a changing world, including emerging methods such as non-invasive genetic techniques and advanced spatial modeling. In addition to synthesizing field and lab methods, we also discuss data interpretation, as well as important guiding questions and principles for students and researchers to consider as they plan research projects in primate ecology and conservation. We incorporate consideration of conservation status and threats to primate populations throughout this volume, where appropriate, encouraging the integration of ecological and conservation methods.

1.1 Organization of the book

Each chapter in this volume is meant to provide a brief overview of a particular topic, which includes the specifics of how to use key methods or refers readers to more detailed sources for the specifics of methods. Methods to assess primate population size, density, and abundance, as well as advice to overcome the difficulties of doing so for rare and nocturnal primates, are examined in Chapter 2. Chapter 3 outlines methods for the capture of non-human primates including darting, anesthesia, and handling, highlighting techniques that previously could only be performed in the lab, but now can be done in the field. Such techniques are crucial for research questions that require unequivocal identification, or for which non-invasive or laboratory methods are insufficient. Multidisciplinary approaches to identify health status and determinants of infectious diseases in wild primate populations are considered in Chapter 4. The tools described in Chapter 4 provide the opportunity for not only basic health monitoring of populations, but also the assessment of the impact of anthropogenic change on non-human primate health, as well as modeling of infection transmission in populations. Chapters 3 and 4 both discuss processes for collecting biological samples as well as issues related to research permitting and the development of ethical protocols.

Methods to understand primate behavioral ecology within groups are considered in Chapter 5, including emerging approaches such as social network analysis and the use of Geographic Information Systems (GIS) in behavioral studies, alongside more traditional but still critically important methods such as how to habituate a primate group. Chapters 6 and 7 discuss how to characterize primate habitats and distributions through direct assessment, advanced spatial modeling, and plant phenology. Chapter 6 is distinctive in its discussion of species niche modeling, which holds considerable promise for enhancing species research and conservation. The methods to assess plant phenology covered in Chapter 7 are essential to characterize temporal and spatial variation in the availability of plant foods, which inform research questions about primate diets, reproduction, grouping, ranging, and sociality.

Ethnoprimatological methods are covered in Chapter 8, including multidisciplinary approaches to measure overlaps between human and non-human primate resource use, and those used to navigate diverse cultural landscapes and to conduct ethnographic interviews. Chapter 9 presents methods to study the social and spatial relationships between primate groups, including inter-group encounters. This chapter is also quite relevant to the study of ranging patterns and space use within individual primate groups due to its focus on GIS including advanced spatial modeling. Experimental methods to study behavioral ecology both in the field and the lab are considered in Chapter 10. This chapter presents an important argument that studies in the field, in the lab, and in free-ranging captive settings can provide complementary insights into the functions and mechanisms of primate behaviors.

Approaches to analyze the mechanical and nutritional properties of the foods that comprise primate diets are discussed in Chapter 11, including methods to examine the diets of elusive primates through stable isotope analysis. Chapter 12 considers diverse techniques for the study of energy expenditure in primates including the measure of heart rate, body temperature, and the use of doubly-labeled water. Advances in non-invasive techniques to monitor and assess hormone–behavior interactions in the wild are discussed in Chapter 13.

Molecular genetic methods for the study of primate ecology, evolution, and conservation are examined in Chapter 14. This chapter highlights emerging approaches including landscape genetics, which integrates GIS with genetic analysis, and next-generation sequencing. Chapter 15 provides methods to build models for the study of demography, life history, and population dynamics, and addresses how these methods can be applied to risk assessment approaches for primate conservation.

Methods for determining the conservation status of your study organism(s) and for contributing to conservation action *in situ* are covered in Chapter 16. These encompass how to identify threats to populations and their magnitude, and the importance of working with local communities on conservation activities. Chapter 17 reviews captive breeding strategies and *ex situ* conservation management for species recovery and reintroduction, including cooperative breeding programs and species survival plans, as well as the important role of zoos in conservation education and primate research projects. Lastly, in light of the wildlife trade's threat to primate populations, multidisciplinary methods used to study populations that are heavily impacted are covered in Chapter 18, including protocols for household, hunter, and market surveys.

1.2 Getting started: key points of advice for planning a research project

Before launching into our chapters, we want to begin our volume by offering some salient points of advice for planning a research project in primate ecology and conservation. These points transcend a variety of research topics and cover: (1) field site selection, (2) permits and visas, (3) digital data collection, and (4) ethical considerations.

1.2.1 Choosing a field site

When choosing a field site to begin a research project, one of the most important considerations is whether to go to an existing, active site, or to establish your own site. Unlike students of primatology 30 years ago, students today may have the luxury of choosing among several active field sites. Although it is crucial to primate conservation as well as to our understanding of primate biology to establish more field sites, the availability of time and funding can limit a researcher's ability to invest in the logistics of setting up a new site. On the other hand, there may not be an existing site suitable to answer the research question(s) that you have in mind. In general, making the "right" choice about a field site will vary greatly depending on the goals of a particular project. Potential field sites vary from simple tent encampments, with little research infrastructure, to sites like La Selva Biological Station in Costa Rica, which has embedded GIS capabilities, many kilometers of groomed trails, and a dining hall that can accommodate more than 100 people.

In almost any case, your choice will be greatly informed by undertaking a short pilot study. A pilot study can either test methodology at a particular site or entail visiting several potential sites. A pilot study can be particularly useful to show funding agencies and academic committees that your project is feasible, since the first-hand experience of visiting a field site before beginning your research will inform you of crucial details that influence a project's timeline and budget: e.g., the extent of the trail system, site infrastructure, specific habitat characteristics, how well Global Positioning System (GPS) units work, relative abundance of the primate population(s), and how long it might take to get them habituated to the degree needed for your project. All of this information allows for more detailed planning of the timeline and budget of your full project; for example, a pilot study can help you to determine if you need one long field season of 1 to 2 years to complete your study, or if several shorter seasons of a few months each will be sufficient.

1.2.2 Permits and visas

Another key issue when designing a research project is lining up research permits and visas for international work. Coordination with international governments and research communities can take time, and we encourage researchers to begin the process of acquiring permits and visas several months ahead of your departure date (or more depending on the country). Specifics are discussed in Chapters 3, 4, and 8, but we want to emphasize here the importance of a head start, especially since many funding agencies require permits and visas in hand before granting funds for any given project. Also, when deciding on a field site for your research, understanding the permitting requirements for the site can be very important as especially challenging requirements can impact the ease of returning to a site year after year.

1.2.3 Digital data collection

A critical consideration when planning a study is how to collect data in the field in a way that will make for more efficient data analysis upon your return. Many primatologists still prefer writing by hand in field notebooks (using waterproof paper), but advances in technology have facilitated the use of digital data collection in the field. In Box 1.1, Michelle Brown offers some key points of

Box 1.1 Digital data collection in the field (by Michelle Brown)

Mobile devices

Advances in technology have facilitated the use of digital data collection in the field, allowing for vastly greater efficiency in data input and processing, and facilitating a faster timeline to analysis after returning from the field. There is an enormous range of options available with regard to mobile devices, including personal digital assistants (PDAs), smartphones, tablets, and handheld computers. New items become available every few months, so I do not attempt to list them all here. However, the following are several issues to consider when selecting a mobile device:

- **Size:** Should the mobile device be small enough to fit into a shirt pocket or cargo pocket? Larger tablet devices are widely available and work with a range of software programs, but generally require two-handed operation and limit the user's ability to simultaneously handle binoculars or other field equipment. Smaller devices, while more portable, have limited screen space that makes it more difficult to enter data.
- **Input:** Do you require a keyboard or a touch screen for data entry? Touch screens dominate the current tablet and cellular phone market, but can be difficult to see clearly under sunny skies. Keyboards or thumb-boards make it easier to enter data without having to spend a lot of time looking at the screen, which means you can spend more time observing your subject(s). Input method is also important when choosing software, as some programs work with only one entry method. Older Palm devices are not only limited to touch screens, but also rely on a unique writing system that takes some time for new users to learn.
- **Battery:** How many hours should the device last between charging bouts? Few devices have batteries that last for a full day of data entry. One way of working around this limitation is to select an older device with removable battery packs, which allows the user to swap an empty battery with a fully charged battery; in some cases, extended-life battery packs are also available. Alternatively, external batteries are available for some of the newer tablets and smartphones, as well as solar or manual hand crank battery chargers.
- **Platform:** Do you need to use a particular operating system, such as Windows Mobile, Android, Apple, Blackberry, or Palm? For instance, if you use FileMaker Pro on your laptop and would like to use FileMaker Go (the mobile version of Filemaker Pro) to avoid switching among software programs, you are limited to Apple mobile devices.
- **Price:** How much are you able to spend on mobile devices, software licenses, external batteries, and other accessories? While some of the older devices are less expensive through eBay and similar sites, they are often plagued with hardware

Box 1.1 *Continued*

and software problems for which manufacturers will no longer offer support. High-end systems provide comprehensive, ready-to-use solutions but require thousands of dollars. Between these options are mid-range solutions, such as smartphones or tablets, that nonetheless cost several hundred dollars.

Data entry software

Table 1.1 lists commonly used data entry software programs, excluding software that only works on desktops and laptops (i.e., which are not compatible with mobile devices). Many of these programs allow you to develop your own data entry forms, and generally do not require a powerful handheld device (i.e., inexpensive or low end units can often work). However, there may be limits on which versions of an operating system the software will work, and many mobile database programs vary in their compatibility with desktop applications.

Table 1.1 *The most widely used data entry software programs. (All websites accurate and accessible as of December 2012.)*

Program	An	Ap	Bl	Pa	Sy	Wi	Reference
FileMaker Go		X					<http://www.filemaker.com/products/filemaker-go/>
HandDBase	X	X	X	X	X	X	<http://www.ddhsoftware.com/>
Pendragon Forms	X	X		X		X	<http://www.pendragonsoftware.com/forms3/index.html>
Open Data Kit	X						<http://opendatakit.org>
Pocket Observer						X	<http://www.noldus.com/the-observer-xt/pocket-observer>
PTab Spreadsheet						X	<http://www.z4soft.com/windows-mobile-pocket-pc/ptab-spreadsheet/>

An = Android; Ap = Apple (Apple iPhone/iPod/iPad); Bl = RIM Blackberry; Pa = Palm OS; Sy = Symbian; Wi = Windows (Mobile Pocket PC and Smartphone).

Accessories

Most primates live in the tropics, meaning plenty of heat and humidity—two conditions that are terrible for the majority of mobile devices, which are generally not waterproof. To protect your handheld device during data entry, a waterproof pouch is a must. A variety of hard cases are also available, but these do not allow the user to protect the device during data entry. Also, consider getting an anti-glare

Box 1.1 *Continued*

screen, which can help protect the screen from scratches, and makes a touch screen much easier to use, by reducing glare under sunny skies.

Many mobile devices now have built-in GPS receivers. Older devices are less likely to work under dense forest canopies, but the signal-to-noise characteristic of GPS receivers are constantly improving and newer models may function quite well even under dense canopy conditions. Some units will allow you to attach an external antenna or a USB GPS receiver to further improve reception.

Some users find it easier to enter data on a touch screen using a stylus. Fingertip models are useful for older devices and are less likely to be dropped or lost than traditional pen-shaped or -tipped styli. More recent tablet devices are not always compatible with point-tipped styli and require a spongier, finger-like stylus.

advice, describing the pros and cons of various digital data collection devices and software used for primatological research in the field.

1.2.4 Ethical considerations

Ethical considerations are fundamental in studies of primate ecology and conservation, especially in those focusing on endangered primate populations or human and non-human primate interactions (see Chapters 8 and 16). Before any study, researchers must get Institutional Animal Care and Use Committee (IACUC) and/or Institutional Review Board (IRB) approvals which ensure the ethical treatment of animal and human subjects in their projects (also see Chapters 3, 4, and 8). Moreover, members of the American Society of Primatologists (ASP) and the International Primatological Society (IPS), and researchers publishing articles in journals by these organizations, must agree to adhere to the set of principles set forth in resolutions and policy statements regarding the ethical treatment of non-human primates. Especially in studies of ethnoprimatology and in conservation management, researchers and practitioners may find themselves in complex situations where human and non-human primate needs and interests must be balanced.

1.3 Primate ecology and conservation in a changing world

Our focus in this volume is on innovative ways to study primates in a world where humans continue to change the habitats in which non-human primates live. The chapters that follow often include case studies that emphasize the use of emerging methods to understand the ecology of primate populations in changing or altered habitats. Thus, a key goal of this volume is to highlight the importance of the human dimension of non-human primate conservation and biodiversity conservation as a whole, as well as the multidisciplinary techniques that will be necessary to incorporate local knowledge, customs, and values into ecological and conservation research and practice. Multidisciplinary approaches are critical not only to improving the effectiveness and sustainability of conservation actions, but also to enhancing our understanding of the ecology of non-human primate communities and the systems within which they live.

2

Primate census and survey techniques

Andrew J. Plumptre, Eleanor J. Sterling, and Stephen T. Buckland

2.1 Introduction

Primates have been counted for more than 50 years as part of scientific research into their ecology, behavior, and conservation. Survey methods vary and have aimed to obtain measurements of either relative abundance (an index that is assumed to be related to abundance, such as encounter rate that is number of animals detected per unit distance), density (no/km^2), or total population numbers. Researchers usually want a measure of density, or its extrapolation to an estimated number of animals by multiplying by the study area, or total population size obtained from total counts, but in some cases this is simply not possible.

Estimating the numbers in a primate community is important for many fields of biology and many studies base their findings on the underlying population density or abundance of the primate species being studied. Understanding what processes limit primate populations, whether ecological or social, depends on knowing the population density and range sizes at any one site. Understanding some of the causes of variation in primate behavior often depends on knowledge of variation in rates of competition, which again is linked to primate density. Conservation of a primate species is very much dependent on knowing how many individuals there are globally as well as within individual populations. Understanding how disease impacts primate populations is again affected by primate density and abundance; there are many other examples. Unfortunately the methods that have historically been used to estimate primate densities and abundance have been varied in quality and the same mistakes have been repeated in numerous studies around the world, making it difficult to compare across studies and sites. This chapter aims to assess the current thinking about methods that can be used to survey primates as well as to point out some of the more common flaws that researchers regularly make.

Primate Ecology and Conservation: A Handbook of Techniques. First Edition. Edited by Eleanor J. Sterling, Nora Bynum, and Mary E. Blair. © Oxford University Press 2013. Published 2013 by Oxford University Press.

2.2 History of primate surveying

Initial survey methods were basic and generally involved trying to estimate a complete count of the population. For example, Schaller (1963) counted mountain and Grauer's gorillas (*Gorilla beringei*) using nest counts of different groups to estimate a total population size. These attempts to make complete or total counts can work in certain situations but are usually impractical. During the 1970s Struhsaker investigated line transect sampling methods to survey primates in Kibale Forest in western Uganda (Struhsaker 1975, 1981). The methods he developed have been built on elsewhere around the world and the "line transect" survey method is now the most commonly used method to count primates.

Struhsaker's work in the Kanyawara study area of Kibale looked at the ecology, ranging, and behavior of several primate species. A network of trails was established at 100 m intervals in a grid across the study area and some of these trails were used to census the primate species. The degree to which trails were selected at random for the surveys is unclear. When walking the trails he (or his assistants) measured the distance from where the observer first sighted a primate group to the group itself—the *animal–observer distance*. Subsequently, as line transect methods were developed in the 1970s, it became clear that the *perpendicular distance* was an important measurement to be taken. Struhsaker had plotted sightings of the primates on maps of the trail system while undertaking the surveys, so he made an estimate of perpendicular distance by measuring from the maps afterwards (Struhsaker 1975). Based on his knowledge of the number of primates in habituated groups in the study area he found that the *animal–observer distance* measurement provided more accurate estimates of the known numbers than the *perpendicular distance* measurement, which he found tended to overestimate the population size. As a result the *animal–observer distance* has been promoted in primatology for the last 30 years as the measurement that should be taken (Struhsaker 1981). Only one other paper published around the same time tended to promote the *perpendicular distance* method and examined several variations of this method (Whitesides *et al.* 1988).

At the same time, the field of what is now referred to as *"distance sampling"* was being developed in other disciplines, notably zoology and wildlife management, with the underlying mathematics and assumptions of the methods being assessed and tested (Burnham *et al.* 1980; Buckland *et al.* 2001). These methods used *perpendicular distance* methods to estimate densities of various animal species and showed that they were fairly robust and accurate when tested. These are now used to estimate the abundance of populations of species as diverse as whales, songbirds, moose, and chameleons. However, primatology has been slow to adopt these methods because of the history of census methods in this field.

There are several reasons why Struhsaker's (1975, 1981) measurement of perpendicular distance may have failed to estimate the "true density" of primates. These include taking measurements from plots on maps rather than measurement in the field, the fact that it was unclear whether he measured to the center of the group or the first animal seen (a high proportion of zeros in his perpendicular distance estimates suggests the latter), whether his estimate of the "true population" was accurate in that it included lone males that move around the periphery of groups, sampling design, and whether his animal–observer distance estimates were accurate because they were estimated by eye. In a later paper, Mitani *et al.* (2000) showed that there can be great variation between observers in their first sighting of primates in Kibale and the animal–observer distances measured, and that you cannot compare observations between different observers because of this. This finding questions the usefulness of the method because you can only compare densities obtained between sites or over time at the same site *if* the same observer has collected the data. However, a more recent paper has also shown that the method used to analyze animal–observer distance measurements in primatology is incorrect and that there are many reasons not to use this method (Buckland *et al.* 2010a). We will therefore focus on the perpendicular distance measure in any reference to the line transect method in the rest of this chapter.

There are still several issues that complicate the use of line transect methods when estimating primate abundance and as a result other methods have also been developed. These include the lure count method (which is based on distance sampling theory), plotting calls of groups to estimate densities for some species that call reliably, and genetic sampling of populations for total counts or capture recapture estimates. The rest of this chapter focuses on the various methods that can be used to survey primates and the issues to be aware of in each method.

2.3 Total count methods

Total count methods are used in very specific situations. These are: (1) when the area to be surveyed is relatively small so that the whole area can be searched, (2) the species can be easily found and identified, (3) the number of animals to be counted is not more than about 500, and (4) individuals or groups can be recognized and separated from others. Mountain gorilla (*Gorilla beringei beringei*) and cross-river gorilla (*Gorilla gorilla diehlii*) surveys have used total count methods because the forest areas are relatively small, the numbers of animals are limited, and each group can be aged and placed into sex categories using dung size and hair found in night nests (Gray *et al.* 2009; Sunderland-Groves *et al.* 2009; Weber and Vedder 1983). In general the method uses a sweep count involving several teams that search for a fresh gorilla trail throughout an area and when they find it follow the trail to count three consecutive groups of night nests. Size of dung in the nests is measured and the nests are searched for signs of silver hairs, indicating an adult male. Nest locations are plotted on a map of the survey area using GIS and any similarly sized groups are compared for their composition; if they are too similar to be separated they are assumed to be the same group. Until recently no independent verification had been made of this method and it suffered from the fact that there was no good measure of the probability of detection of the groups or how well groups are separated or combined. Guschanski *et al.* (2009) compared the total count in Bwindi Impenetrable National Park with the genetic identification of individuals from dung collected in the night nests during the census and showed that the total count method had overestimated the population by 10% because individuals can make more than one night nest during the night and because lone males were double counted.

2.4 Genetic methods of surveying

With genetic analysis costs falling and the ability to identify individual animals improving, the use of genetic methods to survey primates is being tested under two general approaches: (1) the equivalent of a total count of the population where individuals are identified from DNA extracted from feces (e.g., Guschanski *et al.* 2009) or hair samples (Goossens *et al.* 2005) and (2) genetic capture-recapture analysis (e.g., Arandjelovic *et al.* 2011). In the first method many of the prerequisites of the total count method are needed, in that the population must be small in order to be able to genetically identify all individuals, the whole area must be able to be surveyed relatively easily, and it must be possible to find all individuals. Genetic capture-recapture is very new for primate surveys and is still being tested. However, it has been used for the past 10–15 years fairly effectively for elephants (Eggert *et al.* 2003), coyotes (Kohn *et al.* 1999), bears (Mowat and Strobeck 2000; Paetkau 2003), and otters (Dallas *et al.* 2003) amongst other species. For a list of software programs to implement the use of genetic data in population surveys, see Chapter 14.

2.5 Line transect methods

Line transects have been the main method used to survey diurnal primates. This is because they can be used to cover long distances relatively quickly and good methods have been developed to analyze the data. Distance sampling methods are an extension of strip plot methods. In the latter, long thin plots of fixed width are searched during surveys, ensuring that all animals within the plot are found. This method tends to have low sample sizes because the strips need to be relatively narrow (about 10–15 m either side of the transect line for primate surveys in forests) in order to be sure no animals are missed. By measuring perpendicular distances from the transect, a larger strip width is possible and a correction can be made to allow for the decrease in detectability of animals with distance from the transect. The free availability of the software Distance (Thomas *et al.* 2010) has made it far easier to undertake the analyses needed. However, for these methods to work, a minimum of 30–40 sightings of groups are needed to obtain an estimate around the true density and it is better to have 60–80 sightings to minimize the error around the estimate (Buckland *et al.* 2001). If these cannot be achieved with sufficient transect lines (see survey design below), or by repeated walking of the transects, then other methods need to be investigated. It is therefore important to make a pilot test of the method in the study area, before investing too much in the survey, to determine if the encounter rates are too rare to be able to estimate a density. If no method works well it may be necessary to estimate relative abundance or occupancy only (Section 2.7). Line transect methods have four key assumptions:

1. *Objects on the line are detected with certainty.* Therefore, search effort should be concentrated around the line. For primates in tropical forest, which can be 30–50 m high, this means having one observer focus on the canopy above the line while another searches the sides. Van Schaik *et al.* (2005) found that searching plots for nests of orangutans tended to provide more accurate estimates than line transect methods because observers were missing nests on or near the line.

2. *Objects do not move before detection.* If responsive movement occurs, every effort should be made to record the position of detected animals prior to responsive movement. Observers should also endeavor to move quietly, to minimize responsive movement. Non-responsive movement does not generate large bias provided it is slow relative to the speed of the observer. This means that the observers must move at a speed that is relatively fast. Historically survey methods have recommended that observers move at about 1 km/hr to make sure they do not miss primates, but if the average speed of the primates is also around 1 km/hr or faster, substantial upward bias in density estimates can be anticipated.

3. *Measurements are exact.* Many primate surveys estimate distance by sight and the accuracy of these measurements has been shown to be poor (Buckland *et al.* 2010b). There is no excuse, when tape measures and range finders are easily available, to undertake surveys using visual estimations only.

4. *Animal locations are independent of the transects.* This assumption can be effectively ignored if a good sampling design is used when establishing transects (see Section 2.5.1).

There are, however, still several issues that complicate line transect estimates of primate density and which need to be thought about before conducting any survey and we address these here.

2.5.1 Sampling design

A critical issue, which is a flaw in many primate surveys, is the sampling design. When you read the primate literature you will often see surveys conducted along a transect that forms four sides of a square. It is usually assumed that each of the sides is an independent transect and that four transects have been walked. However, they cannot be independent when they are connected in a square shape and effectively only one transect is being walked. There are also problems because at the corners the observer is sampling the same area twice. We refer readers to statistical- and ecological-methods textbooks for more on sampling theory, and to Buckland *et al.* (2001) and Strindberg *et al.* (2004). Basic principles of survey design should address two key issues to infer reliable population size estimates:

1. Transect placement should be random. In the case of simple random sampling, lines have an equal probability of being placed anywhere in the study area. The software Distance can be used to generate a sampling design for a study area and can be used to assess the probability of sampling any point in the study area (Strindberg *et al.* 2004).
2. Replicated transects—at least 10–20 transects (preferably nearer 20)—should be surveyed in each study area. We have seen survey designs where several transects have been surveyed but only one or two walked in each study area and then these different study areas compared for their different primate densities. Effectively the replication is too low in these studies to obtain a good density estimate for each site and therefore the comparisons are flawed.

An example of a good design for surveys of primates in Kalinzu and Maramagambo Forest Reserves in Uganda included surveys of all diurnal monkey species as well as chimpanzees (Fig. 2.1). Here the sampling design was made in Distance with 3 km transects placed using the systematic segmented trackline sampling option. The angle at which the transects were run was varied to assess which direction provided the most even probability of sampling each corner of the park and the north–south orientation proved to be the most balanced design (Wanyama *et al.* 2010). A similar process was used to design surveys for Kibale National Park

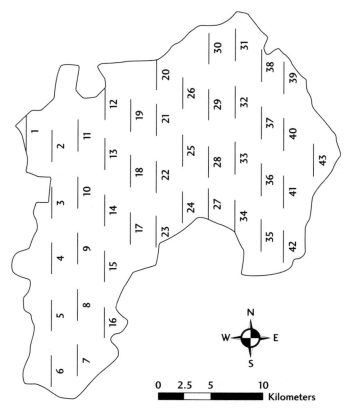

Fig. 2.1 Line transect sampling design for surveys of primates in Kalinzu and Maramagambo Forest Reserves, Uganda. The number of each 3 km transect line is given beside the line.

(Wanyama *et al.* 2010) and this design is now used by the Uganda Wildlife Authority for regular monitoring of populations of primates and other large mammal species.

2.5.2 Group versus individual density

Primates live in groups that can vary from 2–3 individuals up to several hundred individuals. However most primate species live in groups that number between 10–40 individuals. Dealing with groups and what you measure along line transects is tricky when making surveys of primates because different studies have used different methods and there has been no test of which method performs best. Studies measure perpendicular distances from the transect to the first individual seen, to the individual nearest the line transect, or to the center of the group

(Whitesides *et al.* 1988; Plumptre 2000). A complicating factor is that groups of monkeys can spread out over large distances and form subgroups 200–300 meters apart. If encountered along a transect, these subgroups are likely to be counted as different groups. This is fine if the subgroup size is measured at the time, but not if subgroup density is to be multiplied by an average group size in the final analyses, as this will overestimate the primate density (Plumptre 2000; Plumptre and Cox 2005; Buckland *et al.* 2010b). Buckland *et al.* (2010b) looked at various ways of dealing with subgroups/groups and give recommendations that vary depending on what can be observed when a subgroup or group is encountered. Options include: (1) measure to the center of the subgroup and ensure you can count the number of monkeys in the subgroup (all individuals must be counted, at least for subgroups close to the line); (2) measure perpendicular distance to each individual in the group seen. While time consuming, this method might be used if you are unable to detect all the members of a subgroup such that the subgroup size would be underestimated. However it cannot easily be used for species that occur in large groups/subgroups. One way to get around this problem is to truncate the perpendicular distance in the field and only record individuals within this distance of the transect. Other methods proposed by Whitesides *et al.* (1988) of measuring to the nearest individual and correcting by an average group spread, and by Plumptre and Cox (2005) of measuring to the center of the subgroup seen (but probably missing some members of the subgroup), are likely to produce biased results and should not be used.

2.5.3 Indirect sign surveys—nest counts and dung

Some species are very hard to detect from observations on transects as they flee and hide silently when they observe you coming. The great apes are among this category, and despite their large body sizes they can be remarkably elusive, partly because they are at low densities in forests and partly because they spend time on the ground and are hard to observe when there. However, apes build nests, both day and night nests, which can be counted instead of live animals. Dung can also be counted for some species and is often used for ungulate surveys where sightings of a species are rare; however its applicability for primates is poor and it has only been used for surveys of apes (Todd *et al.* 2008). These authors caution that seasonality and differentiation between defecation events and the number of dung piles produced can significantly affect the results. In the case of these indirect signs we need to assess two values: (1) production rate of the sign and (2) decomposition rate of the sign. For example, in the case of chimpanzee nests we need to know how many nests on average are constructed each day and the average time to decay for each nest. Chimpanzees can build more than one nest in a day but also sometimes reuse nests, so this nest decay and reuse needs to be measured at the site rather than assuming every nest is new, as is often the case. An advantage of the use of indirect signs is that they do not move and hence perpendicular distances can be measured easily and accurately.

Decay rates of nests can be highly variable. In Budongo Forest Reserve in Uganda chimpanzee nests decayed between 10–161 days (Plumptre and Reynolds 1996) while orangutan nests could take over 600 days to disappear (Mathewson et al. 2008), and it has been shown that decay rates vary with season and location (Plumptre and Reynolds 1996). The same is true for dung, even elephant dung (Nchanji and Plumptre 2001). This variability increases the 95% confidence limits of the population density estimate. One way to avoid having to calculate and use the decay rate is to use a method of marked nest counts where the transect is repeatedly walked at intervals within which no nest will decay, marking all nests and counting the number of new nests constructed over time (Plumptre and Reynolds 1996). This effectively measures the density of nests produced between the first and last walk of the transects and, by dividing by the time elapsed, can give the density of nest building individuals in the population. It is important to establish an order to walking the transects for marked nest counts so that there is a similar time period between the first and last walk of each transect and so that each transect is walked regularly within 10–15 days. The number of repeated walks depends on the number of nests sighted but one should aim to have at least 60 new nests constructed between the first and last walk of the transects. Although this method takes longer to implement, it is more likely to produce accurate and precise estimates than a method using nest counts that must be corrected for nest decay rates. This is because to truly correct for nest decay, nests should be monitored before and during the census period as well as over the whole study area. To do this will take as much time as carrying out a marked nest count. Most studies that correct using a nest decay rate obtain the decay rate by monitoring nests in one small part of the study area (with easy access) or else "borrow" a rate from another study site altogether. Both methods are unlikely to provide a realistic estimate of the true decay rate in the study area and will not obtain a good estimate of the population density.

2.6 Using group calls to survey primates

2.6.1 Surveys using regular calling by primates

Some primate species regularly call at a specific time of day. For instance gibbons (*Hylobates* spp.) make calls each morning for territorial defense and these calls can be heard over 2 km away (Brockelman and Srikosamatara 1993; Buckley *et al.* 2006). The calls can be used to map the territories of each pair and thereby estimate the number of pairs in a study region. Triangulation of calling pairs is usually made from listening posts by at least three observers at three locations recording simultaneously from dawn to about 10:30 am, at the same site over several days before moving to new areas. It is important that several days are spent at each observation post because some animals may not call on one day and may be missed as a result. Ideally, sightings should be made of calling pairs to assess if they have any young with them at the time. Otherwise the survey is only measuring the density of calling pairs not the total population density.

While point count methods could also be used to census primates, they have not been commonly used instead of the mapping method. Hanya *et al.* (2003) used them to census Japanese macaque groups in mountainous terrain where transects were not feasible. Point count methods measure the distance to any calling/observed individual from randomly placed points throughout the study area. The distances are used in Distance to assess how the probability of detection declines with distance from the point. Assumptions are similar to those for line transects: (1) all animals at the point are detected, (2) measurements are exact, (3) animals do not move during the recording period, and (4) points are established randomly with respect to the primates.

A slight variation on the point count is cue counting (Buckland *et al.* 2006). In this case, each call is recorded, together with an estimate of its distance from the point. An advantage of this approach is that the method is not biased if animals are moving around during the recording period, provided the movement is unaffected by the presence of the observer. To convert an estimate of number of calls per unit area per unit time to an estimate of animal density, an estimate of the mean call rate of animals in the population is required. Distances may be difficult to estimate, and may require multiple observers at different locations to allow triangulation. Given rapid advances in acoustic technology, it may soon be feasible to establish an acoustic array at a sample point, from which distances to calling animals can be recorded; because animal movement does not generate bias, such an array may be left to record calls over an extended period. We are not aware of any primate studies that currently use cue counting methods, but we believe that they have considerable potential for some species.

A related method is to use a line transect design, but instead of walking along the transect, observers are stationed at points along the transect. These observers record calling animals, and their bearing from the point, together with the exact time of each call. Distances of calling animals from the line are then found by triangulation, after matching calls from the same individual recorded by different observers (B. Rawson pers. comm.).

2.6.2 Lure counts

Lure counts differ from simply mapping calling primates by actively luring the animals by playing a call that attracts them. In the case of primates this is usually a call of another group of the species under study (for other species, distress calls of prey animals can also be used). A detection function is developed by testing the playback of the call at different distances from a sample of primate groups whose location is known, from which a model for the detection function is fitted using logistic regression; this function represents the probability that an individual will be detected from the point at which the lure is played. This detection function model is then assumed to hold for the main survey, where a lure is played at each of a number of points systematically spaced through the survey region. The method uses a similar approach to point counts where the detection function is used to estimate an effective radius of response and area searched at each play back point. Once this area is known, the number of responding groups can be converted to a density estimate (Buckland *et al.* 2006). A variation of this method was used by Savage *et al.* (2010) where observers moved along parallel transects playing lure calls for cotton-top tamarins (*Saguinus oedipus*) and luring animals from the strip between the transects. The parallel transects were close enough to be sure that all groups located between the transects would respond, and the direction of approach allowed the observers to determine which of the detected groups were initially located between the two transects.

2.7 Measurements of relative abundance and occupancy

In certain cases it may only be possible to obtain estimates of relative density. This is often for reasons of cost rather than because it is not possible to estimate absolute density, although if one of the assumptions of the method is broken then relative density may be all that can be estimated. Encounter rates of primates along transects can be a measure of relative density but will vary with detectability of the animals, which in turn varies according to many factors, including observer, habitat, animal behavior, group size, and season. Distance sampling methods allow estimation of the probability of detection with distance from the transect (and with other covariates, using the mcds engine of Distance) and so can correct encounter rates for detectability. However, establishing transects can be costly and in some environments attracts poachers. More ground can be covered with reconnaissance walks that aim to follow a compass direction but move around obstacles rather than cut a transect through them. Reconnaissance walks (recce walks) are biased and only encounter rates can be obtained using this method. However, several studies have found good correlations between encounter rates from recce walks and density (e.g., Plumptre and Cox 2005 for chimpanzee nests; Walsh and White 1999 for elephant dung). Encounter rates from recces can therefore be used to estimate density in some areas where resources are limited provided the same correlation holds true. Helicopter surveys of orangutan nests also found a good correlation between the density of canopy nests counted and estimates of orangutan density from ground surveys using line transect techniques, so that aerial counts of relative abundance can be corrected to estimate total abundance (Ancrenaz *et al.* 2005).

Relative measures of abundance assume that detectability does not vary. Hence they should only be used if it is not possible, or if it is prohibitively expensive, to estimate detectability and hence population density.

2.7.1 Occupancy analysis

Occupancy analysis can be used to assess the probability that a primate species is present at a particular site for rare species. This method does not usually allow estimation of primate abundance, although it can in some cases (Royle and Nichols 2003). It is most useful for rare species where the sample sizes would be too small to estimate density using line transect analyses (for some nocturnal species, for instance). In order to undertake an occupancy analysis, several visits to the sites need to be made over a particular survey period. Sites need to be established in a random manner with respect to the primate species, as in any random survey design, and can be points where point counts can take place or short transects. The presence or absence of the species at each site is recorded on each visit. The analysis method calculates the detection probability of the species being surveyed, which makes it a more suitable method to use than encounter rates when monitoring a species. It is not possible to give details about occupancy analysis here but the methods are summarized well in Mackenzie *et al.* (2006) and the accompanying website is useful (Donovan and Hines 2007). We are only aware of the use of occupancy analysis in two primate studies (Neilson 2010; Keane *et al.* 2012) but it has been used widely for other species and would be appropriate for certain rare and elusive primates.

2.8 Nocturnal primates

Line transect methods can be used for nocturnal primates where the assumptions listed in Section 2.5 are met. This method is most often used with lorises (Nekaris *et al.* 2008) and sometimes with galagos as well. For other nocturnal primates that are rarely encountered, the line transect method is not ideal (Duckworth 1998) and in these cases the relative abundance and occupancy methods are suitable (see Section 2.7).

For nocturnal line transect surveys it is best to set and flag the transect during the day. Frequent flagging helps the observer stay on the transect in the dark. Reflective or silver tape is preferable and is relatively easy to purchase. Placing all the flags on the same side of the transect, for instance all on the left-hand side, can help the observer to navigate while also directing attention to the search for animals.

Nocturnal survey timing depends on the activity cycle of the animals, but may need to cover all periods of time from dusk to dawn (Nekaris 2003; Sterling 1993a). As you walk the trail at a slow and steady pace (about 600–800 m/hr), you listen for sounds such as calls, objects dropping, or rustling branches and hold the light source close to your eyes, watching for movement or "eyeshine"—the reflection of light from eyes. It is important to keep the light source near your eyes as primates generally look towards a light source and if the source is at your waist, for instance, you may not catch the eyeshine.

The eyeshine will differ across species depending on whether not they have a *tapetum lucidum*, which picks up available light and reflects it back out. For species with a *tapetum lucidum* white lights will work, but for those without it (like lorises and night monkeys) red lights or filters are best. Red filters may be useful for many species, though they may hinder identification (Nekaris *et al.* 2008). It is best to test different light source options depending on the focal species (Bearder *et al.* 2006).

It is also important to identify the optimal strength of the light source for the species of interest. If the light is too weak, the animal may not look towards you. If the light is too strong, an animal will blink or close its eyes and you may miss the reflection. Blue LED lights may also have the potential to damage an animal's retina; halogen bulbs work well in most forests, providing a softer light.

When you see eyeshine, you should focus the light just below the eyes of the animal (again, to help ensure the animal continues to look towards the light) and use binoculars in order to identify the species. Night vision goggles are not well suited to dark canopy environments. Their resolution is also suboptimal for identifying animals.

It is useful, for both safety and logistical reasons, to have a minimum of two observers. One observer can hold a light on the animal while the other determines the species. Observers record the visibility in the habitat, whether an individual was first detected by sound or sight, the height of the animal above ground, the distance along a transect, the time of the observation, and the group size and structure where applicable and feasible. It is also helpful to note weather and moon cycle information as these may affect activity levels of animals (for instance see Svensson *et al.* 2010 for *Aotus*; Bearder *et al.* 2006 for several species).

Indirect surveys, such as counting feeding remains or nests, have not proved to be useful for nocturnal primates to date. For instance, while aye-ayes (*Daubentonia madagascariensis*) build nests, there are no discernible patterns in the average length of time an animal will use a nest and several individuals may use the same nest at different times (Sterling 1993a,b). While aye-ayes generally nest alone, they have also been observed to share nests (Ancrenaz *et al.* 1994; Sterling 1993a). Researchers have used traps or nets for some nocturnal primates, for instance tarsiers (Neri-Arboleda *et al.* 2002). Trapping could help to establish the presence of animals in an area or be used for capture-mark-recapture estimations (Meyler *et al.* 2012).

2.9 Tools that can help with primate surveys

2.9.1 GPS unit

A Global Positioning System (GPS) unit is very useful for plotting sightings of primates and mapping where observers have passed, as well as establishing line transects. If Distance is used to design a primate survey, it will give the GPS position of the start and end points of the transects. A GPS unit is needed to locate these points and can be used to indicate the general direction to the end point if used while the observer is moving (but see Section 2.9.3). Some GPS units have a high-sensitivity chip (such as the Garmin 60 Csx) that works much better under tree canopies. Although the manufacturers do not often advertise this difference, it is essential that units with the more sensitive chips are used in forests.

2.9.2 Rangefinders or tape measures

There is no excuse for researchers to use visual sighting estimates of perpendicular distances along transects; 30 and 50 m tape measures are readily available, as are range finders that use parallax or lasers to measure distance. Laser range finders are easier to use in forest but do require some thought when being used as they can be affected by small twigs between the observer and the object being measured. If the full distance cannot be measured in one go because the foliage is too dense, then the observer can measure to a tree that is in the same perpendicular direction, then walk to the tree and measure a second distance, and then sum the two measurements to obtain the perpendicular distance. This takes longer than estimating by eye, but it is crucial that measurements are accurate. Tests of measurements by eye typically show that they are biased and have high variability.

2.9.3 Compass

Line transects should be straight and as far as possible should not deviate for barriers or topography, as this may lead to bias in density estimates. Each team of trail cutters must use a compass to sight where the transect should pass and they should be supervised by a member of the research team who will be undertaking the surveys. Make it clear to the trail cutters that metal affects the compass and therefore pangas/knives should not be placed near the compass. Transects should be cut a few days before the census takes place, although nests and dung can be counted along transects where teams are working provided the survey team is not with the cutting team. Do not use a GPS unit to estimate the direction of the transect as the accuracy of the compass direction measurement from a GPS is very dependent on the speed of movement of the observer.

2.10 Monitoring primates for conservation

Many primates are threatened by extinction and therefore it is important that their populations are monitored over time to assess how they are changing. However, studies show that the errors around primate survey data are such that it can be difficult to detect significant changes in the population unless they are fairly large. Plumptre (2000) showed that standard line transect techniques only allow a change of about +/- 10–30% of the population to be detected in subsequent line transect surveys. In the case of surveys of indirect sign, such as nests, only +/- 30–50% of changes in the population could be detected. In other words, the population would have to be almost halved before the results would register as significant at the 5% level. He used surveys that were designed to assess the population density of several primate species at the same time, along transects established randomly along a base line transect. There are several ways in which the ability to detect more subtle changes in the population size could be improved:

1. Focusing on one species would make it possible to stratify the sampling across the study region so that more transects are placed in areas where the density of that species is highest and fewer transects in areas where it is rare. This can only be used on a species-by-species basis unless two species show the same pattern of abundance. (Fig. 2.2).
2. Increasing the number of transects to ensure that at least 20 are sampled in each study area.
3. Repeating the counts along the transects to ensure at least 60–80 sightings of primate groups/subgroups are made. When analyzing repeated counts in Distance, it is important to ensure that the repeated data from the same

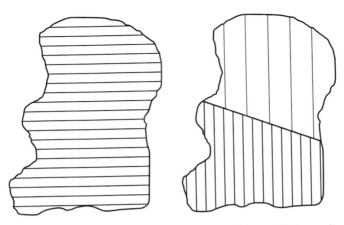

Fig. 2.2 Example of a systematic sampling design (a) and a stratified sampling design (b) for transects. Greater effort is made in an area where the primates are thought to be more abundant to increase the precision of the density estimate in the stratified method.

transect are entered as data from that transect and not as a new transect (the effort of the survey is multiplied by the number of times the transect is walked).

4. If the survey is conducted repeatedly over time, trend analyses can be carried out rather than simple comparisons between successive surveys. Such analyses have greater power, allowing detection of smaller changes.

5. Spatial modeling of line transect data (Hedley and Buckland 2004) allows shifts in the distribution of populations to be identified, and allows these shifts to be related to covariates reflecting, for example, habitat loss or degradation, disturbance, or hunting pressure.

2.11 Conclusion

There are still many issues to be addressed and tested to improve primate survey methods. Ideally, methods would be tested on populations of known size, for example primates on an island where all individuals are known. Aspects of survey methods that still need to be tested include:

1. How to deal with groups in the field—testing methods that treat each individual detected separately versus those that record the location of group or subgroup center and multiply group density by an average group size.

2. New methods of surveying primates, particularly:
   a. Genetic capture-recapture methods—how precise a measure can be obtained?
   b. Double-observer counting methods—as far as we know these have not been tried with diurnal primates but such surveys have become routine in marine mammals (Borchers *et al.* 2006). Double-observer methods were used for some marine mammal surveys as early as the 1980s and are included in the mrds engine of Distance.
   c. Cue or point count methods for species that call.
   d. Lure count methods for species that respond to playback calls.
   e. Occupancy methods for rare species.
3. How the 3D nature of the forest affects detection of primates, particularly what percentage of primates on the line may be missed in different situations. Wich and Boyko (2011) assessed how age of nest, height above ground, distance to the transect, and observer experience affected detection of orangutan nests in Sumatra. They found that all factors significantly affected detection.

We believe that current survey practice can be significantly improved in primatology by observing the recommendations made for line transect sampling in this chapter and in the references we have given. There is a need both to improve the rigor of existing methods and to develop new methods to improve the survey estimates of populations in primatology.

3
Darting, anesthesia, and handling

Kenneth E. Glander

3.1 Introduction—the role of capture in primate field studies

One of the most invasive things that field primatologists do is to dart or trap wild individuals (Wilson and McMahon 2006). Given its traumatic impact, it is vital that each researcher evaluate the pros and cons of capture. Does the research question(s) require capture or is there another less stressful option? What are the benefits and costs the animals being captured and to the researcher (Table 3.1)? These categories have very different goals, but often can be combined for the benefit of the targeted species.

Capture of non-human primates permits standard measurements and samples to be taken from immobilized animals (see Chapter 4). In addition, capture is the

Table 3.1 *Justification for capture*

Unequivocal identification
Management/translocation
Conservation/rescue
Veterinary intervention
Zoonosis
Morphometrics
Collect biological samples
Dental casts
Satellite tracking
Biotelemetry
Re-introduction
Population structure
Gender determination
Dispersal and ranging
Recovery of cadavers via mortality collars
Education/ecotourism

Primate Ecology and Conservation: A Handbook of Techniques. First Edition. Edited by Eleanor J. Sterling, Nora Bynum, and Mary E. Blair. © Oxford University Press 2013. Published 2013 by Oxford University Press.

only way to take advantage of the ongoing revolution in miniaturization that allows the effective exportation of lab-based technologies in a field setting for wild primates. Trapping or darting is required for the application of these formerly lab-based-only technologies: (1) blood glucose levels, (2) core and subcutaneous body temperatures, (3) heart rate, (4) activity levels, and (5) the actual three-dimensional distance traveled. All of these can be continuously recorded and monitored (i.e., in real time) over 24-hour periods and matched with similar continuous recording of environmental parameters of the animal's natural habitat, such as solar radiation, wind speed, air temperature, humidity, and forest canopy temperatures across both space and time.

Collectively, these environmental and physiological data can then be analyzed to better understand how behavior is related to weather conditions, available energy, and food choices, thus relating environmental variation to behavioral and physiological data from free-ranging primates. The behavior and physiology of free-ranging primates results from the interaction of numerous factors that cannot be adequately replicated in the laboratory. These factors include food supply, food quality, predation, social interactions, and three-dimensional variation in forest weather conditions. Thus, existing data from lab studies (e.g., Maho et al. 1981; Müller et al. 1983; Wurster et al. 1985) may not provide an ecologically, and therefore evolutionarily, relevant explanation of most primate behavior or physiology in response to environmental variation.

Additionally, certain research questions can only be addressed with unequivocal individual identification; for example, those relating to individual strategies that require consideration of complete life histories (Merila and Sheldon 2000; Grant and Grant 2000). However, much of the current research on different life stages remains primarily cross-sectional and of short duration. The same marked individuals have rarely been followed through time (but see Alberts et al. 2006; Altmann et al. 2010; Glander 2006). Furthermore, questions about individual strategies often require a large sample size that is longitudinal rather than cross-sectional. The best way to accomplish this type of data set is to have marked subjects that are followed for their lifespan. Visible marking provides unambiguous identification but does require capture, and collars and tags, for example, require replacing for continued accuracy. To ensure that precise, secure, and permanent identification is not compromised, implantable RFID (Radio Frequency Identification) microchips should be used.

Positive identification without marking may be possible when studying terrestrial primates that can be closely approached (Swedell et al. 2011), but unambiguous identification is rarely possible for arboreal primates. In fact, there are justifiable reasons beyond identification for capturing terrestrial primates as well (Jolly *et al.* 2011; see also Table 3.1). Even though technology has advanced to the degree that blood or tissue samples are no longer absolutely needed to perform genetic or health analyses, fecal samples do not yet provide unquestionable genetic results nor do they yield useful data on individuals that may or may not be unequivocally identified, whether they are arboreal or terrestrial. Safe capture of animals for these and other studies does provide better biological samples and allows research questions about individual strategies.

If capture is justified, then every effort must be made to minimize the trauma to the animal and maximize the data gathered. While using the safest capture drug and the proper capture equipment can reduce stress and injury, the most important factor may be the researcher's compassion for the individuals being captured. Human ego must be suppressed to reduce the trauma and stress of capture: the shot not taken is often better than the one taken in haste or pride.

3.2 Permits, licenses, and approvals

The Animal Welfare Act (AWA), passed in 1966, requires that an Animal Use Committee (often referred to as the Institutional Animal Care & Use Committee or IACUC) must approve all proposed research. Each Animal Use Committee generally has forms that can be quite detailed and often have many parts that focus on laboratory rather than field studies. Regardless of their complexity or parts that are not relevant, they must be completed and approved before any research may take place. Some institutions, such as Duke University, have modified their IACUC form to include a shortened "main form" plus a separate form specifically for "field capture/field studies." Fedigan (2010) lists several other institutions that also have "field forms" (McGill, York, Stony Brook). In addition to IACUC approval, many institutions require training in safety, animal handling, and working with primates plus an annual negative TB test. It is advisable to seek institutional help and allow ample time to complete the necessary Animal Use forms and to determine what shots and training may be required by your institution.

Many countries have specific requirements and approvals that must be obtained before they permit research. These requirements range from host country forest or wildlife ministry permits, to entry fees for national parks and reserves. In most instances, permits are required even if the research will be done on private property. Working on private property adds another layer of permission needed from the owner. It is advisable to find local assistance to legally, economically, and expeditiously obtain the necessary permits and/or pay the fees.

Working with most capture drugs requires a Drug Enforcement Agency (DEA) license. Application can be made online (<http://www.deadiversion.usdoj.gov/>). The DEA may also require the researcher to acquire a state license. Check with your local state. There are fees for both of these licenses.

Permits from the private property owner and the local government agency to carry out the research do not include the collection or exporting of biological samples. Additional permits from local agencies are often needed to collect the samples, and an export permit is then required to take them out of the country of origin. Many local government agencies demand a copy of the permit to import the samples to the United States (US) before issuing the export permit. This means, at minimum, a Center for Disease Control (CDC) <http://www.cdc.gov/od/eaipp/importApplication/agents.htm>, and US Fish and Wildlife document <http://www.fws.gov/permits/applicationforms/ApplicationIJK.html> giving the researcher permission to import biological samples into the US. A CITES export permit from the host country is also required for all samples from live animals <http://library.fws.gov/IA_Pubs/CITES_permits-certs.pdf>. The application process should be initiated well before the fieldwork is expected to commence as it may take as many as 9–12 months to obtain these permits.

3.3 Public relations

Establishing a positive relationship with the local people is equally important to the success or failure of a research project. This is true whether the research occurs inside or outside the US and it is vital if animals are going to be captured. Making contact and hiring local individuals to assist in the research provides an avenue to establish trust and understanding. Ignoring this step in the process of doing any research will inevitably lead to misunderstanding and likely failure. Making the effort to involve the local population often results in them becoming invested in the animals and habitat to the extent that they protect both in the short- and long-term.

3.4 Trapping vs. darting

The method chosen for capture depends almost entirely on whether the target is terrestrial or arboreal. Trapping does not succeed for arboreal species unless they can be baited to the ground or to platforms in the canopy (Garber *et al.* 1984; Aguiar *et al.* 2007; Rocha *et al.* 2007). Even then, trapping of arboreal species is limited to situations where preferred foods are in limited supply and there are few if any other species present to "steal" the bait or to keep the targeted species away from the trap (Aguiar *et al.* 2007). Advantages are that traps yield multiple animals at a time, there is no danger from falling, and traps can be the only effective method for capturing terrestrial primate species (Jolly *et al.* 2011). The primary disadvantages of trapping are stress from the close proximity of other individuals, the length of time the primates are held in the trap before being tranquilized, and possible wounding from individuals outside or inside the trap. The use of multi-chambered traps and frequent monitoring can significantly reduce these negative effects (Garber *et al.* 1993; Savage *et al.* 1993).

An alternate method of capture via trapping is the use of netting that is either: hand-thrown or rocket-propelled over individuals attracted to a baited site; spread on the ground after felling isolated trees; or set out in a sheltered area such as a cave. Nets also are used to rescue primates and other animals from rising water due to dam construction (de Thoisy *et al.* 2001). Netting is not applicable for arboreal primates, unless they are in the above-mentioned situations.

Camera traps capture an image rather than the individual and are used to verify presence of a secretive or nocturnal species. However, they are not effective in the arboreal environment and the images they capture are random, providing no opportunity for obtaining data other than establishing presence.

3.5 Darting methods and equipment

Darting, the delivery of a drug-carrying projectile, is the primary method of capturing arboreal species, but it cannot be used on large primates such as chimpanzees, bonobos, or gorillas because their fall from the trees is likely to be lethal. Darting also is used to anesthetize individuals in traps or at zoos without squeeze-cages. A pole-syringe is another way to anesthetize animals in traps (Jolly 1998).

In darting with either syringes or explosive darts, the preferred injection site is the hindquarters. The hit must be perpendicular to the target surface to ensure injection of the entire drug dose. Since the chest, thorax, abdomen, shoulder, neck, and head are vulnerable, unsuitable target areas, a shot should only be taken when the subject is facing away from the darter. Thus, if a shot misses the targeted hindquarters, it will miss all vulnerable areas. The use of a jab stick brings other areas such as the shoulder into consideration as target areas because of greater control and lessened impact of the jab.

The darted animal may fall as much as 30 meters. The fall must be cushioned in some manner to prevent injury or death. An effective method is to catch the falling animal in a mesh net or "camper's hammock" held by two or three people (Fig. 3.1). When the darted animal does not fall, the branch on which the animal is hanging is either shaken or cut down with a saw attached to the end of an aluminum pole. The pole comes in six-foot sections that can be bolted together

Fig. 3.1 Camper's hammock held in preparation to catch a falling mantled howling monkey in Costa Rica.

until it is long enough to reach the hanging animal (Aazel Corporation, <http://www.aazelcorp.com/>). Catching a falling primate requires an experienced individual to properly place the net and the net holders to prevent injury to both falling primate and each other.

Darts carrying a small explosive charge or syringes with compressed air are used to deliver the anesthesia. The explosive charge detonates on impact, quickly injecting the drug while a small plastic collar on the needle of the syringe is pushed back on impact thus exposing a hole through which the drug is injected. Both darts and syringes usually bounce out and away from the target upon impact, and it is important to try and recover these in order to avoid harm to other animals. A barb on the dart or needle would prevent it from immediately falling out, but the resulting tissue damage is unacceptable.

Table 3.2 lists the types of darting equipment. The choice of equipment must be based on which system results in the least injury and is the most effective based on the highest capture rate per hit and the least mortality. Based on 39 years of personal experience, I have found that the safest and most effective choices are CO^2 powered projectors with explosive darts, for distances of 5–20 m, and a blowpipe with explosive darts for distances of 4 m or less. Cartridge-fired projectors are useful

Table 3.2 *Projectors and other equipment (all websites checked and accessible as of September 2012)*

Company and location	Products	
Pneu-Dart™ (Williamsport, PA)	Rifles, pistols, blowpipes	<http://pneudart.com/>
Telinject™ (Agua Dulce, CA)	Rifles, pistols, blowpipes	<http://telinject.com/>
Dan-Inject™ (Knoxville, TN)	Rifles, pistols, blowpipes	<http://dan-inject.com/>
Cap-Chur™ (Powder Springs, GA)	Rifles, pistols	<http://www.palmercap-chur.com/products.html>
Animal Capture, Inc. (League City, TX)	Jab stick or pole	<http://www.ace-cap.com/>
(Army/Navy stores)	Camper's hammock	local

for large-bodied mammals at greater distances, but are much too powerful for the smaller-bodied primates at any distance. Pistols are effective for close-up or zoo-capture, but the blowpipe is just as effective when employed properly and eliminates the sound of the pistol.

The brand choice of darting equipment is often based on expense or personal preference. I prefer the Pneu-Dart® system for its ease of use, safety factors, reliability, and cost. Its reliability comes from the fact that the explosive darts never fail to expel the drug, which frequently happens with syringes. This is because the syringes rely on air pressure to inject the drug when the hole in the needle is exposed and, unless used immediately, the air pressure can bleed off without the researcher knowing, resulting in failure to inject when the syringe successfully hits the target. Under field conditions darts or syringes are frequently not used immediately. This is not an issue with the explosive darts since the explosive cap is inactive until igniting on impact. Even though every explosive dart reliably expels the drug, not every hit may successfully bring the target down because the dart can hit at an angle causing most or the entire drug amount to be expelled into the air. This is due to the darter's skill and not to the reliability of the dart or syringe.

Pneu-Dart® and Cap-Chur® use explosive darts while Telinject® and Dan-Inject® use air pressure syringes. All of these systems have been successfully used in primate capture. Also, they can all be used to administer medication and vaccinations to either wild or captive animals without handling.

Following capture, the marking, measuring, and collection of biological samples proceeds. Since part of the reason for capture is to provide the researcher with a visible way of identifying an individual, brightly colored collars with uniquely colored or shaped tags are used. Nylon webbing with indelible colors is most effective for larger primates, with keychain-like ball-and-chain collars being most effective for marmoset and tamarin-sized primates. Most pet stores carry nylon dog collars or leads of different sizes and colors. The leads can be cut to appropriate size and a D-ring used to attach a tag, or the buckle cut from the collar and the D-ring retained to display the tag. The cut ends of either the lead or the collar are melted and two rivets are used to secure the collar around the animal's neck. It takes experience to provide enough slack to prevent injury while preventing the animal from slipping off the collar. Tags are attached to the D-ring with an S-hook. Tags are available from farm suppliers such as Nasco™. D-rings and S-hooks are available in most hardware stores.

A more permanent method of identification is needed to prevent loss of information should the visible collar and/or tag disappear. RFID microchips serve this purpose. They are easily implanted under the skin on the back between the shoulder blades and last longer than a primate's lifespan since they have no battery or moving parts. Microchips may not be a viable solution for those primate species that are dextrous groomers. The best known microchip suppliers are AVID™, HomeAgain™, and Pro ID™.

3.6 Drug type and dose

An understanding of the differences between anesthetizing laboratory and wild primates is critical in choosing the best drug and using it safely. Table 3.3 lists the various drugs and dosages. When used under laboratory conditions, an effective

Table 3.3 *Drugs used for capture and/or sedation*

Brand name and drug for capture	Dose for intra-muscular injection (IM)
Telazol®/Zoletil® (tiletamine HCL & zolazepam HCL	20 mg/kg
Ketaset®/Vetalar® (ketamine HCL)	20 mg/kg
Tubarine® (tubocurarine)	Not effective or safe
Sernylan® (phencyclidine HCL)	No longer available
Rompun® (zylazine HCL)	Not effective or safe
Brand name and drug for sedation	
Dexdomitor® (medetomidine HCL)	0.1–0.5 mg/kg
Atipamezole® (atipamezole) Medetomidine reversal	0.1–0.5 mg/kg
Valium® (diazepam)	Not effective for capture
Drug for protection of heart	
PromAce® (acepromazine)	0.5–1.0 mg/kg

dosage is much lower than is required for the effective and safe capture of wild primates; for example, 3–6 mg/kg in the laboratory compared to 20–25 mg/kg in the wild.

The most commonly used drugs for capture are Ketaset® and Telazol® (Vetalar® is a different brand name for the drug known as Ketaset® and Zoletil® is the European/Latin American brand name for Telazol®). I will discuss Ketaset® and Telazol® only, but all descriptions and dosage hold for their other brand names as well.

The recommended dosage for Ketaset® is 10–15 mg/kg IM for immobilization and 25–30 mg/kg for surgery. Since there is no distinction made between the laboratory or field, the assumption is that these doses are effective for immobilization under both circumstances (surgery is seldom done in the field and if it is, it should be done using isoflurine which is much safer and more effective). In reality, the lower dose is not effective for darting of wild arboreal primates, while the lower dose is effective for terrestrial primates (Brett *et al.* 1982; Phillips-Conroy *et al.* 1991).

Ketaset® is not effective for howler monkeys, spider monkeys, or woolly spider monkeys (all have prehensile tails). It is partially effective for *Cebus* who have semi-prehensile tails. The reason it is not effective for primates with prehensile tails is the same reason that makes it a poor choice for use in the capture of any wild primate or for manipulation of laboratory primates, that is, it causes muscle rigidity. The limbs become stiff and non-flexible which greatly increases the possibility of broken bones if the falling animal hits branches. This rigidity also causes the muscles of the prehensile tail to lock onto a branch, preventing an animal from falling even though the drug has completely immobilized it. These animals will hang by their tails without falling until they recover enough to regain an upright posture and move away. Its continued use in both laboratory and field situations is possibly linked to it being much cheaper than the more effective and safer Telazol®.

The recommended dosage for Telazol® is 2–6 mg/kg IM for macaques and 1–20 mg/kg IM for other species. The lower doses are not effective for wild arboreal primates. In my experience, a dose of 20 mg/kg IM is required for the safe and effective capture of wild primates (see Glander *et al.* 1991and Glander *et al.* 1992 for different drugs' use and effectiveness). A dose of 20 mg/kg IM will keep darted individuals immobile for about 30 minutes with subsequent doses of 2–6 mg/kg IM being safe and effective to maintain immobility. Unlike Ketaset®, Telazol® produces totally relaxed muscle tone with completely flexible limbs and tails, resulting in more easily manipulated individuals, greatly reducing the probability of broken bones from falling, and of prehensile-tailed monkeys remaining hanging by their tails. The margin of error for over-dosing is much greater with Telazol®/Zoletil® than any other drug. All of these factors make it the obvious drug of choice for safe and efficient handling of primates in either the laboratory or field.

Animals that recover from the capture dosage before the research procedures are complete are given injections of 1–3 mg/kg of Telazol, repeated as often as needed. After all procedures are complete the animals are placed in burlap bags until they recover enough to walk or climb unaided. The bags are kept in the shade and are the best means of holding an animal until it recovers because they reduce visual stimulus and stress.

The other drugs in Table 3.3 are either not safe, not effective, or not available for use in capturing arboreal primates. Sernylan® is no longer available nor is it safe for use with primates (Glander *et al.* 1991). The others have been used, or are recommended to be used, in combination with Ketaset® to reduce muscle rigidity. These combinations may be useful in the laboratory, but are not safe or effective for capture of arboreal primates, though they may be safe and effective for immobilizing trapped primates.

Dexdomitor® is safe and effective for maintaining immobilization after capture but cannot be used for capture. Its effects can be reversed by Atipamezole® which greatly shortens the recovery time for captured animals. Neither Dexdormitor® nor Atipamezole® interact negatively with Telazol®. Valium® is not effective for capture, but is recommended in combination with all the other capture drugs in Table 3.3 except Telazol® which already has a tranquilizing agent.

Acepromazine is highly recommended for all wild captures as it is antidysrhythmic, antiemetic, and antispasmodic (Stepien *et al.* 1995). It protects the heart by lowering blood pressure, a factor that is impacted by capture (Muir and Mason 1993).

3.7 Safety considerations while handling a captured animal

Handling an animal carries possible dangers for both the animal and handler. These include the transmission of disease and parasites in both directions (see Chapman *et al.* 2005b for a list of possible exchanges between human and non-human primates). Common sense dictates that the handler should take universal precautions in terms of dress and protective gear such as gloves and masks. Humans should be particularly cautious if blood or saliva samples are being collected from Old World Primates as SIV and herpes viruses are common among these non-human primates. Given those basic preparations, the primary concern has to be for the safety of the animal. The following issues are critical. The order does not indicate relative importance or danger.

1. The animal's body temperature must be closely monitored, as many of the capture drugs interfere with thermoregulation. If the body temperature

increases beyond 39.5 °C, the animal should be immersed in water until the temperature returns to 37.5 °C. Conversely, any temperature below 34 °C must be treated by placing the animal on a heating pad or placing plastic bottles filled with warm water in contact with the animal's inguinal and auxiliary regions. The duration of immobilization and the ambient temperature are the major factors impacting the animal's body temperature.

2. Keep the mouth free of any regurgitate to prevent inhalation. Use a spatula or tongue depressor and not hands or fingers, as spasmodic jaw-closing reflexes are often still intact with dissociative drugs. Carry an anesthetized animal with its head down to prevent either saliva or regurgitate from being inhaled. Holding the animal with its head down and shaking can be effective if no stick, spatula, or tongue depressor is available to clear the airway.

3. Make sure that breathing, heart rate, and oxygen levels are normal. This is best done with a pulse oximeter. Monitoring these should occur during the entire immobilization period. Without this type of monitoring, there may be little or no warning as an animal returns to almost full awareness, resulting in possible harm to itself as well as the handlers.

4. Attend to any bleeding from the dart wound or from recent injury. The treatment should be limited to external care without the administration of antibiotic shots. Wild primates have incredible recuperative abilities and antibiotics will destroy more beneficial gut flora and fauna than bad, thus resulting in further harm rather than assistance.

5. Seizure or severe muscle contractions may occur if using ketamine or phencyclidine. These can be life-threatening and must be treated with diazepam. Telazol® use does not carry this risk as it contains zolazepam.

6. Broken limbs from the fall are possible, particularly with Ketaset® and similar drugs. Unless the break is compound, the best treatment is to ensure that no further damage is done during handling and to release the individual without splinting or casting. This seems counterintuitive, and goes against most veterinary training, but splinting and casting are effective only if the animal is held in a captive situation. In the wild, animals with broken bones resulting from natural events (i.e., not capture) have been observed to reduce their travel for about 2–4 days and then move only enough to acquire food. A similar response occurs with breaks due to capture. Within a period of 6–8 weeks the broken bones heal with no visible evidence of the break.

3.8 Recovery

At the conclusion of all procedures the animal is allowed to recover before being returned to its group. This does not and cannot include being left on the ground near the group. Since it takes about 3–5 hours for an anesthetized animal to gain full control of motor and cognitive abilities, it must be protected from predators as well as other group members who may take advantage of a disoriented and uncoordinated individual to gain an advantage because the recovering animal is acting strangely. The best method is to put the recovering individual in a burlap bag that is then placed in the shade or inside an occupied building if it is necessary to keep the animal overnight. The bags are the best means of holding an animal until it recovers because they reduce visual stimuli, prevent hypothermia, and allow free-flow of air. Cages are not effective as they allow the recovering individual visual access to the surrounding environment resulting in over-stimulation and stress that can cause self-inflicted injury and diarrhea or both.

3.9 Release

It is imperative that the animal be fully awake and have full control of motor and cognitive facilities before being released. Burlap bags again are ideal for this as it is easy to determine whether or not an animal is sitting upright. When that is evident, the bag can be held upright, untied, and opened just enough to evaluate the animal's awareness. This viewing does not appear to stress the animals as the bag apparently gives them a sense of security that is not provided when a human checks a primate in a cage (personal experience). Burlap bags have been used for 27 primate species and may have contributed to a low mortality rate (Glander *et al.* 1991, 1992).

Release is simply a matter of placing the bag on the ground at the capture site, untying it, pointing it away from the holder, and gently shaking it if necessary to encourage the animal to exit. The unobstructed view of the opened bag is almost always the only thing needed for the animal to quickly leave the bag without looking back. The bag serves to hide the human(s) who are not visible from the opened end until the animal is out and running toward the trees. An open space with nearby small trees is the ideal release location as this focuses the animal's attention on rapidly climbing the trees. I have found it unnecessary to release individuals near their group as they are much more efficient at finding their group than I am.

3.10 Safety concerns for researchers

In addition to the danger of contracting a disease or parasite from handling non-human primates, there is the danger of injury due to an animal falling from the trees, a bite from a partially anesthetized animal, or personal injury from a fall or accident. All non-human primates have parasites and some of them can and do survive on or in humans. Non-sterile surgical gloves should be worn and hands thoroughly washed after handling the animal or any biological sample. Particular care should be taken when working with blood or saliva of Old World monkeys and apes as they may harbor potentially deadly viruses such as B-virus, SIV, or Ebola. Beware of accidental "sticks" while recapping or disposing of all needles and have a "sharps" container.

Extreme care must be taken when handling the capture drugs, especially when initially loading the darts or filling a new syringe under time constraints during a partially successful darting. It is imperative that strict firing protocols are used before, during, and after a darting procedure. These include never pointing the projector horizontally or putting anyone in the line of fire. Strict protocols are even more critical when more than one darter is employed. These are common sense precautions, yet they are often not practiced; particularly in the moments of disappointment of a failed capture attempt or the celebration of a success. Insufficient respect is given to projectors used for capture because they are not "real" guns. They may not contain bullets, but the darts and syringes that they do contain can cause real damage and a so-called "empty" projector is just as dangerous as an "empty" gun.

3.11 Accountability

Whether trapping or darting is used, there should be an accountability to report the overall success or failure of any methodology employed. This accountability must include mortality rate, impact on the subject's behavior and/or ecology, and local public relations. The reporting of these factors (certainly mortality) are required by most Animal Use Committees, but should be available to the general research community.

3.12 Conclusion

Capture, whether by darting or trapping, has considerable consequences in terms of the dangers to the subjects and humans as well as with respect to alterations in the subject's group, or misunderstanding by the local community which may lead to restriction of researcher access. Thus, it is imperative that there are clear communications with national and local governments; there is a well thought out rationale for capture; all permits are in place; and sufficient and appropriate preparations and contingencies have been made for the safety of the subjects and associated humans.

Given these caveats, a successful capture can produce a range of information that is extraordinary and can be obtained in no other way.

4

Health assessment and epidemiology

Michael P. Muehlenbein and Cari M. Lewis

4.1 Introduction

The interests of primatologists, veterinarians, immunologists, and disease ecologists, when taken together, create a multidisciplinary approach for identifying the ecological, physiological, and behavioral determinants of infectious diseases in wild non-human primate populations. The fruits of these combined interests are a number of published field studies of intestinal parasites and other pathogens in wild primates ("primates" to be used in lieu of "non-human primates" for the remainder of this chapter). The variety of methods employed in these studies continues to grow. Our goal here is *not* to provide a comprehensive list of protocols for sample collection and analysis, (see Gillespie 2006; Gillespie *et al.* 2008; Greiner and McIntosh 2009; Leendertz *et al.* 2006; Unwin *et al.* 2011). With this chapter, we rather outline the importance of health assessment in wild primates, suggest how this information may be used for research and conservation purposes, and focus on topics that have been stressed less in previous publications. For the novice, this discussion may help you choose appropriate methods; for the expert, this discussion may help to expand your existing toolkit or point your research in a new direction by providing additional options for consideration.

4.2 Primate zoonoses and anthropozoonoses

The following material describes health assessment as it primarily relates to infectious organisms. This of course does not negate the importance of assessment of injury and chronic conditions, although they are beyond the scope of this paper. Infectious organisms include viruses (and bacteriophages), bacteria (including rickettsiae), parasitic protozoa and helminthes (nematodes, cestodes, and trematodes), and fungi. These parasitic organisms live all or part of their lives in or on a host from which biological necessities are derived. This state of metabolic

Primate Ecology and Conservation: A Handbook of Techniques. First Edition. Edited by Eleanor J. Sterling, Nora Bynum, and Mary E. Blair. © Oxford University Press 2013. Published 2013 by Oxford University Press.

dependence usually results in host energy loss, lowered survival, and reduced reproductive potential. Disease or illness is the impairment of host body function due to a pathogen.

There is remarkable variation in the transmission dynamics of infectious (communicable) organisms (Anderson and May 1992; Combes 2004; Poulin 2006). The primary infection transmission routes for wild animals include fecal–oral (ingestion of contaminated food, water, or other objects), respiratory, vector-borne (e.g., mosquitoes, ticks, flies, etc.), blood-borne, sexually-transmitted, and congenital. Zoonotic pathogens are those transmitted from non-human animals to humans; anthropozoonoses are those pathogens transmitted from humans to non-human animals. Primates can serve as the reservoir host for a number of zoonotic diseases (Chomel *et al.* 2007). Given that over half of all human infections are zoonotic in origin (Cleaveland *et al.* 2001; Woolhouse and Gaunt 2007), it is important to monitor the health of wild primate populations for our own health purposes.

For example, the current HIV pandemic, with an estimated global prevalence of 33 million people, appears to have originated from non-human primate simian immunodeficiency viruses (SIV). Through the hunting and butchering of chimpanzees (*Pan troglodytes troglodytes*), western lowland gorillas (*Gorilla gorilla gorilla*), and sooty mangabeys (*Cercocebus atys*) in West Africa (Gao *et al.* 1999; Santiago *et al.* 2005; Van Heuverswyn *et al.* 2006), SIV likely entered into the human population several times and became established as HIV around 1900 AD (Worobey *et al.* 2008).

Primates are hosts for a number of malaria species (Prugnolle *et al.* 2010), and likely serve as a reservoir for some human infection, particularly from *Plasmodium knowlesi* (Cox-Singh *et al.* 2008). Malaria is a mosquito-borne disease caused by protozoa of the genus *Plasmodium* (phylum Apicomplexa, suborder Haemosporidiidea, family Plasmodiidae), with 172 named species that parasitize reptiles, birds, and mammals (Garnham 1966; Coatney *et al.* 1971).

Some pathogens transmitted to humans via contact with wild primates can be particularly deadly. For example, Cercopithecine herpesvirus 1 (B virus) is common in wild and captive macaques (*Macaca* sp.). When actively shed from the macaque mucosal epithelia, transmission to humans can, unlike infection in the reservoir hosts, result in fatal encephalopathy (Huff and Barry 2003). Nearly all cases of Ebola in humans can be traced back to the handling or consumption of infected wildlife carcasses, particularly those of apes (Leroy *et al.* 2004). *Ebolavirus* infection has wiped out several non-human primate populations, particularly great apes in Gabon, Cameroon, and Democratic Republic of Congo over the past 20 years (Leroy *et al.* 2004).

Primates can also be highly susceptible to human pathogens (Kaur and Singh 2009), due in part to phylogenetic relatedness between human and non-human primates. Wild populations are usually immunologically naïve to human pathogens, and ape populations in particular can be quickly decimated because of slow reproductive rates. For example, outbreaks of polio, measles, anthrax, and respiratory pathogens have been responsible for major mortality in wild ape populations (Leendertz *et al.* 2006). Contact between human and wild primate populations is the likely cause of several of these epidemics, particularly that of respiratory syncytial virus and metapneumovirus in chimpanzees in Côte d'Ivoire (Köndgen *et al.* 2008), as well as intestinal pathogens *Giardia* and *E. coli* in mountain gorillas and chimpanzees in western Uganda (Graczyk *et al.* 2002; Goldberg *et al.* 2007; Rwego *et al.* 2008). Primates can be very susceptible to *Mycobacterium tuberculosis, M. bovis, M. avium,* and other bacteria (Burgos-Rodriquez 2011). Monitoring the health of primate populations is obviously important for conservation purposes and for human health concerns. But these wild populations can also offer unique opportunities for addressing interesting questions about the influence of host and ecological conditions on susceptibility to infection and complex transmission dynamics. In this case, a variety of epidemiological methods can be employed.

4.3 Methods for collection and analyses

4.3.1 Study design

Several introductory texts on infectious disease epidemiology are available: Friis (2009), Heymann (2008), Merrill (2009), and Rothman, K. J. *et al.* (2009). Epidemiology is the study of the distribution (frequency, time, hosts, locations, etc.), determinants (biological, behavioral, ecological, etc.), course (etiology, spread throughout a population, effects of control measures and interventions, etc.) and outcome (morbidity and mortality) of disease in a population. There are a number of epidemiological study designs, and the most common ones used to document primate health are observational, descriptive, cross-sectional ones. These "ecological" (or correlational) studies measure exposure and effect simultaneously, and document occurrence of disease by analyzing groups of hosts in different places or in a time series. This simple study design can lead to difficult interpretation of results (as there are many potential explanations for findings). And of course these studies are subject to bias (the "ecological fallacy") because associations observed between variables at the group level may not exist at the individual level (Freedman 1999). Studies of primate disease ecology should attempt to utilize more sophisticated epidemiological designs such as case-control analysis, in which the occurrence of disease is monitored longitudinally over time in exposed versus unexposed groups. Cohort studies could measure the same disease-free animals multiple times over a given period of time to observe development of disease in different groups. Relative risk of infection can be calculated by comparing rates of occurrence in exposed versus unexposed groups in cohort studies. But remember, the two groups ideally need to be comparable in all other ways.

Regardless of the study design, basic projects in primate disease ecology will begin with describing the *prevalence* of infection, which is the frequency of existing cases of disease in that population. These values are obviously influenced by a number of variables, including duration of disease, diagnostic abilities, and migration of hosts in and out of the population (so narrowly define your population at risk to the greatest extent possible). *Incidence* represents the number of new cases of disease in a specified population during a given period of time, which can be calculated if the host population is sampled multiple times across a given period of time. Finally, *richness* represents the number of unique pathogenic species that infects a given host at a given period of time. Richness is a measure typically reported in studies of primate disease ecology.

4.3.2 Permits

Researchers are usually required to obtain several different permits for animal sampling and specimen transport. Permit requirements vary by host and home countries and species of animal. Researchers must inquire about their specific requirements. Some more common issues include:

1. *Local (host country) permissions*: In most cases, foreign researchers must identify local counterparts and submit applications for research permits (also see Chapters 3 and 8). The local collaborators may also provide researchers with important letters of support, such as letters to customs officials describing how the researcher is traveling with necessary research supplies. Likewise, they might supply letters to airlines and shipping companies describing the context upon which the samples were collected, and confirming that they are diagnostic specimens for research purposes only.

2. *Ethical permissions*: Primate research most often requires adherence to institutional guidelines, like those imposed by research/educational organizations. In the US, these policies are usually enforced by local Institutional Animal Care and Use Committees, under the guidance of the US National Institutes of Health, Office of Laboratory Animal Welfare. Similar policies are enforced by the Academy of Medical Sciences, Royal Society, Medical Research Council and Wellcome Trust in the UK (Weatherall 2006).

Permits to ensure adherence to these guidelines are often required for release of funds from granting agencies.

3. *Exportation/importation of specimens*: Appendix I (dated 27 April 2011) of the Convention on International Trade in Endangered Species of Wild Fauna and Flora (CITES) lists 55 primate genera or species as endangered, and therefore requires both an export permit from the host country and an import permit from the final repository of samples. These import permits for the US are granted by the US Fish and Wildlife Service, International Affairs, Division of Management Authority. For the export permit, it is necessary to identify a CITES management authority (this is different from the scientific authority) officer in the host country and supply this person with information regarding species utilized, numbers and volumes of samples, and intended uses.

For any samples that may contain etiologic agents (including vectors of disease), a CDC Etiologic Agent Import Permit is required in the US. Currently, the US Fish and Wildlife Service does not require a permit to import animal feces, as they consider feces a byproduct, not a product or derivative. Similarly, the US Department of Agriculture, Animal and Plant Health Inspection Service, Veterinary Services, National Center for Import and Export does not require an import permit from non-human primate samples as long as they do not contain any potential or actual zoonotic pathogens. Therefore, samples that may contain such pathogens (including those discussed throughout this chapter) must be rendered inactivated, as through the use of preservatives. If the researcher is specifically diagnosing a biological agent determined to have the potential to pose a very severe threat to human or non-human animal health, they may also have to conform to Select Agent regulations in the US. Importation of soil (i.e., for microbial analysis) into the US is regulated by the US Department of Agriculture, Animal and Plant Health Inspection Service, Plant Protection and Quarantine program, and usually requires a permit.

When permits are not required for various reasons, it is still a good idea to obtain letters from the above-mentioned organizations stating that permits are not required. Permits and letters from the various organizations should be affixed to both the outside and inside of shipping containers.

4.3.3 Biosafety

Specimens from primates must be handled carefully. Researchers should be trained in both biological and chemical safety, and follow good general laboratory practices even in the field. While most investigators are careful to utilize personal protective equipment in the laboratory when working with primates (Morton *et al.* 2008), fewer are inclined to do so in the field. Exceptions, of course, include long-running projects like the Mountain Gorilla Veterinary Project that regularly monitors the health of their researchers (The Mountain Gorilla Veterinary Project Employee Health Group 2004).

Institutional Biosafety Committees may require applications for research involving animal handling or collection of biological specimens in general. Many of these regulations are set forth by the US Centers for Disease Control and Prevention's Office of Safety, Health and Environment and the World Health Organization's Collaborating Centre for Applied Biosafety Programmes and Training.

4.3.4 Sample collection

The pathogens or other measures of health you choose will be determined by your research question, which should be influenced by what you know about the natural history of your primate host. However, field conditions will determine the types of specimens you can collect.

4.3.4.1 Non-invasive samples

At a minimum, clinical signs like decreased activity, visible weight loss, overt injury, hair loss, and loss of appetite can be recorded. Feces, collected immediately following defecation, can be sealed in polypropylene tubes, kept on ice, and frozen back at camp. Fecal samples should be macroscopically inspected immediately, and consistency, the presence of blood, and the presence of cestodes or nematodes should be noted. Samples should be free of water, urine, soil, or other contaminants, and so should be collected from the center of the fecal mass. Feces can be mixed with RNAlater® for genetic analyses, or even desiccated for later hormone analyses (Muehlenbein 2006; Muehlenbein 2009). Fecal samples can be preserved in 10% neutral buffered formalin, polyvinyl alcohol, Schaudinn's fixative or SAF fixative for later parasitological analyses, or analyzed immediately using a direct wet smear to identify motile protozoa. Samples can also be placed in 10% glycerine or various transport media over the short term for bacterial culture. Several companies (e.g., Meridian Biosciences) provide excellent collection materials with reagents pre-aliquoted. Samples are usually preserved using a ratio of three parts preservative to one part feces (approximately 2 grams of feces). Tubes should be thoroughly labeled, sealed with ParaFilm, shaken to mix the contents well, and kept in a cool dry place until analysis.

Urine can be frozen, mixed with RNAlater®, and even spotted and dried onto filter paper (Simon & Schuster #903). Other non-invasively collected samples include saliva from food wadges, sperm plugs, and even blood found on foliage.

Ticks, fleas, mites, lice, mosquitoes, and leeches can be collected from nests or from the surrounding area using nets, suction devices, traps, combs, sheets, and sifters. Excellent resources on specific methodology can be obtained from various agencies such as the US Department of Agriculture, Agricultural Research Service, Insect and Mite Identification Service. Soil sampling for microbial identification is also described in detail by the US Department of Agriculture's National Soil Survey Center and other resources (Carter and Gregorich 2008).

4.3.4.2 Invasive samples

If it is feasible to anesthetize the animals, several samples should be taken. You can record weight and body temperature and note the presence of any injuries. Blood should be collected from the femoral, tibial, or brachial veins using a sterile technique and materials (like BD disposable needle sets and EDTA Vacutainer tubes). Note that there are limits on how much blood can be collected from an animal at any one time, and this is based on weight of the animal, the frequency of sampling, and your own Institutional Animal Care and Use Committee's regulations.

Blood can be frozen whole, mixed with RNAlater® (500 µl blood to 1.3 ml RNAlater®), or spotted and dried onto filter paper (Simon & Schuster #903). Serum can be separated via centrifugation and frozen. Thick and thin blood smears (films) should be made following techniques described in Houwen (2000). Blood for smears and spots can also be collected via finger or ear stick. Saliva can be collected and mixed with Qiagen's RNAprotect® saliva reagent (150 µl saliva to 1 ml reagent). Swabs can be desiccated or placed in a variety of bacteria or virus culture media. Tissues from necropsies (Travis 2009) can be preserved in ethanol, RNAlater® or 10% formalin. Ectoparasites can be collected from the host, killed using various liquid and solid substances inside killing jars, and preserved via freezing, desiccation, ethanol, or RNAlater®. In all cases, tubes and other containers should be properly labeled using a solvent-proof pen. The more analyses you want to make, the more sample aliquots you should take.

4.3.4.3 Transportation

Airlines have strict restrictions regarding liquids, flammables, and dangerous substances. Whereas some of us remember bringing blood samples back from the field in a cooler in the overhead compartment of an airplane, we are now largely restricted to shipping samples back or placing them in checked luggage (although proper packaging must be utilized). Dry ice and liquid nitrogen are usually not allowed on airplanes. Acceptable products include industrial-strength freezer packs (for −20 °C) and "dry shippers" which contain an absorbent material for keeping samples at cryogenic temperatures ($-150\ °C$). Researchers must bring these things with them, or have them available in the country (including liquid nitrogen for dry shippers). In any case, packaging conditions are outlined by the International Air Transport Association, and require appropriate primary and secondary containers and labels.

4.3.5 Sample analyses

Methodological standardization is necessary if results from different studies are to be even remotely comparable (Muehlenbein 2005). That said, comparisons will always be limited by an enormous range of inter-observer error when more than one laboratory is involved in the diagnostics.

4.3.5.1 Feces

If you have decided to analyze the fecal material for parasites yourself (but See Section 4.4), Ash and Orihel (1991) and Garcia (2007) are invaluable references. Proper identification of intestinal parasites usually requires the combination of different concentration techniques. Sedimentation via centrifugation helps to recover all oocysts, eggs, and larvae, and flotation procedures using high-specific gravity liquids allow the separation and detection of all but the heaviest larvae and eggs. The most common sedimentation techniques utilize ethyl acetate or ether, and work via centrifugation to extract fat and debris from the stool and concentrate the parasites in a lower sediment suspension. The most common flotation techniques utilize zinc sulfate, sugar, or sodium nitrate to cause lighter parasites to rise to the surface while debris and heavier parasites sink. Although others have suggested that zinc sulfate flotation and formal-ether concentration methods are inadequate for detecting primate parasites (Gillespie 2006; Gillespie et al. 2008), these methods have been used extensively in veterinary medicine. Furthermore, there has been no adequate comparison of these techniques using primate samples.

Examination of a permanent stained smear via oil immersion will also aid in the correct identification of a variety of intestinal protozoa. Examination of a permanent stained smear can be performed when it is convenient for the examiner, and the smears represent a permanent record of the parasite. Depending on the difficulty of the preparation and the amount of time available for staining, the examiner has a wide variety of staining techniques to choose from, including the trichrome stain, iron hematoxylin stain, and acid-fast stains. However, before the permanent stains can be performed, the smears must be prepared depending on how the fecal sample was preserved.

Different techniques can be used for quantification of helminth eggs including the Kato thick smear and the Stoll technique. However, correlating egg counts with the predicted number of parasites present is not very effective. Estimates of worm burdens are highly vulnerable to variability in fiber content, consistency, parasite factors, and host characteristics (Stear *et al.* 1995). Additionally, egg production by parasites also varies with the age of the parasite population, the presence of co-infections by other parasites, and the level of the parasite population.

Nematode infections may be diagnosed via specific culturing techniques, known as coproculture, that concentrate larvae that typically hatch in soil or tissue. Culturing of feces allows for the detection of larvae in feces when they cannot be detected via concentration methods or distinguished from other first-stage larvae. Culturing also allows for the development of filariform larvae, which aids in the differentiation process of diagnosing species (Garcia 2007). Common techniques include Harada–Mori filter paper strip culture and the Baermann procedure.

Whichever combination of methods you choose, the entire area of the coverslip on the slide should be examined with a low-intensity light and a low-power objective (10x). A higher objective (40x) should be used for identifying specimens. You should measure the size of the parasites using a calibrated ocular micrometer, and photograph questionable specimens if possible. Iodine or buffered methylene blue can be used to stain the specimens.

In addition to classical parasitology, other techniques like immunofluorescence (Kowalewski *et al.* 2011), pyrosequencing (Yildirim *et al.* 2010), and RT-PCR (Johnston *et al.* 2010) can now be employed on feces. These tests are more accurate and rapid, but also much more expensive. Use of these and other techniques have facilitated the identification of SIV (Santiago *et al.* 2003), malaria (Prugnolle *et al.* 2010), and metapneumovirus (Kaur *et al.* 2008) from primate fecal samples.

For bacteria, ideally, samples should be cultured immediately after collection on media appropriate for detecting the microbe of interest. This can be done in the field if refrigeration and incubation are available. Various culturing techniques are described in the US Food and Drug Administration's Bacteriological Analytical Manual, available free online. The shelf life of samples can be extended through the use of an appropriate transport medium, although such media only preserve the infectivity of bacterial cells for time periods on the order of days to a week. Freezing samples for future analysis is possible, but this likely reduces infectivity, even in the presence of such preservatives as glycerol. In situations where infectivity cannot be preserved, molecular detection methods can be used. These are highly sensitive but preclude further analysis of the biological properties of cultured organisms. Techniques include multilocus sequence typing, and restriction fragment length polymorphism analysis of PCR-amplified genes encoding 16S rRNA, allowing researchers to characterize the epidemiological relationships between pathogens (Goldberg 2003; Goldberg *et al.* 2006).

4.3.5.2 Blood

Thick and thin blood smears (films), usually stained with Giemsa, are helpful in diagnosing protozoan parasites, like malarias, babesias, and trypanosomes, based on morphologic features. A minimum of 200 microscopic fields of the thin smear (at 100x magnification) and 100 fields of the thick smear should be evaluated. If blood parasites are suspected but still not found in the thick or thin films, blood samples can be concentrated using several techniques such as Knott concentration, membrane filtration, and density gradient centrifugation.

Hematocrit, packed cell volume, mean corpuscle volume, hemoglobin, and erythrocyte sedimentation rate can all be determined using field-friendly supplies. Red blood cells can be lysed, the peripheral blood mononuclear cells separated and fixed in formaldehyde, and a white blood cell differential performed (much more accurately than via microscopy) using flow cytometry, but samples must be analyzed within a very short timeframe of days. However, physiological ranges for normal blood parameters in many primate species are available through the International Species Information System.

Antibodies against specific infections can be determined in whole blood and serum using enzyme immunoassay and western blots (Goldberg *et al.* 2009; Khan *et al.* 2006). Blood samples can be used on microarrays or "virus chips" (e.g., Greene chip) as well as with mass spectrometry (Abbott's PLEX-ID and other systems by AnagnosTec and bioMerieux) for casting a wide diagnostic net.

PCR has the ability to detect very low levels of parasitemia. Additionally, PCR represents a diagnostic test that can identify parasites to the species level, even in mixed infections. This is particularly the case with primate malarias. Whereas a good morphological characterization is always desirable, malarial parasites are identified using few morphological characteristics, and even experts must take the host into consideration as a context to interpret such traits. The gene encoding the mitochondrial Cytochrome b (cytb) is now frequently sequenced for identification of closely related malaria species (Escalante *et al.* 1998).

PCR inhibitors, genotypic variants, and DNA degradation can affect PCR results (see Chapter 14 for more details). As PCR techniques become more utilized in the future, researchers must keep in mind the sensitivity of PCR detection, which can be influenced by the collection mode and storage conditions of blood samples. These factors are especially important in cases of low parasitemia, mixed infections, and when comparing data from multiple laboratory and field settings

(Farnert *et al.* 1999). Avoiding sample contamination is another reason to wear personal protective equipment when collecting samples.

4.3.5.3 Rapid tests

There are a number of field-friendly rapid tests for parasites, bacteria, and other indicators of infection status. These include urinary dipsticks (e.g., Bayer Multistix for assessment of leucocytes, ketones, glucose, and other factors in urine), rapid culture supplies for coliform analysis of water (e.g., ColiComplete from BioControl Systems and ColiScreen from Hardy Diagnostics), and various lateral flow immune-chromatographic antigen-detection tests (e.g., PrimaTB STAT-PAK assay from ChemBio Diagnostic Systems, VetScan Giardia rapid test from Abaxis, ImmunoCard Stat! products from Meridian Bioscience, and the Malaria Pan/Pv/Pf test by Core Diagnostics). Before employing these tests, it is important to consider whether they can in fact detect your organism of interest (i.e., they may not detect the species of parasite that is commonly found in your host population), and the sensitivity of the test, as many of these tests may have unacceptably high rates of producing false negatives.

4.4 Some modest advice

Experience allows us the luxury of hindsight. And many publications in primate disease ecology suffer from a number of methodological flaws. Here we offer some modest advice we hope will be helpful, so that someone might also learn from our mistakes.

1. *Find an expert*: Just as tests must be reliable (precise, repeatable), valid (accurate; measures what it is intended to measure), sensitive (ability to correctly identify those with disease), and specific (ability to correctly identify those who are healthy), those conducting the analyses must too be proven. Measurement error is caused not only by the use of inappropriate techniques, but also by untrained individuals. No-one is qualified to conduct the majority of the tests described above just because they have read a book on parasitology, or have taken a course in parasitology. New researchers should find an expert with an advanced degree in tropical medicine if they are going to use parasitological techniques.
2. *Involve local colleagues beyond sample collection*: Local colleagues would benefit from being trained in whatever techniques they may be able to perform in the host country. Not all samples need always be exported from a host country to a foreign lab; in many cases, local laboratories exist or can be

developed (Cheesbrough 2005). Contributing to the education and economic development of our host countries is vital to sustained collaborations.

3. *Choose the research questions before the research methods*: Too many students become enamored with some romantic topics like parasitology or endocrinology. They decide they want to use these methods, sometimes for the wrong reasons. They then work backwards to identify a research question so that they can use these methods. In most of these cases, the research questions turn out to be mediocre. Choose an important question, generate a list of predictions and potential confounding variables, and then seek advice about the best methods to evaluate these questions. It may be the case that you should be using a measure of animal health rather than infection status (as the two are not the same). Additionally, conducting a pilot project will minimize the potential for serious problems with techniques after the data are collected.

4. *Do not collect samples randomly from unknown individuals*: This will produce an inaccurate measure of prevalence by basing analyses on parasites per sample instead of parasites per animal. Huffman *et al.* (1997) demonstrated that significant reporting bias is created when one uses the number of samples instead of the number of individuals to calculate the prevalence of parasitic infection in a population. Furthermore, fecal samples are not randomly distributed in the environment, as ill individuals defecate more frequently than healthy hosts. Random collection of feces would then result in sampling bias (i.e., some animals sampled more than others). This is particularly problematic if your pathogen of interest is not evenly distributed throughout the host population, which is usually the case. Samples should be collected from known animals immediately following defecation, and prevalence estimates should be based on number of positive animals, not number of positive samples. It is possible to genotype the samples in order to determine sample identity, although this is an extremely expensive option.

5. *Multiple samples should be collected from each animal across time*: Intestinal parasites are shed intermittently. Therefore a minimum of three samples, collected on non-consecutive days, is necessary for accurate diagnosis of intestinal infection. For example, Muehlenbein (2005) demonstrated that not one of the twelve parasitic species recovered from the chimpanzee Ngogo group in Kibale National Park was found in *all* samples from any one animal, and the most commonly occurring parasites were found in all of the serial samples of only a fraction of the chimpanzees sampled. Cumulative parasitic species richness for the chimpanzee hosts significantly increased for every sequential sample (up to four samples) taken per animal during this study.

While a single sample may be adequate for determining antibody levels indicative of past infection, current infection usually requires multiple samples.

6. *Try to sample as much of the population as possible*: Sampling error can happen when only one group of animals is sampled, and if that group of animals is not representative of the entire population (perhaps because they are better habituated than others). Furthermore, sample size must be large enough to have statistical power to detect important differences. We recommend using the free software G*Power (Erdfelder *et al.* 1996). Only a handful of published studies on primate disease ecology have ever reported results of their sample size or power analyses.

7. *Try to sample across seasons*: There is likely seasonal variation in pathogen transmission in wild primates (Gillespie *et al.* 2010; Huffman *et al.* 1997). And many factors such as rainfall, temperature, sex, age, reproductive state, body condition, dominance rank, host population density, and daily range could influence these transmission dynamics. It is important to try to account for as many of the variables that may influence susceptibility to pathogens as possible as well as contact with them.

8. *Be conservative in your interpretations*: It is difficult to determine whether observed associations are causal, as the guidelines for judging causation are difficult to establish in wild primate studies. Among other considerations for causation, the cause must precede the effect, there must be a strong relationship between the cause and effect (the cause should covary with the effect), the results must be consistent with other studies, and the mechanism of action must be plausible (Rothman 1988). In field studies involving primates, it is usually impossible to demonstrate that increased exposure is associated with increased effect, nor to illustrate that removal of the cause leads to reduction of disease risk. There can be much confounding (when another factor is associated with both exposure and disease, and it is not accounted for in analyses and hypothesis generation) in ecological studies guided by inductive reasoning.

9. *Realize that not all infections are virulent in primates*: We need to know how naturally occurring infections impact the survivorship and reproduction of wild primates. Morbidity and mortality is to be expected under some, but not all, conditions. Some animals may be resistant to infection, while most may develop a tolerance for chronic infection. Many infections are thought to impair nutrient absorption, cause anemia, and influence energy expenditure. This is certainly the case for some infections like *Oesophagostomum*. Low levels of chronic infection matched with a relative absence of clinical

signs most likely indicate that many infections are well tolerated. This is probably the case for most intestinal parasites of wild primates. These infections may only cause overt disease under certain circumstances, such as following seasonal fluctuations in food or during breeding seasons. Too many recent publications assume that these intestinal infections cause high levels of morbidity, when in fact there is little evidence for this assumption at this time. Research should be focused on determining the effects of parasitism on primate host morbidity and mortality. Research should also be focusing more on pathogenic viral and bacterial infection rather than basic intestinal parasitism, although the latter will continue to be important.

10. *Consider measuring immune function*: Nearly all measures of primate health have been limited to determination of exposure (i.e., baseline parasite levels). It is feasible under certain circumstances (usually involving collection of invasive samples) to go beyond reporting parasitemia and signs of illness, to include immunological measures such as immunophenotyping (e.g., white blood cell differential), lymphocyte proliferation (ability of T and B cells to undergo mitosis *in vitro* following exposure to a mitogen), hemolytic complement activity of serum, *ex vivo* bacteria killing abilities of blood and urine, antibody and cytokine levels in blood and urine via enzyme immunoassay, multiplexing, western blot, immunofluorescence, or haemagglutination. Gene expression of immune effector pathways can be determined using quantitative PCR or microarrays. Concentration and activity of complement proteins, lysozyme, antimicrobial peptides, NK cells, nitric oxide, and macrophage phagocytosis can all be determined using a variety of assays. Which assays you choose should depend on your model system and availability of blood. Many of these assays represent functional, integrated, biologically relevant measures of different immune pathways (Boughton *et al.* 2011; Demas *et al.* 2011).

But remember, baseline immune measures in wild animals do not necessarily represent a "disease absent" state, as some upregulation of immunity due to the ubiquitous presence of parasites is to be assumed. Immune responses are influenced by more than just infection, including factors such as reproductive state, energy balance, stress, and season. Furthermore, it is advised not to rely solely on a single immune measure, as this cannot adequately represent the functioning of an entire system. Multiple measures would ideally be combined with assessment of animal infection status.

4.5 Going beyond basic health monitoring

Primate epidemiology should attempt to go past basic health monitoring to answer questions regarding risk assessment (i.e., likelihood of changes in disease patterns; Travis *et al.* 2006) and most importantly the causal factors of changes in disease dynamics within and between populations.

4.5.1 Behavioral ecology and disease risk

Population density, mating system, geographic range size, body mass, terrestrial substrate use, diet, and group size are all likely related to risk of infection in complex ways in wild primates. In general, arboreal primates may have fewer parasites than terrestrial ones due to less potential contact with feces and soil containing parasites (Muehlenbein *et al.* 2003). Howler monkeys may defecate in the peripheral areas of the canopy in order to avoid contamination of food with intestinal parasites (Gilbert 1997; Henry and Winkler 2001). Nest building and alteration of sleeping sites may reduce parasite exposure (including infection from malaria via mosquitos) in primate hosts (Hausfater and Meade 1982). And chimpanzees and baboons may dig wells to specifically filter water and decrease the likelihood of water-borne infections (Galat *et al.* 2008).

Pathogens may influence primate sociality, and social behaviors can influence risk of infection (Nunn and Altizer 2006). If contact or proximity occurs frequently among individuals in dense populations and elevates the rate of disease transmission, then higher levels of gregarious behavior or sociality should result in greater infections within social systems (Altizer *et al.* 2003). Parasites may influence primates to emigrate out of a larger group in search of a smaller one, or dominant individuals may force subordinate animals to leave larger groups (Freeland 1976). Furthermore, if pressure from pathogens increases emigration from a primate social group, then subordinate, younger, or elderly individuals, as well as the dispersing sex, are likely to experience the most pressure (Nunn and Altizer 2004).

Primates could potentially reduce their risk of disease by avoiding group members displaying visible cues of infection, including physical, behavioral, and olfactory signs of infection. This could influence grooming patterns and distances between individuals. An increase in territoriality might result in greater use of a group's home range, which may lead to exposure (or re-infection) of pathogens (Ezenwa 2004; Stoner 1996). Despite these interesting predictions, there is little evidence to definitively support these ideas. Testing these predictions in wild primates will prove difficult, but certainly rewarding.

4.5.2 Modeling infection transmission in primate populations

Traditional models of susceptible, infected, and recovered hosts in a fixed population underestimate the complexity of host ecology, transmission dynamics, pathogen virulence, spread of resistance/tolerance, and much more (Brauer and Castillo-Chavez 2001). In addition to detailed comparative analyses (see Nunn and Altizer 2006), agent-based modeling is being more frequently used to model these complex ecological systems (Nunn 2009). These models attempt to account for spatial structure as well as interaction patterns of individuals (also see Chapter 9). For example, dominance rank may influence the likelihood of infection transmission due to variation in diet, contact between animals, and physiological stress (Muehlenbein and Watts 2010).

A variety of computer programs are available for agent-based modeling depending on the programmer's level of experience. Software that is most often used for the social sciences includes LSD, MAML, MAS-SOC, Repast, NetLogo, FAMOJA, SimBioSys, StarLogo, Sugarscape, and VSEit. The primary step in constructing an agent-based disease model is building the host population. Here, the researcher needs to decide group composition, size, structure, dispersal, sex, and spatial organization (Nunn 2009). Decisions need to be made regarding the number of neighbors the host population may have, how often they come into contact, and if there are any geographical or social barriers that may exist between the two. The researcher needs to consider life history variables, such as mortality: Which agents die of natural causes and how many? Which agents die from the disease? These types of questions are important, since those killed by natural causes most likely represent a population with a constant size, while populations with a large number of agents that die from disease most likely have a decreasing population size; any changes in population size can affect subsequent spread of disease as the model progresses. Additionally, the researcher needs to make decisions about which variables will fluctuate and those that will remain static, as in models demonstrating the evolution of certain traits. Taken together, these parameters provide vital clues as to how a disease might spread in real time populations.

4.5.3 Assessing the impacts of anthropogenic change on primate health

Increased contact between humans and wild primates will likely lead to changes in infection profiles in all populations concerned. Given the rapid nature of anthropogenic modifications of our physical environment, the likelihood of pathogen transmission to wild primate populations in the future is significant. Human and livestock populations continue to grow rapidly, increasing the number of hosts potentially susceptible to novel infections. Mass transportation of people, products, livestock, and vectors of disease bring each of these closer to one another and more quickly at that. Population movements due to war, social disruption, and rural-to-urban migration, in addition to general urbanization, increase the densities of non-immune human hosts and pose significant sanitation problems. Changes in water usage, such as during the construction of dams, culverts, and irrigation systems, can increase the potential breeding sites of vector species like mosquitoes and snails. Human encroachment into previously undisturbed areas increases remote area accessibility and introduces more vectors and reservoirs of infection to new hosts. Encroachment, extensification of agricultural land, and urban sprawl all alter population densities and distributions of wildlife, which changes disease dynamics. Forest fragmentation can produce an "edge-effect," increasing the flow of organisms across ecotones, novel species contact, and the likelihood of infection transmission between organismal populations. Biodiversity loss due to global climate change, deforestation, the spread of invasive species, overexploitation, and other causes increases the likelihood of cross-species transmission. Increased inbreeding and decreased genetic diversity in remaining wildlife populations could even facilitate further outbreaks due to impaired immune functions in host animals.

Forest fragmentation has been associated with a variety of infection patterns in wild primates (Gillespie and Chapman 2006; Gillespie and Chapman 2008; Goldberg *et al.* 2008). In these and other examples, it remains to be seen whether or not changes in infection patterns are due to alterations in host susceptibility a the result of altered distribution of nutritional resources, altered transmission dynamics (such as increased contact with humans and livestock), or some other factor. Forest habitats may differ in complex ways other than just by degree of fragmentation, including amount of human contact, population density of hosts, fruiting and flowering schedules, amount of ground water, and so on.

Recreational use of natural areas can also affect the risks of zoonotic and anthropozoonotic pathogen transmission. Habituation of animals to human presence can increase the likelihood that animals will actively seek out contact with humans, particularly in the form of crop raiding and invasion of garbage pits and latrines. Despite the fact that primate ecotourism is increasingly perceived as a venue for promoting awareness about conservation issues, tourist–wildlife contact has the potential of producing devastating health and economic outcomes (Muehlenbein and Ancrenaz 2009). Understanding the risks of pathogen transmission from tourists to wildlife is a necessary, but overlooked, aspect of wildlife conservation; information regarding risk of transmission in these different locations could help to optimize responsible ecotourism for animal well-being and economic development.

The relative contribution of tourists to the spread of pathogens to wildlife is largely unknown, but the number of tourists visiting wildlife sanctuaries worldwide is increasing dramatically. The majority of tourists who visit wildlife sanctuaries arguably underestimate their own risk of infection, as well as their potential contribution to the spread of diseases themselves. Using the largest survey ever of ecotourist health behaviors at Asia's most frequented wildlife tourism destination, the Sepilok Orangutan Rehabilitation Centre in Sabah, Malaysia, we have demonstrated relatively low levels of vaccination in visitors (Muehlenbein *et al.* 2008). Many tourists had animal contact immediately before coming to Sepilok, and many of them had at least some basic knowledge about infection transmission (i.e., medical-related occupation) (Muehlenbein *et al.* 2010). Despite their interests in environmental protection, these ecotourists very likely create unnecessary risk of infection transmission to wildlife; despite the fact that 96% of respondents believe humans can give disease to wild animals, 35% of them would still try to touch a wild monkey or ape if they had the opportunity (Muehlenbein 2010). Clearly this problem deserves more attention, specifically by comparing risk factors at primate-based ecotourism destinations that differ significantly by visitor characteristic and degree/type of human–wildlife interaction. The tools described above may help equip researchers working with wild primates to begin to address this and other related research questions, including those which can contribute immediately and directly to primate conservation.

Acknowledgments

Mary Blair, Nora Bynum, Eleanor Sterling, Tony Goldberg, Ananias Escalante, and one anonymous reviewer contributed to the improvement of this manuscript.

5

Behavior within groups

Beth A. Kaplin and Apollinaire William

5.1 Introduction

Almost all non-human primate species live in some form of a social group. The main factors believed to be responsible for the evolution of group living include predation pressure (van Schaik and van Hooff 1983; Dunbar 1988; Janson 2000), food competition and access to resources (Krebs and Davies 1993; Janson 2000), and avoidance of male harassment and infanticide (e.g., van Schaik and Kappeler 1997; Wrangham 1979; Janson 2000). Individual differences in physiology and morphology related to body size, age, sex, or reproductive status can create challenges for social living, demanding that individuals coordinate their behaviors and activities to accommodate one another (Conradt and Roper 2005). The need to resolve conflicts within a group that arise between individuals with differing needs and constraints can lead to the emergence of complex patterns such as coordinated movements, fission–fusion dynamics, or specific communication signals (Conradt and Roper 2005; Sueur *et al.* 2010). The study of behavior within primate groups is focused on understanding the evolution of social living and the mechanisms behind these observed patterns, including maintenance of group cohesion, social organization and structure, communication, predation avoidance, culture, and reproductive behavior. In addition, studies of habitat use and ranging behavior, foraging behavior, diet composition, and activity budgets help us understand patterns in primate social systems.

In this chapter, we review approaches to the study of behavior within groups, including a discussion of techniques and methodologies. We begin with a discussion of what is a primate group, including the costs and benefits of group living, and review the diversity of social systems found among primates. We describe the process of habituating a primate group in order to study behavior. This is followed by a discussion of sampling methods. We then discuss some of the types of research questions posed in the study of behavior within a group of primates. We also

Primate Ecology and Conservation: A Handbook of Techniques. First Edition. Edited by Eleanor J. Sterling, Nora Bynum, and Mary E. Blair. © Oxford University Press 2013. Published 2013 by Oxford University Press.

review two exciting approaches and techniques that complement within-group behavior studies: GIS and network analysis. Lastly, we discuss the conservation of primates and how within-group behavioral information can contribute to conservation efforts.

5.2 What is a primate group?

A primate *group* (also sometimes referred to as a *troop*) refers to a relatively stable association of individuals ranging in number (from 3 or 4 to 30 or more), which is relatively cohesive in time and space. These groups are in contrast to very large associations of individuals of the same species that periodically form subgroups or small parties in a fission–fusion system, or some form of multilevel social system composed of subgroups. A large, often less cohesive group is called a *community*, especially in reference to chimpanzees (Goodall 1973). The term "community" used in this way can be confusing, as it is not used in the ecological sense of community, which refers to the interactions of populations of different species within a particular ecosystem. A primate community usually forms temporary subgroups in a fission–fusion dynamic with each subgroup often called a *"party"* (Pepper *et al.* 1999; Lehmann *et al.* 2007a). Fission–fusion systems are highly flexible and can respond quickly to environmental changes with fluctuations possible from day to day or even hour to hour (Asensio *et al.* 2008), possibly due to group member responses to food availability and efforts to reduce competition or aggression (Chapman 1988; Lehmann *et al.* 2007a; Di Fiore *et al.* 2011).

Researchers have also identified *multilevel societies* composed of consecutive levels of social groupings. At the most basic level, these are usually one-male groups, which cluster into increasingly higher levels of organization. Depending on species, one-male units may cluster into *clans* in which social interactions occur more frequently within a clan than between clans, and clans can be organized into *bands*, then *troops* or *herds*, the highest level (Stammbach 1986). Some examples of multilevel societies include snub-nosed monkeys (*Rhinopithecus* spp.; Ren *et al.* 2012) that form large (>100 individuals) highly cohesive groups with the behavioral flexibility to form subgroups when high-quality food resources are scarce. Geladas (*Theropithecus gelada*) and hamadryas (*Papio hamadryas hamadryas*) also form large, non-cohesive, multilevel social systems (troops) at sleeping sites that disaggregate into *bands* which are spatially cohesive during the day (Kummer and Kurt 1963; Kummer 1968; Swedell 2002). Mandrills (*Mandrillus sphinx*) congregate in very large, stable aggregations referred to as troops or herds; extremely large congregations (>500) have been called *hordes* or *supergroups* (Rogers *et al.* 1996;

Abernethy *et al.* 2002), which still appear to be fairly cohesive and stable over time (Abernethy *et al.* 2002).

Primate groups can be described in terms of their *social organization, mating system*, and *social structure* (Sterling 1993b; Kappeler and van Schaik 2002). The first of these attributes, social organization, is typically described in terms of *group size*, spatio-temporal *cohesion*, and *sex ratios*. While the benefits of group living include an increased ability to detect and defend against predators, and an enhanced ability to locate and guard food resources, costs involve increased conspicuousness to predators, and intra-group competition for food (Terborgh and Janson 1986). These costs and benefits lead to trade-offs that help determine *group size* which varies from 2 individuals in some species to more than 300 for black and white colobus (*Colobus angolensis ruwenzorii*; Fashing *et al.* 2007b) and up to 845 documented for mandrills (*Mandrillus sphinx*) in Gabon (Abernethy *et al.* 2002).

The distribution of food resources and predation pressure are the two most-studied factors influencing group size (e.g., Chapman *et al.* 1995; Janson and Goldsmith 1995), but cognitive abilities have also been proposed as a constraint on size in socially bonded groups, as well as the time available to maintain individual social bonds (Dunbar 1992). The cognitive abilities hypothesis comes from the finding that brain size (neocortex size relative to total brain size) is highly correlated with group size and this is assumed to limit the number of social relationships an individual can maintain; if group size becomes too large, an individual can no longer maintain close social bonds with all group members (Dunbar 1991; Lehmann *et al.* 2007b).

Group cohesion, another parameter of social organization, refers to the degree to which individual group members are associated with one another in time and space, and varies within and among primate species, influenced by factors including availability of food, predator threats, and reproductive requirements. Cohesion is maintained by behavioral attributes including social grooming (Lehmann *et al.* 2007b) and vocalization (Snowdon 1993), and is especially important in socially bonded groups. Groups may be relatively cohesive, meaning that the individuals spend the majority of their time and space together, or relatively non-cohesive, meaning that they exhibit temporal and spatial variation in cohesion (Kappeler and van Schaik 2002). Temporal variation is manifested in fission–fusion groups, where subgroups of varying size and composition form temporarily (van Schaik and van Hooff 1983; Strier 1992), as well as in multilevel societies where small breeding units usually composed of a male and one or more females and offspring, and small social subgroups, comprise the social group (Rubenstein and Hack 2004). A group that can form frequently changing subgroups is thought to be

better able to regulate feeding competition, maintain flexibility in resource exploitation, or allow males to maximize monitoring of reproductive females (Dunbar
1988; Lehmann *et al.* 2007a). The exact social and ecological factors that encourage fission–fusion and subgroup behavior in primates with typically large cohesive
groups are not fully understood and this is an area where more research is needed,
although many hypotheses exist (Ren *et al.* 2012).

The diversity of *sex ratios* or *sexual composition* in primate groups is exceptionally
wide, ranging from adult male–female pairs to multi-male–multi-female groups
(Kappeler and van Schaik 2002). Groups of multiple individuals may have a single
adult male (e.g., a dominant breeding male with a multi-female harem) or be
composed of multiple adult males. Some species form all-male groups, including
mountain gorillas (*Gorilla beringei beringei*; Robbins 1996), patas monkeys (*Erythrocebus patas*; Ohsawa 2003), and other old world primates. The reasons why all-
male primate groups tend to be rare remains unclear, as do details about relationships within these groups (Kappeler 2000), and insights from field research would
be beneficial here. Another interesting configuration is found in some callitrichids,
where some groups are composed of a single breeding female with multiple males
(Strier 2003) and some of these males exhibit helping behavior, or care of offspring
(Goldizen 2006).

Our understanding of social organization among nocturnal primate species is
rudimentary by comparison with diurnal species. Nocturnal primates are difficult
to observe and they are often detected solo, but research during the last 15–20
years has documented complex social organizations in nocturnal primate species,
which were previously thought to be solitary with only rare social interactions
(Sterling 1993b; Muller and Thalmann 2000; Weidt *et al.* 2004). The lack of an
observed cohesive structure may not automatically imply a lack of complex social
networks in nocturnal primates (Charles-Dominique 1978; Bearder 1999) and this
is another area in need of further investigation.

5.3 The habituation process

Primates are among the most amenable of vertebrates to field studies: they are
generally large and can usually be readily habituated to close observation, often to
the point that the observer can be spatially integrated into the study group,
allowing for detailed observations (Terborgh and Janson 1986). It is difficult to
study primates in the wild without some degree of habituation, and there are limits
to what questions one can ask with unhabituated primates. The goal of habituation
is to have the animals comfortable with the researcher's presence so that the
researcher can observe behaviors and collect data, while the animals are aware of

the researcher but continue their behaviors and activities, tolerating or ignoring the presence of the human observer. Williamson and Feistner (2011) present a thorough discussion of the specific steps in the habituation process including use of trail system, techniques for finding primates during the habituation process (e.g., auditory clues and waiting at key sites), and suggestions for ways to approach the study animals non-intrusively. The authors also discuss ethical issues surrounding habituation, which include, but are not limited to, the potential for significant long-term changes to normal activity and behavior patterns, increased disease risks due to the contact with humans, and increased vulnerability to hunters and poachers due to loss of fear of humans.

The duration required to habituate a primate group for a field study varies, depending on the species' natural history and its previous exposure to humans. For instance, primates exposed to poaching or human harassment will take longer to habituate (reviewed in Williamson and Fiestner 2011). Pepper *et al.* (1999) reported that after 1 month some individuals in a chimpanzee community that had been studied already for approximately 20 years allowed observers to follow them within 15 meters along the ground, and after 4 months almost all adults and adolescent males and a few females tolerated observers within ≤ 15 meters. On the other hand, it took one year of contact with researchers before Hamadryas baboons (*Papio hamadryas*) could be approached to within 60 m during daily movement, and two years before researchers could walk among them (Kummer 1995). Primates that spend significant amounts of time on the ground in dense understory (e.g., *Cercopithecus hamlyni*, B. A. Kaplin, pers. ob.; western lowland gorillas, Remis 1997) will take longer to habituate or may never fully habituate to human presence.

Nocturnal primates present an interesting case. Most nocturnal primates (e.g., bushbabies, *Galago* spp.; sportive lemurs, *Lepilemur* spp.; woolly lemurs, *Avahi* spp.; and dwarf lemurs, *Cheirogaleus* spp.) with the exception of the aye-aye (*Daubentonia madagascariensis*), readily habituate to observers (Williamson and Feistner 2011). However, nocturnal species have remained largely unknown to science until fairly recently. This has mainly been due to the challenges of locating and observing the animals in low light conditions, as well as the cryptic behavior of some of the species. Advances in technology, particularly radio telemetry, have greatly facilitated field studies of nocturnal primates (Sterling *et al.* 2000). For example, to obtain data on sleeping site behavioral ecology of the nocturnal golden-brown mouse lemur (*Microcebus ravelobensis*) Thoren *et al.* (2010) captured individuals using Sherman live traps and attached radio collars to facilitate locating individuals during focal observations with a telemetry receiver.

Many research questions require that individuals in a group be identified during observation and data collection and this is an additional level of detail that may take time to achieve. Subcutaneously injected transponders (e.g., Trovan Small Animal Marking System, Telinjects) can facilitate individual identification, which requires trapping to insert a unique transponder into each captured individual (see Chapter 3). Other options include paint, hair dye, haircuts, ear tags, or collars. For some research questions age-sex class identification is adequate and individual identification is not necessary, although this will not affect the degree of habituation needed to be able to observe individuals well enough to classify age-sex category and record behaviors.

5.4 Techniques and approaches in sampling behavior within primate groups

The quality of a behavioral study will depend critically on the sampling methods used. There are several techniques available to sample and quantify animal behavior, and since Altmann's (1974) foundational paper reviewing these techniques, some studies have presented useful comparisons and evaluations of sampling methods (e.g., Bernstein 1991). Here we present an overview of the standard sampling approaches.

Focal sampling refers to the recording of behaviors of interest while observing a focal individual. Focal animal sampling allows for a record of all acts for which the focal animal is the actor or the receiver. Typically the observer will create a protocol for the selection of the focal individual for each sample period to avoid biases; the assignment of individuals to sampling periods and the scheduling of sampling periods will depend on the question being asked (Altmann 1974). When individuals cannot be identified, the focal animal should be chosen at random from the visible animals in the group. The focal animal can be sampled during a predetermined period of time, or until it moves out of view. Sample session duration should be set at the upper limit by observer fatigue—when sample sessions are too long the observer's attention and focus may begin to wander, jeopardizing accuracy of the data. If duration of behaviors is the main interest in the research, sample sessions need to be long enough to allow for adequate sampling of the distribution of durations (Altmann 1974).

Focal data may be recorded using continuous focal sampling or focal interval sampling. *Continuous focal sampling* (also called continuous time sampling) uses a set time period to follow and observe the focal animal and record all occurrences of behaviors together with information about the time and duration of occurrence.

This type of sampling gives an estimate of frequencies, durations, and latencies of behaviors. *Focal interval sampling* also uses a set time period within which the observer records the focal animal's predominant behavior at the start of each session and at set time intervals, usually ranging from every minute to every 3 or 5 minutes until the set time period ends (e.g., Rose 2000). This sampling yields estimates of percentage of time spent in various behaviors.

Focal animal sampling is usually preferable to other sampling techniques because it provides relatively unbiased data applicable to a wide variety of behavioral research questions (Altmann 1974). When the collection of focal animal data is a challenge due to thick vegetation, when the study animals are not fully habituated to human observers, or when trying to observe fast-moving arboreal animals, instantaneous scan sampling may be preferable (Gilby *et al.* 2010). *Instantaneous scan sampling* uses the group as the unit of observation rather than the individual as in focal sampling, and involves the observer recording the predominant activity of each individual in view during preselected time periods, or scans (e.g., every 5 minutes throughout the day). The first behavior of each individual in view at the start of the scan should be noted, and each individual in the group is scored only once in each scan. Every effort should be made to view as many individuals as possible within the group for the same brief or instantaneous amount of time during each scan, avoiding the temptation to focus on readily visible animals doing behaviors that attract the observer's attention. If the interval between consecutive scan samples is short, the resulting data are essentially equivalent to rate and frequency estimates derived from focal animal sampling but are not accurate for estimates of duration (Altmann 1974).

In scan sampling conducted at constant intervals, the percentage of intervals in which a state (such as resting) occurs is a good estimate of the amount of time spent resting only if each scan is independent. However, Gilby *et al.* (2010) pointed out that scan sampling may overestimate the total time an average individual spends feeding, and could under-represent individual food intake, thus missing intra-specific variation in feeding behavior (such as variation associated with age or sex class) while not representing a meaningful average (across ages or sexes) of activities within the group. Independence of consecutive scan samples for statistical purposes should also be considered. To ensure that the activities scored in one scan were not dependent on those scored 15 min previously, Poulsen *et al.* (2001) statistically tested the monthly proportion of scan samples that each activity was observed in using 15- and 60-min scan intervals and found that for all activities except resting 15 min intervals maintained independence (i.e., the study species almost never maintained an activity for more than 60 min except for resting). If the researcher cannot ensure that individual scans are independent of one another, scans should be combined into monthly units or daily time blocks (i.e., 6–9am, 9am–12pm, 12–3pm, 3–6pm).

All-occurrence sampling involves recording all instances of certain behaviors by all individuals in a group during each observation period. This type of sampling is best when observation conditions are excellent and the behaviors to be recorded are easy to detect and do not occur so frequently as to make them hard to record. The results provide a frequency record or rate of occurrence of a behavior. Data from all-occurrence sampling can provide an accurate estimate of the rate of occurrence of behaviors in the group as a whole, for example the frequency of all agonistic vocalizations of a group within a particular time period, or the frequency of sexual mounting between adults (Altmann 1974). However, Bernstein (1991) pointed out that it is difficult to sample brief or subtle responses accurately with all occurrence sampling, and the larger the group or space occupied by the group, the less reliable the results.

One-zero sampling, also called time sampling, involves scoring occurrence and non-occurrence of an action during a sample period. The observer makes a single entry during a predetermined time period if any instance of the behavior category is seen during that time period, regardless of its duration or the number of times it occurs (Bernstein 1991). If the same state persists into the next time period, another entry is made. For each individual or class of individuals the number of sample intervals and the number or percentage of intervals that included an occurrence of the behavior of interest can be calculated (Altmann 1974). One-zero sampling thus represents the frequency of intervals in which a behavior occurred in any amount of time, not the frequency of that behavior. Although Altmann (1974) suggested that alternative methods be used for sampling frequency and duration of behaviors and percentage of time spent in various activities, Bernstein (1991) recommended that one-zero sampling may be appropriate when behaviors occur in non-independent flurries, and where the probability of an event occurring is more important than the frequency or duration of that event. For example, play behavior or copulations may be effectively sampled using one-zero sampling, which provides data on the probability of the behavior happening within a given period of time. Zinner *et al.* (1997), in a comparison of the estimation of true values and power for different sampling methods, recommended that one-zero sampling is appropriate for estimating frequencies (but not estimations of time budgets) of short-interval rare events, but is not effective for more frequent events.

Ad libitum sampling does not follow implicit systematic sampling rules, and the hazard is that the observer will be drawn to recording more readily visible behaviors that grab attention (Altmann 1974; Bernstein 1991). Bernstein (1991) found that *ad libitum* scoring may produce absolutely more data than focal animal sampling in a given time period, but the data do not provide reliable measures of absolute frequencies, they underestimate the relative contributions of many responses, and observations are biased towards larger or noisier age-sex classes. However, *ad libitum* techniques can be used effectively when the sole objective is to obtain large data sets without regard to absolute or relative frequencies, such as for development of directionality matrices to determine which member of a dyad is more likely to groom, mount, or aggress against (Bernstein 1991). *Ad libitum* scoring is also useful for detecting rare events and in initial descriptive work and pilot studies (Bernstein 1991). For example, Rose (2000) recorded *ad libitum* observations (also called *opportunistic sampling*) of events such as inter-group encounters, mating, fights, predator encounters, and other rare events during and between focal sessions, even interrupting focal sampling when necessary. Finally, *ad libitum* sampling is effective for collection of data for *ethograms*, which are catalogs of behavioral repertoires for a given study species (see Petrů *et al.* 2009).

Selecting a sampling technique depends in part on the types of research questions being asked. It is important to determine the extent of temporal and individual variation in activity and the observability early on in a field study (Fragaszy *et al.* 1992); this information will inform sampling method decisions. In some cases, employing a variety of methods may be desirable, sequentially or as appropriate. Based on a comparison of focal interval sampling and group scan sampling, Fragaszy *et al.* (1992) found that a combination of sampling approaches in repeating sequences can be an effective field sampling strategy for understanding behavioral events and states. Lastly, ancillary ecological or environmental data may be very important to collect before, during, or after any of the above sampling techniques. Examples include data on weather, habitat type, presence of other species or conspecifics, or GPS waypoints taken at set intervals (see Section 5.6 and Chapter 9).

5.5 What kinds of questions are asked about behavior within groups?

Many questions in primate behavior stem from an interest in understanding the evolutionary function of particular behaviors. We want to know why primates behave in certain ways under particular conditions, as well as questions about the underlying or proximate causes for certain behaviors (Strier 2003). Here we present a few of the common research areas pertinent to within-group behavior, highlighting some of the current theories, questions, and challenges associated with their study.

5.5.1 Activity budgets

An enduring question in primate studies is how primates allocate their time between activities including eating, resting, traveling, and social interactions, and these activity budgets are highly variable among species and among members of the same group (Strier 2003). One approach to quantifying behavior in a group of animals is by observing them over an extended period of time, systematically collecting data on their activities, and creating an "activity budget" from these data. An activity budget shows how much time an animal or a group spends engaged in various activities. Typical activity categories might include traveling or moving, foraging (including searching for food, handling food, and inserting food into the mouth), resting, and grooming; categories can be added or refined depending on the specific research questions in order to gather data to construct the activity budget. Activity budgets are presented as the percentage of time spent in each activity. Many factors are hypothesized to influence activity budgets including internal factors such as individual energy requirements, age, sex, reproductive state; external or environmental factors such as weather, spatial and temporal distribution of food, food quality; and presence of other species, conspecifics, predators, or tourists. For example, the activity budget of pregnant and lactating females is predicted to be different from other females, reflecting different nutritional requirements (Altmann 1980).

Studies of a group's activity budget can reveal differences between groups in different locations, and patterns within groups such as synchrony, which may allow individuals to maximize the benefits of social living. For example, individuals can increase opportunities for obtaining information about food location and quality by foraging together and monitoring neighbors, and may also benefit from coordinating antipredator scans as long as information is shared rapidly (King and Cowlishaw 2009). However, behavioral synchrony can be costly, for instance when optimal activity patterns of individuals of different age or sex within a group differ, when groups forage on scattered resources or move through highly heterogeneous habitats, or when distance between neighbors within a group causes a visual or auditory break in communication (King and Cowlishaw 2009). A study of chacma baboons (*Papio ursinus*) found that while behavioral synchronization, as measured by diversity in activities among group members, can be affected by variation in activity budgets among individuals, habitat constraints were also found to affect patterns of synchrony, which have to date received little attention in the literature (King and Cowlishaw 2009).

Comparative studies of activity budgets have found differences between groups exposed to disturbances such as tourists, agriculture, and fragmentation. In a study of tourism impacts on Tibetan macaques (*Macaca thibetana*), Matheson *et al.* (2006) found differences between more- and less-habituated macaque groups in responses to humans, and tourist presence affected interactions within groups and between monkeys and humans. L'Hoest's monkeys (*Cercopithecus lhoesti*) living along the edge of Bwindi forest, Uganda, spent relatively equivalent amounts of time on major behavioral activities such as feeding, traveling, and resting as did an interior group (Ukizintambara 2010). However, the group living near the edge spent time outside the forest attracted to the agricultural fields where they were exposed to significant threats, and the forest edge in this case is considered an ecological trap due to the high injury and mortality rates experienced by these monkeys (Ukizintambara 2010). Studies are needed that focus on the effects of anthropogenic activities on changes in activity budgets, fitness, and population health in order to better understand the long term impacts of human activities.

5.5.2 Diets and foraging behavior

Temporal and spatial variation in food abundance and quality influence primate diet, habitat use, social structure, mating, and life history strategies (Strier 2003; Hemingway and Bynum 2005). Data collected on the feeding ecology of primate groups allows us to understand the role of specific food items in the diet and how a group of primates interacts with and is influenced by these resources. For example, some primate species have flexible foraging behaviors allowing them to adapt to changes in food abundance and quality. Sometimes this flexibility means they use *fallback foods*, typically defined as foods consumed during periods of overall low food availability (Marshall and Wrangham 2007). Fallback foods are often relatively low quality and used in inverse proportion to the abundance of high quality, preferred foods (Wrangham *et al.* 1998; Marshall and Wrangham 2007; Harrison and Marshall 2011). They may also be high quality foods that are rare in the environment (Lambert 2007). Quality refers to ease of energy extraction including processing time required before putting the item into the mouth and digestibility or structural carbohydrate content (Watts *et al.* 2012). Leaves, pith, and other plant parts high in cellulose and hemicellulose are usually low-quality foods, while fruits are usually, but not always, high quality (Wrangham *et al.* 1998; Marshall and Wrangham 2007; Harrison and Marshall 2011).

Many Old World frugivorous primates use leaves and other nutritionally poor resources as fallback foods during lean periods, while smaller-bodied new world frugivorous primates fall back on insects and gums. For example, leaves are a fallback food for chimpanzees when fruit abundance is low (Watts *et al.* 2012), while the callimico (*Callimico goeldii*) consumes exudates as a fallback food when other resources, such as ripe fruits, are scarce (Porter *et al.* 2009). Identification of fallback foods for specific primate populations requires sampling phenology (systematic sampling of plant food availability, such as fruits, new leaves, and flowers; see Chapter 7) to document availability and abundance of food resources. Nutritional analyses of food quality will also contribute to identification of fallback foods (see Chapter 11). Identifying fallback foods for primate populations and protecting these resources has important conservation implications, especially when related to differences in habitat quality and anthropogenic impacts (Marshall *et al.* 2009b). For example, in a small, completely isolated forest fragment in Rwanda, Chancellor *et al.* (2012) found that during periods of fruit scarcity, fallback foods were critical to chimpanzees with limited options to find foods outside their forest fragment. A study of diademed sifaka (*Propithecus diadema*) found that a group resident in a forest fragment fed on a typical fallback food item year round (Irwin 2008).

Foraging behavior studies can show us when a primate species plays an ecological role in the community, such as seed dispersal or pollinator services. The effectiveness of a seed disperser refers to the contribution of that disperser to the future reproduction of a plant and is divided into quantitative and qualitative components (Schupp 1993). Quantitatively, seed dispersal involves the number of visits made to a plant by a disperser and the number of seeds dispersed per visit. Valenta and Fedigan (2008) provide a valuable framework for standardizing the quantification of seed dispersal so that values can be compared within and among species. Reliable and comparable measures of the quantity of dispersed seeds include percentage of fruit in the diet (measured as percentage of time spent feeding on fruit per total feeding time), number of plant species consumed, percentage of plant species whose seeds are passed intact, number of seeds and species per defecation, number of defecations per animal per day, animal density, and percentage of defecations containing seeds (Valenta and Fedigan 2008). The quality of seed dispersal depends on seed handling behaviors, treatment given to a seed in the mouth and gut, and aspects of seed deposition that determine the probability a deposited seed will survive and become an adult (Schupp 1993). The seeds in fruits consumed by primates may be chewed and digested, swallowed, and defecated intact, or removed from the flesh and spat out (Corlett and Lucas 1990). Seed handling is dependent on fruit and seed type and will vary depending on what food resources are available. For example, a study of *Cercopithecus* monkeys covering several seasons showed they will alternate between acting predominantly as seed predators, seed spitters, or seed swallowers depending on fruit availability (Kaplin and Moermond 1998).

Primates that feed on nectar may also play an ecological function as pollinators (e.g., Overdorff 1992; Gautier-Hion and Maisels 1994; Kress *et al.* 1994). To assess effectiveness three main kinds of data are needed: (1) observations of regular and non-destructive visits to flowers, (2) observations that pollen has been transported to conspecific flowers, and (3) experimental evidence that flowers visited by the primates have produced a seed set (Carthew and Goldingay 1997). More research is needed to assess the importance of primates as pollinators including field observations and experiments to manipulate timing of nectar and pollen presentation, for example. Carthew and Goldingay (1997) present a useful framework for establishing effectiveness of non-flying mammals as pollinators, and for testing floral traits associated with primate pollination. The identification of specific roles that primates play in maintaining ecosystem functioning and provisioning of ecosystem services, such as pollination and seed dispersal, can provide important arguments for establishing the conservation priority of these species.

5.5.3 Competition, aggression, and cooperation

All primates experience some level of within-group competition, mainly over access to food and safe sites (Janson 1988). Competition can be in the form of direct contests over the resource, or indirect scrambling for the resource, and the type of competition will influence the social relationships among primates that maintain cohesive groups (Wrangham 1980; Isbell 1991). Contest competition is often associated with aggressive interactions over a resource and is typically considered to be dependent on resource abundance, distribution, and quality (van Schaik 1989; Koenig 2002). In some cases, contest competition is correlated with the formation of dominance relationships in which dominant individuals obtain relatively more of the defensible resource (Mathy and Isbell 2001; Vogel *et al.* 2007). Scramble competition occurs with non-defensible resources, where an individual competes for food resources without interacting with others directly by getting to the resource before the others get it. Scramble competition is often difficult to measure directly because it is essentially based on the assumption that individuals lose access to resources when other group members have already used them (Sussman *et al.* 2005).

The competitive regimes related to food distribution have been used to explain spatiotemporal variation in the distribution of females and female–female social relationships. Under this framework, predation risk forces females of most diurnal primate species to live in groups, while the strength of the resultant contest competition for resources within and between groups is then the main determinant of female social relationships (Wrangham 1980; van Schaik 1989; Isbell 1991; Sterck *et al.* 1997). Sterck *et al.* (1997) presented a review and test of models addressing variation in female–female social relationships that focused on ecological factors, notably predation and food distribution, and alternative models (i.e., between-group competition, forced female philopatry, demographic female recruitment, male interventions into female aggression, male harassment) and found the original model (Wrangham 1980), based on predation risk leading to group living—with strength of the contest competition for resources within and between groups determining social relationships between females—quite robust. However, Koenig (2002) has argued that due to the indirect nature of most tests of feeding competition and social structure, the results of predicted interrelations are inconclusive, and ultimate approaches using mechanistic correlates of fitness (net energy gain) or lifetime reproductive success are the more effective ways to measure consequences of feeding competition.

Several authors have suggested that more data are needed to understand the role of competition or reconciliation in structuring social systems in primates (Pruetz 1999; Chapman and Chapman 2000a; Sussman *et al.* 2005; Newton-Fisher and Lee 2011). Sussman *et al.* (2005) argue for more attention to quantifying the advantages to kin and non-kin group members in developing affiliative and cooperative behaviors in which partners receive collective benefits. The authors call for a new framework in which affiliation, cooperation, and social tolerance form the core of group-living in primates in contrast to the competition–aggression/affiliation–reconciliation paradigm. Alternatively, Koenig *et al.* (2006) have argued that affiliation and cooperation are no more important than agonism in structuring social relationships within primate groups, and call for more rigorous tests of the costs and benefits of affiliation and agonism using correlates of lifetime fitness. This is a rich area for further research. An understanding of an individual's abilities to maintain affiliative and coordinated behaviors while minimizing agonistic interactions is likely to provide critical insights into the evolution of sociality and group-living in primates.

Biological market theory has been gaining in popularity as a framework for understanding cooperation and reciprocity among individuals of a group (Fruteau *et al.* 2011). Biological market theory provides a model for understanding the evolution of "altruistic" behavior in social groups (Noë and Hammerstein 1995). The model is predicated on the idea that individuals (or classes of individuals) differ in what they offer one another through their interactions, that the interactions have an inherent value, and that choice of social partners is contingent on what partners offer and what value the recipient attaches to the interaction and to the participant (Newton-Fisher and Lee 2011). In biological market theory, the action of natural selection on the behavioral strategies of animals is modeled as a marketplace in which animals (or the strategies they express) are considered traders with goods (behavioral interactions) to offer and exchange (Newton-Fisher and Lee 2011). Grooming, for example, has been considered a low-cost good in the sense that it can be used to compensate for imbalances in the trading of other commodities (Barrett *et al.* 2000). Grooming is thus either reciprocated in kind by the partner, or exchanged for another commodity such as tolerance at food sources, agonistic support, or mate compliance (Fruteau *et al.* 2011).

5.6 Geographic Information Systems (GIS) and within-group behavior

A geographic information system (GIS) integrates hardware, software, organizational structure, and data for the purpose of retrieval, analysis, synthesis, and display to promote understanding of relationships, patterns, and trends related to spatially located objects and associated non-spatial information (Kennedy 2009). Currently, there are more than a hundred GIS software products including ArcGIS, qGIS, ENVI, ERDAS Imagine, and ILWIS (see summary at <www.spatialanalysisonline.com/SoftwareBrief.pdf>). ArcGIS (ESRI) is one of the more popular products, particularly in the environmental and natural sciences fields.

A global positioning system (GPS) is a navigation system consisting of a network of satellites. GPS receivers (units) are devices that provide data about latitudinal and longitudinal coordinates for any geographic area on Earth and near the earth (for more detail on GPS units see Chapter 9). GIS and GPS units have become important analytical tools in wildlife biology and management, including mapping and analyzing space-use patterns of within-group primate behavior such as in wild orangutans (Wartmann *et al.* 2010; also see Bentley-Condit and Hare 2007; Fashing *et al.* 2007b). Specifically, GIS provides tools and techniques that help researchers monitor primate ranging behavior (see Chapter 9 for methods to define the home range and measure home range overlap) as well as many other aspects of behavioral ecology such as foraging behavior. Furthermore, hydrologic models within GIS can help monitor seasonal and inter-annual water scarcity and its effects on primate ranging behavior (e.g., Scholz and Kappeler 2004).

Habitat characteristics can be mapped using a GPS unit and downloaded into a GIS program to be used in the creation of maps showing habitat features. Such information has a variety of applications for ecologists and primatologists (see Chapter 6). For instance, these data can be used to identify the location and utilization of flowering and fruiting trees, and to examine habitat use by animals (Phillips *et al.* 1998). Data on the spatiotemporal distribution of resources such as food can also be collected and plotted using GPS and GIS. The use of both technologies allowed Campbell (1994) to determine that the patterns of home range use by spider monkeys (*Ateles geoffroyi*) and capuchin monkeys (*Cebus capucinus*) are related to dispersion and availability of food resources.

Most studies involving an evaluation of the status and trends of ecological systems require the use of remote sensing technology (Horning *et al.* 2010). Remote sensing refers to the use of satellite or aircraft-based aerial sensor technologies to detect and classify and interpret objects and phenomena on Earth (Joseph 2005). Remote sensing methods are invaluable to detect natural features in the landscape such as effects of storm events, fires, and vegetation change. Remotely sensed data work most accurately if used in combination with field surveys (e.g., Pozo-Montuy *et al.* 2008).

Horning *et al.* (2010) provide a detailed review of how remote sensing can be used in ecology and conservation science. As an example, GIS together with remote sensing technologies can be used to model plant phenology and help assess seasonal and inter-annual availability of food resources for primates in studies requiring data on resource availability in space and time (see Chapter 7). By combining such data on the spatiotemporal distribution of resources with other information on the foraging and ranging behavior of primate groups, based on field observations, one can analyze the influence of food resource distribution on social organization, group cohesion, feeding competition, and reproductive success (e.g., Phillips *et al.* 1998; Knott *et al.* 2008). GIS can also be used in combination with remote sensing technology to monitor landscape fragmentation. One can investigate the effect of fragmentation on the distribution of rare or endangered primate species and how groups use such landscapes, for instance (see Chapter 6).

The Normalized Differential Vegetation Index (NDVI), which is derived from remotely sensed data, has emerged as an effective tool for assessing habitat productivity. NDVI is calculated from reflectance patterns from the Earth's surface in the red and near-infrared regions of the electromagnetic spectrum (Tucker 1979). NDVI remains the basic vegetation index most widely employed for global monitoring of vegetation. It is defined by the following ratio:

$$NDVI = \frac{\rho^{NIR} - \rho^{red}}{\rho^{NIR} + \rho^{red}}$$

where ρ^{NIR} and ρ^{red} are reflectances for visible (red) and NIR spectral bands (Trishchenko *et al.* 2002).

Time series analysis of NDVI allows long-term monitoring of vegetation indices. For example, canopy leaf cover can be modeled using remote sensing, based on high or medium resolution imagery such as the Moderate Resolution Imaging Spectroradiometer (MODIS), the Advanced Very High Resolution Radiometer (AVHRR), Quickbird, or Landsat ETM imagery. There are many useful applications of this index. For example, this approach could be particularly useful for predicting locations of primate groups in landscapes not yet surveyed for primate presence. Furthermore, behaviors such as vigilance and foraging can be correlated to canopy leaf cover by analyzing seasonal and inter-annual variations in NDVI. See Chapters 7 and 9 for further discussion of the use of NDVI in studies of primate ecology. Another application of GPS and remote sensing involves Digital Elevation Models (DEM), which can be used to assess the relationship between altitudinal gradients (derived from the DEM) and ranging behavior. In fact, the overlay of high-resolution DEM with land cover allows exploration of the correlation between land use, slope, and ranging patterns (e.g., Sueur *et al.* 2011b).

GIS and remote sensing also have conservation applications. These techniques can be used to classify satellite images for various land uses and support site selection for high priority conservation areas or assisted primate migration, for example. GIS has been used to help design corridors to facilitate movement between animal groups as a way to manage the risk of inbreeding (Epps *et al.* 2007; Li *et al.* 2009; see Chapter 14). GIS tools can also be used to assess risks of extinction due to climate change (see Box 19.1).

5.7 Social network theory and network analysis

Because primates are generally highly social animals, theories about social networks have strong applications to the study of primate within-group behavior. A social network is essentially a set of social units such as groups (nodes) and the relationships (ties or edges) between them. The network itself is usually defined as a set of individuals that are directly or indirectly connected to each other via particular interactions (Voelkl *et al.* 2011). *Social network analysis* (SNA) is a framework used to study the structure of societies or social groups (Brent *et al.* 2011) by quantifying aspects of social structure that have long been part of animal behaviorists' understanding but have often not been quantified in depth (Sih *et al.* 2009). Social network analysis is not a new concept, but its application to primate social interactions and dynamics has grown dramatically in recent years. For a thorough history of SNA, see Brent *et al.* (2011).

The strength of the SNA framework is in the way it allows the researcher to view and explore relationships that define members of a group (Jacobs and Petit 2011). A key feature of SNA is its focus on interactions between individuals, rather than on the individuals themselves (Sueur *et al.* 2011a). Sih *et al.* (2009) identified four general aspects of social network structure that are familiar to animal behaviorists: (1) individuals differ in their social experiences; (2) indirect connections matter; (3) individuals differ in their importance in the social network; and (4) social network connections in one context carry over to influence social dynamics in other contexts. SNA uses *social network metrics* to quantify social structure. An example of a metric is an individual's *degree*, or number of social partners, which is a major focus of studies of sexual selection—a male's mating degree, or his number of mating partners, reflects mating success (Sih *et al.* 2009). Other metrics include an individual's *reach*, or the number of friends of friends, the number of individuals connected to that individual via two steps, and the *clustering coefficient* which is the tendency for a focal animal's friends to also be friends with each other (Croft *et al.* 2008; Sih *et al.* 2009).

SNA encompasses a number of different graphical tools to visualize networks (e.g., sociograms), as well as tools for mathematical modeling (e.g., matrix algebra and permutation-based analyses) that allow detection and quantification of patterns in social networks (Freeman 2004). Over the last 15 years there have been large advances in SNA, resulting in new perspectives on the structure of complex networks, development of new network metrics, and the creation of several readily available software packages and scripts for the implementation of network analyses (Brent *et al.* 2011). SOCPROG, Ucinet, Pajek, and NetDraw are among the most commonly used software packages for analyzing social networks. SOCPROG, for example, can be downloaded for free at <http://myweb.dal.ca/hwhitehe/form. htm> and interfaces with Excel worksheets to incorporate observational data. The program can handle data sets containing large numbers of individuals, and it interfaces with other specialized network analysis programs available such as UCINET and NetDraw (Whitehead 2009).

The great value of these computer software programs lies in part in what they can do with sociograms. Sociograms have been popular with primatologists as a way to visualize a variety of interactions among primates including grooming, proximity, vocalizations, copulations, aggression, and play (Brent *et al.* 2011). Programs such as Netdraw (Borgatti 2002) readily apply algorithms to interaction matrices to create sociograms based on specific criteria; for example, the spring-embedding algorithm places individuals with the highest centrality scores toward the center of the sociogram (Brent *et al.* 2011). These graphical representations consist of nodes, which represent actors, and edges, which represent relationships. Using these programs, nodes can be assigned different colors to distinguish males from females or age ranges, for example (Sueur *et al.* 2011a). The graphical representation offers flexibility; edges can indicate the direction of an interaction and its thickness can show information about the importance of the interaction (Sueur *et al.* 2011a). In addition to sociograms, hierarchical clustering analysis and multidimensional scaling, both common visualization techniques in SNA, have been used to identify subgroups within larger groups of primates, based on the interactions between pairs of individuals, and graphically represent them (Brent *et al.* 2011).

An interesting application of SNA is in the exploration of social group or network instability. For example, fission–fusion societies alternate between a single large group and several small subgroups (e.g., Kappeler and van Schaik 2002; Lehmann *et al.* 2007a). SNA can support analyses to explore the connections between subgroups, and allows identification of a keystone individual who serves as a bridge between subgroups, thus playing a critical role in maintaining subgroup connection. This example highlights an important insight from social network theory for stability of animal groups: not all individuals are equal in their influence on group cohesion (Sih *et al.* 2009). In another interesting application of social network analysis, King *et al.* (2011a) explored social feeding, which can provide individual benefits including reduction in predation risk, collective defense of resources, and efficient information sharing, but can also lead to contest competition, aggression, and possibly injury. Individuals may attempt to reduce such costs by selecting foraging partners who are likely to be most tolerant and willing to share food resources with them (King *et al.* 2011a). Partners may share a strong social bond with the individual ("social bonds hypothesis") and/or may be closely related to the individual ("kinship hypothesis"). Co-feeding behavior may thus be mediated by social bonds or kinship, or a combination of the two, and King *et al.* (2011a), using SNA, tested these two hypotheses by exploring whether individual chacma baboons (*Papio ursinus*) feed with individuals with whom they share a particular social (grooming) or kin relationship. Results showed that baboon co-feeding was significantly correlated with grooming relationships but not genetic relatedness, and dominant individuals were also found to be central to the co-feeding network through their sharing of food patches with multiple group members. SNA allowed for analysis of multiple foraging associations between individuals and results underline the importance of dominance and affiliation to patterns of primate social foraging.

Interesting questions that one can test using social network analysis include the following: Do cooperators tend to interact primarily with other cooperators? Within a group, is there a correlation between an individual's social network position and its tendency to give or receive favors? Are high degree individuals less (or more) cooperative (Sih *et al.* 2009)? SNA has also been used to study disease transmission, and advances in this area have shown that transmission does not occur randomly among primates, due to the nature of the social ties within groups and the associated spatial proximity which correlates with disease transmission risk (Jacobs and Petit 2011). The use of social network analysis has been gaining in popularity (Brent *et al.* 2011) and will probably only grow in sophistication and use as an effective approach to investigate social interactions such as aggression, cooperation, competition, and mating behavior as well as phenomena such as dominance hierarchies, disease transmission, and social learning.

5.8 Conclusion

The study of within-group primate behavior remains a rich and exciting area of research with many recent innovations that improve our ability to collect data and understand the mechanisms underlying the patterns we observe in the field. Our review of techniques used to collect behavioral data in the field emphasizes the importance of selecting a systematic approach, and an understanding of the temporal and spatial variation in behaviors to be observed.

We have described the tension that exists between the costs of group living, including intra-specific competition, and the benefits and advantages of group living, including reduced predation. This tension is mediated by resource distribution, predation, infanticide, and other variables, and has led to a breathtaking diversity in primate social organization. Considering this diversity in the context of global and regional environmental change, we can begin to make predictions about how changing temperature regimes, along with changes in plant phenologies, distributions, and rainfall patterns will influence particular social groupings. Furthermore, it is possible that habitat disturbance and alteration may affect stress levels and rates of aggression in primates, as well as other behaviors. Research testing the impact of environmental changes on social organization through measures of fitness will help us understand potential effects. This is an urgent area of research given current patterns of global environmental change.

Much of the behavioral research conducted today on primates may be applied to conservation issues and mitigation steps. For example, data on activity budgets can help us monitor changes in energy intake and expenditure with changing environments and disturbances, identification of diet composition and critical fallback foods can help us develop effective habitat conservation and land use plans, and research on the ecosystem services (e.g., seed dispersal and pollination) provided by primates can support conservation planning and priority-setting. We now know that certain primates may have a significant effect on forest structure or composition. For example, seed dispersal behaviors of chimpanzees and L'Hoest's monkeys increase seedling survival in some large-seeded, late successional tree species in tropical montane forest in Rwanda (Gross-Camp and Kaplin 2011). As a further example, the critically endangered black-and-white ruffed lemur (*Varecia variegata*) was shown to be capable of dispersing large seeds of canopy tree species, and may thus play an important role supporting carbon sequestration (Moses and Semple

2011). Few other forest animals disperse such large-seeded fruits. These finding suggest that the loss of primates in certain ecosystems has the potential to profoundly impact vegetation distribution and abundance.

Two relatively recent newcomers to primate studies that offer interesting approaches to studying within-group primate behavior, Geographic Information System (GIS) and social network analysis (SNA), also have the potential to contribute to primate conservation planning. For example, Buckingham and Shanee (2009) created a Habitat Suitability Model for the critically endangered yellow-tailed woolly monkey (*Oreonax flavicauda*) using inductive GIS modeling to predict current geographic distribution and to create an ecological risk assessment model of the study region, which contributed to site selection for priority areas for conservation action, including identification of corridors and new protected area designation. Further examples of the kinds of contributions that GIS can make to primate conservation are discussed in Chapter 6 and Box 19.1.

Social network analysis (SNA) offers a new approach to exploring complex interactions involving multiple individuals in a group, and it can facilitate the quantification of relationships that were previously only qualitatively evaluated (Jacobs and Petit 2011). SNA can also be valuable in the management and conservation of primates (Sueur *et al.* 2011a). For instance, group size may increase under certain conditions causing increased within-group food competition and aggression, possibly resulting in permanent fission. Movement and immigration may become limited due to physical barriers created by human activities, preventing natural fission and potentially escalating human–primate conflicts. SNA can facilitate development of mitigation steps based on an understanding of individual behaviors and responses within networks (Sueur *et al.* 2011a).

Acknowledgments

We thank the editors for inviting us to participate in this volume and for the constructive criticism on the earlier versions of this manuscript that greatly improved the manuscript. We also thank the external reviewer for comments that helped us revise this manuscript. Lastly, thanks to Jim Jordan for editing revisions of this manuscript.

6

Habitat assessment and species niche modeling

E. Johanna Rode, Carrie J. Stengel, and K. Anne-Isola Nekaris

6.1 Introduction

More than half of the world's primates remain unstudied. While many of the well-known primates live in accessible "quantifiable" landscapes, less-known taxa remain so, at least partially, because they live in difficult, inaccessible habitats that defy typical habitat assessment techniques. This chapter will present an overview of classic habitat assessment techniques and how such studies can be carried out on various budgets and with a range of equipment from traditional to modern. This chapter also addresses the important issue of how to decide when and where to look for elusive species. In our view, species niche modeling (SNM)—computer-based models that predict potential species distributions, futures, and fates from existing distribution and habitat data—holds considerable promise and we discuss how it can complement field data to enhance species research and conservation.

6.2 Habitat assessment

6.2.1 Habitat classification depends on scale

Habitat descriptions can range from large scale to local microhabitat. Primates live mainly in forest habitats such as subtropical or tropical dry, moist lowland, mangrove, swamp or moist montane forest, but may also live in grassland or human altered landscapes such as plantations, rural gardens, and urban areas (Campbell *et al.* 2011a). On a finer spatial scale these habitats may be further classified based on different criteria such as level of disturbance, floristic composition, vegetation structure, or environmental conditions (see Section 6.2.2).

Primate Ecology and Conservation: A Handbook of Techniques. First Edition. Edited by Eleanor J. Sterling, Nora Bynum, and Mary E. Blair. © Oxford University Press 2013. Published 2013 by Oxford University Press.

Habitat selection is based on factors important to the animals and thus happens at the spatial scale of microhabitat (Krebs 2001) (see Section 6.2.3).

6.2.2 Classification of habitat types

Researchers must first become familiar with their study area. Often it is necessary to stratify the habitat in order to discriminate between different conditions. Stratification in this sense means the separation of the studied area into different habitat type classes or strata. Habitat classification can be based on different abiotic or biotic factors (Ganzhorn 2003; Table 6.1). Appropriate classification must be decided on site and within the context of research goals. If the project investigates feeding behavior, classification might be based on the floristic composition of the area, while conservation-related projects might favor a classification by habitat disturbance level. Characteristics of classification systems are not mutually exclusive and may even be combined (e.g., open eucalyptus forest, Nicholls and Goldizen 2006). For studies on a particular species, the same classification should be implemented across studies to enable direct comparison of results, and the classes used should be clearly defined. For example, the terms "primary forest" and "secondary forest" are often used, but rarely defined. As many habitats are influenced by humans to some degree,

Table 6.1 *Examples of habitat type classifications, based on abiotic and biotic factors.*

Criteria	Examples
Abiotic criteria	
Disturbance (human/natural)	• Primary forest, secondary forest
	• Degraded forest, logged forest, (un-)disturbed forest
Human modification	• Natural forest, plantation, rural cultivation, gardens
Altitude	• Mountain forest, lowland forest
Topography	• Plane forest, cliff, mountainous forest, steep forest, gallery forest, karst forest
Environmental factors	• Swamp forest, mangrove forest, deciduous forest
	• Dry forest, wet forest
Biotic criteria	
Vegetation structure	• Tall forest, small forest
	• Dense forest, open forest (undergrowth, climbers, tree density)
	• Open forest, closed forest (canopy)
	• High canopy, low canopy
Dominant/indicator species; floristic structure/composition	• Bamboo forest, eucalypt forest, oak forest
Age of forest	• Old forest, re-growing forest

definitions based on time left undisturbed might be most appropriate. Forest types may be determined from current data and management records, for example in the case of National Parks. If no previous data exist it is up to the researcher to divide the habitat into different strata. If the research period allows, one should thoroughly assess habitat structure and vegetation (see Section 6.3) in order to base habitat classification on ground-truthed data. Unfortunately for most short projects this is not an option. In these cases, classification must be based on visual inspection and short pilot visits to the area, perhaps combined with satellite imagery or aerial photos. Asking local people about the forest characteristics and history may also be helpful in the classification process. For further information on classifying habitat types based on satellite imagery, see Horning *et al.* (2010).

6.2.3 The microhabitat

In all research design the researcher has to ensure that samples are collected randomly. Randomness means that each sample unit has an equal chance to be included in the sample. Most techniques to identify a random sample assign a number to each potential unit. Random numbers between a lower and upper bound can then be calculated to identify the units included in the sample. In the field other easy methods can be used like using the second hand from an analog wristwatch to select a random orientation. If random points in a forest must be identified for plotless sampling, the minimum and maximum northing and easting can be determined from a map and random coordinates between these bounds selected. If points fall outside the study area they have to be omitted and the next point should be chosen.

6.2.3.1 Methods

6.2.3.1.1 Plot based methods

(See also Chapter 7, Section 7.2.) Plots can be quadrate, circular or rectangular shaped (for a visual overview see Hill *et al.* 2005 p.256). The shape must be consistent throughout the study. A long rectangular shape is considered a strip transect. Belt transects are quadrates that are laid continuously spaced along a line. Choosing an appropriate shape depends on the study objectives. Quadrates tend to represent the habitat of a large forest area in general especially if it is homogenous, while transects may give a better impression of the zonation of a forest. For example, Enstam and Isbell (2004) used transects of several hundred meters length and five meters width placed at a perpendicular angle to a boundary between two different forest types in order to assess the abruptness of change between them. All trees or features of interest should be sampled within a plot. Care must be taken at the layout of plot systems. Usually plots are placed randomly, however different plots need to be spaced sufficiently to prevent autocorrelation; computer models are available to aid in random placement of plots (Buckland *et al.* 2001). The size of the quadrates depends on the entities to be sampled and the size should be at least larger than the average distance between entities. For detailed information on plot design, refer to Hill *et al.* (2005). Permanent plots need durable markers, as they are prone to more destruction over time. For example, buried metal has been used to demarcate plots with researchers employing the use of a metal detector to re-find them in the future. While permanent plots are able to detect changes in time, such as forest regeneration, temporary plots are a snapshot in time and are more useful to show changes over a spatial scale.

6.2.3.1.2 Plotless sampling

Plotless sampling is mainly used to investigate tree density, but characteristics of the trees or entities of interest may be recorded at the same time. The three commonly used plotless sampling techniques described below (Greenwood 1996; Hill *et al.* 2005) are quick to perform and require little equipment. All three measures start with the location of random points and involve measuring distances to the estimated center of a tree (see Hill *et al.* 2005 p.233). Be aware that in the following three methods the result is given in trees per m^2 and has to be transformed into trees per hectare if needed. Methods 1 and 2 are appropriate if trees are randomly distributed. This is mostly not the case, as trees tend to occur in a clumped or regularly spaced manner. If trees are aggregated, the nearest neighbor method overestimates the true density while the point-centered quarter method underestimates it. If trees are overly dispersed the bias is the opposite for each method. The T-square method includes both of the above named techniques and thus cancels out the effects of bias.

1. Nearest individual method.

The distance D_i between a random point and its nearest tree is measured.

$$Density = \frac{1}{(2D_m)^2}$$

where mean distance for all samples is $D_m = (\sum \frac{(Di)}{n})$ and n is the number of points.

2. Point-centered quarter method.

Two perpendicular lines cross each other at the center point and create four quadrants. The compass directions of the lines have to be the same for each sample point. In each quadrant the distance D_i to the nearest tree is measured and the four distances averaged.

$$\text{Density} = \frac{1}{D_m{}^2}$$

where the mean averaged distance for all sample points is $D_m = \frac{\sum D_{avg}}{n}$, the averaged distance between four trees is $D_{avg} = \left(\sum \frac{(D_i)}{4}\right)$, and n is the number of points.

3. T-square sample.

First, the distance (x) to the nearest tree from a random point is measured. Then, a line is placed perpendicular to the connection between point and first tree. Finally, the distance (y) to the nearest neighbor of the first tree is measured.

$$\text{Density} = \frac{n^2}{2.2828 \sum x \sum y}$$

where n is the number of points.

Greenwood (1996) presents a formula to test a T-square sample for random tree distribution in the plot.

6.2.3.1.3 Transects

Transects are fixed lines placed within a study area, typically parallel to one another (for further details see Chapter 2). Ideally they should be straight; however this might be difficult in certain habitats. Therefore while it may be necessary to bypass obstacles, in general a straight line should to be followed in order avoid generating bias (Buckland *et al.* 2001). In contrast to strip and belt transects, line and point intercept transects are composed of single, one-dimensional lines. When using line intercept transects, entities touching the line are counted over its entire length. Alternatively the distance on the line covered by certain features can be recorded. Point intercept transects record the absence or presence of certain features at pre-defined distances. The results can be transformed into percentage of cover, relative abundance, and frequency.

6.2.3.1.4 DAFOR and Braun–Blanquet scale

A researcher can either create percentage classes for estimating vegetation type cover, or follow traditionally used scales like the DAFOR or the Braun–Blanquet scale. In the DAFOR method, the researcher assigns the category dominant, abundant, frequent, occasional, or rare to a species (= DAFOR), plant type or other feature within an area or plot (Bullock 1996). The scale can be transformed into numbers for statistical analysis, and z-transformed for more sophisticated modeling (MacKenzie *et al.* 2006). Ruperti (2007) used the DAFOR scale to assess the abundance of disturbance indicators (e.g., tree stumps, cattle feces) in each $10 \times 10 \text{ m}^2$ subplot of a 1 ha large plot. In the Braun–Blanquet scale method, the researcher assigns different percentage coverage classes to species (Kent and Coker 1994; Bullock 1996; Table 6.2). Both methods are quick and simple, but relatively coarse. They may be biased towards more obvious plant types or aggregations. When employing multiple observers in data collection, cross-checks might be useful to ensure inter-observer reliability (Bullock 1996).

6.2.3.2 How many samples?

If one aims to establish, for instance, the mean height of trees in a forest, it is not enough to measure all trees in one 20×20 m plot. This would lead to an imprecise estimate. Instead, samples need to be repeated to be representative. As the number of sampled plots increases, the precision of the estimate increases and it is possible to attach confidence intervals to the estimate. The precision of the overall estimate depends on the precision in each plot measured and on the variation between plots. The maximum number of plots would cover the whole area and would essentially represent the true mean if all trees are measured precisely. The sample size should be chosen based on a trade-off between obtaining the precision needed and resource constraints (time, money, person power). Greenwood and Robinson (2006) present two methods of choosing a sample size, based on precision needs and fixed costs. These calculations are based on the variability between samples, which have to be estimated in a preliminary survey. As the sample size also determines what kind of statistical tests can be performed, taking future analysis

Table 6.2 *The DAFOR and Braun-Blanquet scales for visual estimation of cover.*

Value	DAFOR	Braun–Blanquet
+	A single individual, no measurable cover	< 1% cover
1	1–2 individuals, no measurable cover	1–5% cover
2	Several individuals, but less than 1% cover	6–25% cover
3	1–4% cover	26–50% cover
4	5–10% cover	51–75% cover
5	11–25% cover	76–100% cover
6	26–33% cover	
7	34–50% cover	
8	51–75% cover	
9	76–90% cover	
10	91–100% cover	

into account is strongly suggested when planning data sampling, which is commonly referred to as power analysis.

6.2.3.3 Variables of microhabitat

The features and variables to be measured depend on the study aim and focus. The habitats of all living beings consist of abiotic and biotic factors. Depending on the lifestyle of primates, certain factors may be more important than others. For example, while all primates need food and habitat, highly specialized species may be more vulnerable to changes in these factors while generalists may be better able to cope with disturbance. The formulation of study objectives and the subsequent selection of variables to be recorded depend on many aspects. The sections below introduce biotic and abiotic factors, the resources used by primate species, and what variables can be sampled if little to no forest is present and the study species live in a human-dominated environment.

6.2.3.3.1 Biotic factors

As most primates live or at least spend a majority of their time in forested environments, many measurements concern trees. Undergrowth measurements may also be important for some species. Although a measuring tape is always useful, various kinds of equipment can be used and are mentioned within the following descriptions.

Density: Density is the number of individuals of a tree species per unit area. Density of trees is calculated simply by dividing the number of trees by the area sampled when using plot sampling. Calculation of density using plotless sampling has been described above. Density of undergrowth or other vegetation type can be estimated in percentage classes per plot or on a DAFOR scale. Laying a grid over the plot allows for an estimation of vegetation presence, calculation of frequency or presence/absence on the grid.

DBH (diameter of breast height): DBH is usually measured at 1.3 m height (Hill *et al.* 2005), on the upper side of the tree if the area has a slope.

$$\text{DBH} = \frac{c}{\pi}$$

where c = circumference.

Calipers can be used to measure the DBH of smaller trees but this can be a time consuming process. Usually the circumference is measured and the DBH is calculated from it (if the cross-section is relatively circular). A (tele-) relascope, a multifunctional tool used in forestry for more than 50 years, calculates the DBH, basal height, tree height, and distance. Trees with buttress roots should be measured 20 cm above the buttress roots. Trees with branches lower than 1.3 m are considered as two trees and both are measured. Many researchers group trees into tall trees (> 9.9 cm DBH), small trees (e.g., 2.5–9.9 cm DBH), and saplings (e.g., < 2.5 cm DBH; Ganzhorn 1989; Bellows *et al.* 2001) and may decide to use only one of these categories. Other variables are then only measured for the respective DBH category.

Basal area: Basal area gives an impression how much of a given area is occupied by the cross-section of tree trunks (if necessary for a certain species). Basal area as calculated by πr^2 where r is the radius of the tree or $\frac{1}{2DBH}$ respectively, and divided by the total area.

Importance value: If a primate species is known to utilize a specific tree species it may be useful to calculate the tree species' importance value, which is the product of relative density, relative frequency, and relative mean basal area (Mueller-Dombois and Ellenberg 1974; Raynal *et al.* 1998). The percentage of the species in question can be calculated, for instance, by dividing the density of one tree species by the average tree density.

Biomass: Biomass is a useful variable for studying folivorous primates. Direct measurements may be work-intensive and destructive. It is easier to use DBH or basal area as these variables have been found to be useful indicators for predicting biomass (Gholz 1982; Niklas 1993).

Height: The height of different variables can be measured, such as tree height, bole height (height of first branch), or canopy height (first leaves). Height can be estimated by the observer; for instance to the nearest meter. If there are two observers the mean may be taken between the two estimates. Previous training of all observers improves reliability. To obtain more precise measurements, a range finder can be used, which measures distances by pointing a laser beam at an object. The Pythagorean Theorem can be applied if the observer stands away from the tree and measures the distance to the tree and the distance to the tree top. Equally, the viewing angle when looking at the top and the distance to the tree can be measured and the height calculated using trigonometry.

$$\text{Height} = (\tan\alpha \times D) + x$$

where α is the viewing angle, D is the distance to tree, and x is the height of the observer's eyes.

A clinometer measures angles and slopes in the field and is based on trigonometry. Alternatively, a compass with mirror can be used to determine slope and angle.

Crown class, crown diameter, crown area, crown volume: Crowns can be categorized into dominant, co-dominant, intermediate, and suppressed (Gill *et al.* 2000). The crown dimension is directly related to the availability of food for primates.

As crowns are not typically circular, one may decide to use maximum crown diameter if results are only compared within a study. Gill *et al.* (2000) measured the radius of trees towards the center of their plot and a radius perpendicular to the first radius. If trees overlapped, the radius was measured to the line produced by the intersection points of the two adjacent tree crowns, although this measurement does not represent the full extent of the crown. The quadratic mean of both radii was used for further analysis. Crown area is $2\pi r$, where r is the radius (or mean of measured radii from the same tree). Crown volume is $2\pi r y$ where r is the radius and y is tree height minus bole height (Ganzhorn 1989).

Canopy cover, canopy closure: The canopy influences the light penetrating to the lower levels of the forest as well as the microclimate. The activity of many nocturnal primates and their prey species is influenced by light (Nash 2007). Canopy cover is often used as a measure of disturbance and often leads to the assignment of habitat types such as primary or secondary forest. As direct measurements are difficult to perform and often do not measure the light regime influencing plants and animals, indirect measures such as the intensity of light penetration obstruction by the canopy can be measured (Jennings *et al.* 1999).

Jennings *et al.* (1999) defines canopy cover as the "proportion of the forest floor covered by the vertical projection of the tree crowns." In contrast, canopy closure is the "proportion of the sky hemisphere obscured by vegetation when viewed from a single point." Canopy cover has to be measured from single points, which should be sufficiently spaced to prevent autocorrelation. A sighting tube (with an internal cross-hair) may help to ensure vertical projection. Although this is a quick and easy method, a large sample size is necessary. Alternatively, line transects can be used with the length of the line covered by canopy recorded. Finally, photographs can be used but one should take care not to introduce an angle as a viewing angle of less than $5°$ is the maximum acceptable deviation from the vertical (Bonnor 1967). Canopy closure might be more useful if the influence of light on the animal is examined as it reflects the condition at one point in all directions. A camera with fisheye lens takes hemispherical photographs with an angle and can be analyzed with computer programs like Adobe Photoshop, whereby the proportion of pixels covered by foliage or sky is divided by the total pixels to get a proportion of cover or openness. An appropriate threshold between light and dark pixels must be determined, and then the picture manipulated to account for pinhole effects (light passing through small canopy holes) or enlightened vegetation (Jennings *et al.* 1999). Special equipment such as moosehorns or spherical densiometers can be useful. Moosehorns are sighting tubes that include $45°$ mirrors and show equally spaced dots that have to be recorded as lying inside or outside the canopy. Spherical densiometers have concave or convex shaped mirrors with dots and basically work the same way. Both instruments often include bubbles to ensure vertical projection but may be stabilized using a tripod. Finally, canopy density can also be estimated visually.

Connectivity: Arboreal primates that do not habitually leap wide distances are dependent on tree connectivity at their preferred travel height. Canopy connectivity can be assessed by counting the numbers of other tree canopies touching the canopy of the tree in question. The researcher may decide to record categories like "standing alone," "connected to one tree," and "connected to two or more trees." As it is often difficult to identify different individuals of bushes, undergrowth connectivity may be better represented as density (See *Heterogeneity and complexity* below).

Lianas: The amount of lianas within a study area may be interesting for several reasons. Lianas increase the microhabitat structure and may offer shelter as well as forage sites. Lianas often can simply be counted or recorded using a DAFOR scale. The researcher may decide for a minimum thickness of lianas, which can be based on the ability to support the weight of the study species (Ruperti 2007). The size of lianas and branches can be visually gauged on a scale from small to large by comparing them with the size of the study species' hands and feet (Nekaris 2001).

Height diversity: As diversity in habitat structure creates a variety of ecological niches and has been shown to have an effect on species diversity, many studies investigate the height diversity of foliage or the canopy. Foliage height diversity (FHD) measures the lowest leaf at different points in a plot grid and uses the standard deviation as a measurement of foliage height diversity. Many primates live in very tall forests, thus vertical transects might not always be feasible. However, Malcolm (1994) used vertical transects to measure foliage density per strata for up to 40 meters. Ganzhorn (1989) recorded trees in plots by using different height classes and applied a Shannon–Wiener diversity index to his data. Aber (1979) used a camera to measure the height of foliage above different points, but a range finder or estimates will work as well.

Complexity and heterogeneity: Complexity is the vertical development of strata, while heterogeneity or patchiness describes horizontal variability (August 1983; Williams *et al.* 2002). Arboreal primates use vertical structure for moving, feeding, and sleeping. Although this has an important daily impact on their travel, energy expenditure, and home range it is rarely measured due to difficulties in accessing the canopy. August (1983) and Williams *et al.* (2002) measured standard variables, or the estimated vegetation density (measured in classes), in different heights within an area surrounding trapping grid points. They expressed complexity via data reduction techniques, using a single Principle Component to represent multiple ecological variables per grid. Both studies express homogeneity as the standard deviation of their reduced variables. As August (1983) used one grid for each of his five habitats, he applied Spearman rank correlation to compare complexity and heterogeneity with species abundance, richness, diversity, and evenness. He also showed that heterogeneity and complexity may have independent effects.

Floristic composition and invertebrate communities: Floristic composition means the combination of plant species or other chosen taxonomic level within the study area. In order to identify species correctly, it is advisable to cooperate with a local botanist or botanical institute. They may advise you on which plant parts will be needed for identification. In the field a pruning pole may be used to take samples of leaves, fruits, or whole small branches. To dry and preserve the plants, they should be pressed to make a herbarium specimen. Pressing can be done by placing samples between newspaper and cardboard, fixed between two wooden boards using straps or belts. The whole press can be additionally dried over a fire (careful with fires in dry forests!) or with a bright light in a box lined with tin foil. Furthermore, the composition and characteristics of invertebrate communities are an important habitat aspect for primates and may be recorded in the same way as floristic composition. Compositions can be reported descriptively, but several ways are commonly used to quantify the characteristics of communities, including species richness, diversity indices, evenness, and measures of similarity between two populations (Stiling 1999; Magurran 2004).

Species richness is simply the number of species for a certain sample of individuals. If the species richness of two communities is compared, a standard sample size must be used, as the number of species is dependent on the number of individuals sampled. If 100 individuals are sampled you will probably find more species than if only 50 individuals are sampled. This can be demonstrated in species accumulation curves (Stiling 1999). When the curve asymptotes one can assume that most species in the community have been sampled and equal sample size is not important any more. If one compares two communities where the curve does not reach an asymptote, the method of *rarification* can be used. Rarification calculates the species richness expected from each sample if the higher sample size would be reduced to the lower sample size (see Stiling 1999).

In contrast to species richness, *species diversity* indices take the relative abundance of each species into account. Two indices are presented below in Box 6.1 but more exist (Magurran 2004; Stiling 1999). The Shannon–Wiener Index H' (also called Shannon Index or Shannon–Weaver Index) assumes that all species are present in a random sample of a large community. Although it has the advantage of not favoring rare or abundant species, H' tends to increase relative to the amount of species sampled. Unless the researcher can safely assume all species are likely to be present within a sample, this may not be the index of choice—despite its

---

Box 6.1 Diversity indices, evenness, and similarity index. Working examples can be found in Magurran (2004)

---

1. Shannon–Wiener index

$$H' = -\sum_{i=1}^{s} p_i \ln p_i$$

where p_i is the proportion for the i-th species and S is the total number of species. Shannon–Wiener's measure of evenness $E = \frac{H'}{H_{max}} = \frac{H'}{\ln} S$.

2. Simpson diversity index

$$D = \frac{1}{\sum_{i=1}^{S} p_i^2}.$$

Simpson's measure of evenness $E_{\frac{1}{D}} = \frac{D}{S}$.

3. Jaccard index of similarity

$$C_J = \frac{a}{(a+b+c)}$$

where a is the number of species common to both sides, b is the number of species in site A, but not B, and c = the number of species in site B, but not A.

popularity. The Simpson Index gives the probability that any two individuals in a random sample belong to the same species. Although this index is weighted towards the most abundant species in the sample (adding a few animals of a rare species might not even change the index, Stiling 1999), Magurran (2004) calls it the most robust and meaningful index. For each index, the *evenness* of different diversity indices can be calculated; allowing the researcher to compare the actual diversity value with the maximum possible diversity. There are several *measures of similarity* that express the overlap between two communities. One of them is the Jaccard Index (Stiling 1999).

Jack-knifing can be performed to improve the estimate of any statistics, including diversity indices, and attach approximate confidence levels (Stiling 1999; Magurran 2004). The method recalculates diversity n times by always leaving out each sample in turn. The mean of the recalculated "pseudo values" becomes the new, jack-knifed estimate. This way a standard error and confidence limits can be attached.

Undergrowth cover: Apart from canopy cover, other vegetation cover or resources may be important for primate species. Cover of all vegetation forms can be estimated using the cover classes and predefined scales such as DAFOR or Braun–Blanquet. Line and point intercept transects can be useful as well. To improve the precision of an estimate, paper boards of a certain surface area (e.g., 25×25 cm^2) can be used for comparison (Corbin and Schmid 1995).

6.2.3.3.2 Abiotic factors

Abiotic factors such as temperature and humidity can be measured with a standard thermometer and hygrometer. Self-made rain gauges are a cheap option to measure rainfall but require near daily monitoring and emptying. Various measurements can be recorded in classes. For instance light can be indirectly recorded by estimating cloud cover. Temperature, humidity, rain, light, and wind are precisely recorded and stored with a data logger if information from professional weather stations is obtained. These are expensive but are advisable for long-term projects and field stations. Slope and inclination can be recorded by using clinometers and/ or compasses. Standard GPS devices include altitude and moon phase. Moon phase is especially important when studying nocturnal primates. While some nocturnal primates become more active during moonlit nights, for instance due to better foraging efficiency (e.g., *Tarsius tarsier*; Gursky 2003), other species become less active (e.g., *Nycticebus coucang*; Trent *et al.* 1977), possibly due to their cryptic defense strategy (Starr *et al.* 2012). Moon phases may be placed in different classes (Gursky 2003) or measured as a continuous variable using an online database (Starr *et al.* 2012).

Disturbance: As many primate species are affected by human disturbance, conservation-related studies often try to measure the amount of, or presence/ absence of, disturbance. Often disturbance signs are measured in plots (Merker *et al.* 2004, 2005) and include variables such as the number of cut tree stumps, exotic plants, human excrement, evidence of recent fire, litter, tree or vine tapping, and others (Muoria *et al.* 2003; Merker *et al.* 2004, 2005; Ruperti 2007). Natural disturbance may include wind damage or erosion (although the latter may be human-caused). Presence or abundance of the signs of a healthy forest, such as tall trees, may be used as well to assess disturbance (Merker 2003). Merker *et al.* (2004) introduced a disturbance index that can be used to compare plots within a study.

6.2.3.4 Other resources

Other resources and ecological parameters such as tree holes used as sleeping sites (e.g., sportive lemurs, mouse lemurs), termite mounds (chimpanzees), gum licks (lorises, marmosets), or clay/salt licks (indris) and others are important for primate species (Campbell *et al.* 2011a). To measure abundance, they can simply be counted in plots or recorded as present or absent from an area. Further characteristics can be sampled. Characteristics of tree holes may be the tree species, height of the hole, size of the opening, volume, wall thickness, and orientation of opening (Biebouw *et al.* 2009; Radespiel *et al.* 1998); although rarely recorded in primate studies, it would also be valuable to record materials used to construct and line the nests.

6.2.3.5 When there is no forest left?

Some primates are capable of existing in human-modified areas such as plantations or gardens (Moore *et al.* 2010). In plantations, habitat assessment is similar to that used within the forest. Variables like distance to forest, frequency of human presence, diversity of food resources, or others may also be investigated. Moore *et al.* (2010) studied purple-faced langurs living in small home ranges in gardens and rubber plantations, enclosed by rice paddy fields and a lake. They plotted feeding and non-feeding trees with DBH ≥ 10 cm and houses on a map, measured DBH, basal area, and density of trees, and calculated the importance value and species diversity of trees. Apart from trees, roofs, fences, walls, and electrical wires and posts were considered as potential substrate while observing the langurs' behavior.

6.2.4 Fragmentation and patchiness of habitats

Patchiness of a landscape can have anthropogenic but also natural causes, such as tree fall zones created by storms. However as human-induced deforestation continues, primate habitat is decreasing and becoming increasingly fragmented. An isolated forest patch is called a fragment while the environment around those fragments is referred to as the matrix. Chimpanzees are able to disperse between two isolated patches of a given distance more easily than a mouse lemur. Gillespie and Chapman (2006) for instance defined 50 m as the minimum distance between fragments for red colobus monkeys.

Edges of fragments are dynamic zones characterized by the penetration of environmental conditions from the matrix into the forest interior (Lehmann *et al.* 2006) and may have very different microhabitats compared to the forest interior. Depending on the spatial and resource needs of species as well as their flexibility they may not be able to persist in small fragments as the ratio between forest edge and forest interior is negatively correlated to patch size (Kapos 1989). While some species like insectivorous, folivorous, and dietarily flexible species may be tolerant towards forest edges (Corbin and Schmid 1995; Lehmann *et al.* 2006), in general, overall species richness and diversity decreases in smaller fragments.

Microhabitat in relation to distance to forest edge can be sampled by using transects leading into the forest interior at a right angle from the forest edge. Lehmann *et al.* (2006) used transects placed perpendicular to the forest edge reaching 1250 m into the forest. They also took vegetation samples along the line and estimated densities of different lemur species for every 100 m increment of the transect. They found several lemur species to be edge-tolerant, which may explain why many lemur species have survived extreme habitat loss and fragmentation in Madagascar. If a researcher is undertaking a study involving the effect of fragmentation and forest edges, several of the above described measurements (e.g., DBH, canopy cover) may be used to describe the fragment. The following measurements can be added to account for the size and position of the fragments.

Fragment size: If a fragment is approximately regularly shaped, the perimeter or the length of the edges can be used to calculate area. In other cases GPS points can be taken along the perimeter and the size of the resulting polygon calculated in a GIS (see Section 6.3.2 below). The track function of GPS devices takes track points at different time or distance intervals and some GPS even include a function to calculate the area enclosed by a track directly.

Distance to nearest fragment: In order to assess the dispersal opportunities for the species of interest, the distance to the nearest fragment can be measured using a range finder, a tape measure, or a GPS. These distances have to be estimated for individual species as they will differ depending on a species size and movement style.

6.3 Species niche modeling

6.3.1 Species distributions and conservation

Comprehensive primate distribution data can help researchers examine relative levels of threat and protection, define conservation status, and set up effective conservation strategies. Although these data are needed, they are often unavailable and difficult to collect (Anderson *et al.* 2002). The methods we describe above and those described in Chapter 2 (on surveys) often require intensive fieldwork in hard-to-reach areas. Many of the primates for which few data are available are relatively small, located in areas of political dispute, or are nocturnal or simply do not attract conservation funding. For these species where complete distribution data are unavailable, species niche modeling (SNM), including ecological niche modeling (ENM) and predictive niche modeling (PNM), can be useful solutions. These terms (SNM, ENM, PNM) are often used interchangeably but their use may vary depending on context; for example, PNM may refer more specifically to using modeling algorithms to project species distributions while ENM may refer to clipping habitat layers in GIS to show a species' general niche. These models have great utility for conservation, such as directing where researchers should spend often limited time and resources searching for understudied or cryptic species, as well as providing foundational information for making landscape management decisions.

6.3.2 Geographical Information Systems (GIS)

Geographical Information Systems (GIS) can be utilized to construct, image, and describe SNM as they enable large-scale data analysis of many variables over a spatial area. GIS is used in many fields and has several applications within conservation biology including environmental classification and mapping, forest fragmentation analyses, population viability assessments, and predictive modeling for selecting priority conservation areas (Hughes 2003). Several recent studies illustrate the wide-reaching and useful application of GIS for primate conservation (Irwin *et al.* 2005; Serio-Silva *et al.* 2006; Buckingham and Shanee 2009; Shekelle and Salim 2009; Thorn *et al.* 2009; Peck *et al.* 2011). Most of these studies were first attempts at assessing the distributions of understudied species (Peruvian yellow-tailed woolly monkeys, Sangihe Island tarsiers, brown-headed spider monkeys, Indonesian slow lorises, spider and howler monkeys in the Yucatan Peninsula).

6.3.3 Species niche modeling and GIS

SNM is a relatively a new method for depicting and studying species ranges. Previously, distributions were portrayed by conservative dot or shaded outline maps, which did not specify underlying data and were highly dependent on subjective knowledge of the study species and/or region. Comparatively, SNM is a significant step forward in the imaging and study of species distributions. Further general information can be found in Franklin (2009) and Peterson *et al.* (2011).

A benefit of SNM is that it allows for the production of highly detailed and testable maps, increasing both the reliability and potential impact of the modeled results (Rodriguez *et al.* 2007). Using GIS to image, analyze, and manipulate distribution information makes it possible to build complex maps of species occurrences, prioritize areas for conservation, and assess the completeness of existing protected area (PA) networks (Box 6.2).

6.3.4 Predictive niche modeling and the ecological niche

PNM combines known species occurrence information with environmental variables to create a model of the ecological conditions within which a species may exist (Franklin 2009; Peterson *et al.* 2011). This methodology is based on the concept of the ecological niche (EN), a critical determinant of species distributions (Hutchinson 1957). An EN is defined in two ways: the fundamental niche and the realized niche. The fundamental niche is the conjunction of ecological conditions (temperature, precipitation, altitude, vegetation, etc.) within which a species is able to maintain populations without immigrations (Grinnell 1917). This can be thought of as the range of theoretical possibility, or a species' potential distribution without boundaries or competition. Although suitable conditions may exist within an area, species rarely occupy an area completely due to biological and historical restrictions. This reduces the fundamental niche to an exploited subset, the realized niche (Hutchinson 1957).

Only the realized niche can be observed in nature, but by modeling species across their entire geographic distributions using a set of specific environmental variables, the fundamental niche can be described (Peterson 2001). In modeling the fundamental niche it is important to note that some areas identified by models may be occupied by closely related species, or may represent areas the study species failed to disperse into, or be areas in which the species in question has become locally extinct (Anderson *et al.* 2003). Therefore, care is recommended in choosing a PNM algorithm and interpreting results.

6.3.5 Predictive niche modeling algorithms: GARP and Maxent

Ideally the best predictive system is one that converges quickly to the greatest accuracy. A variety of modeling methods can produce different predictions, even when using the same data sets (Loiselle *et al.* 2003; Hernandez *et al.* 2006). Therefore it is imperative to select, compare, and test PNM methods cautiously. Currently the most commonly used algorithms are GARP and Maxent. These algorithms are relatively easy to use, do not require absence data, and have produced useful predictions.

GARP has been tested and applied widely in conservation biology. GARP functions as a genetic algorithm that can scan over a large search space to find solutions that give high values for a selected set of criteria (Anderson *et al.* 2003). This method acts as a super-set of bio-climatic envelope (i.e., the upper and lower bounds for each environmental variable), atomic (i.e., specific values for each variable), and logistic regression (logit) rules (Anderson *et al.* 2002; Pearson *et al.* 2007). GARP uses an iterative process of rule selection, evaluation, testing, and incorporation or rejection to search for non-random associations between the environmental characteristics of known species occurrences and the overall study region.

The standard GARP program produces many results, requiring the researcher to select a best subset and synthesize these results into a final prediction. A best subset is the portion of GARP results with the strongest prediction. There are many ways to determine and synthesize a best subset (Anderson *et al.* 2003); this process can be simplified by using a version of GARP "with best subsets option" in open-Modeller, an open source niche modeling library (<http://openmodeller.source-forge.net/>). Within a set of researcher-specified parameters, openModeller analyses results and selects a final best model, a process more comparable to that of Maxent. The algorithm has been modified to deal with presence only data (See Section 6.3.6) by selecting pseudo-absence localities from the background sample at random (Phillips *et al.* 2006; Pearson *et al.* 2007). For a complete description of the GARP method see Stockwell and Noble (1992) and Stockwell and Peters (1999). Morales-Jimenez *et al.* (2005) used GARP to model the distribution of Colombian spider monkeys, and were able to assess the conservation status of all three Colombian spider monkey species. Morales-Jimenez found *Ateles hybridus* in particular was not covered well by the protected area system and subsequently garnered conservation action for this species (Morales-Jimenez and Link 2006).

Maxent is a general-purpose algorithm for predicting distributions from incomplete information (Phillips *et al.* 2006) and has been used successfully to direct conservation priorities. For a thorough definition of Maxent see Phillips *et al.* (2006). Maxent estimates distributions by using a set of environ-mental variables and species localities to find the probability distribution of maximum entropy (i.e., that which is most spread out or closest to uniform), relative to a set of constraints imposed by the inputted data sets. The program computes a probability distribution based on environmental variables spread over a study region using a random sample of background pixels (Phillips *et al.* 2006; Pearson *et al.* 2007). Maxent has proven to be valuable for highlighting potentially unknown species populations and areas of high habitat suitability (Pearson *et al.* 2007; Thorn *et al.* 2009). In some comparative studies, Maxent was statistically more accurate than and generally outperformed GARP, especially with modeling smaller sample sizes (Elith *et al.* 2006; Hernandez *et al.* 2006; Phillips *et al.* 2006; Pearson *et al.* 2007).

Box 6.2 Species niche modeling anthropogenically altered habitats: a case study of two critically endangered primates in Sri Lanka's fragmented rainforests

Sri Lanka is home to two highly endangered primates: the Horton Plains slender loris (*Loris tardigradus nycticeboides*) and the Western purple-faced langur (*Trachypithecus vetulus nestor*) (Nekaris 2006; Parker *et al.* 2008). The distributions of both species are highly fragmented and basic ecological and field data are lacking for both. The rate of natural forest loss is considered to be one of the highest in South Asia; combined with one of the highest population growth rates in a biodiversity hotspot, Sri Lanka's habitats will likely continue to degrade (Cincotta *et al.* 2000). In such a scenario, SNM can be an indispensable tool to predict areas of species' presence and highlight remaining areas of habitat suitability. Results can be used with GIS to quantify the amount of remaining habitat within and outside PA boundaries, and analyzed to make conservation recommendations for PA extensions, connecting corridors and survey sites.

In 2009, we modeled the distributions of *L. t. nycticeboides* and *T. v. nestor* using presence only locality data (5 and 22 final data points, respectively) from natural history museum collections, published accounts, and personal field notes. These data sets and 20 environmental data layers were entered into Maxent. A receiver operating characteristic (ROC) curve was used to test for model significance. Following Phillips *et al.* (2006), ROC curves were developed to distinguish presence from random, where random prediction AUC values are $n = 0.5$, and strong prediction AUC values are closest to $n = 1$. Maxent models were projected into ArcGIS 9.2 to perform a Gap Analysis and Risk Assessment. A Gap Analysis was carried out by comparing remnant habitat fragments to current PAs and protected vs. unprotected areas were quantified. A Risk Assessment was carried for unprotected areas using a set of developed criteria. Unprotected fragments were highlighted as Priority 1 or 2 for conservation action.

Maxent developed distribution models that were significantly better than random for both *L. t. nycticeboides* ($n = 0.99$) and *T. v. nestor* ($n = 0.93$). As locality data for this study were carefully selected with the aid of expert opinion and therefore considered highly reliable, this study used a "lowest presence threshold" (LPT). LPT works to identify areas that are at least as suitable as those where a species' presence has been recorded.

Our research suggests that although a considerable amount of habitat of *L. t. nycticeboides* is currently protected, the species could benefit by enlarging one sanctuary (Fig. 6.1a). Additionally our research highlighted five areas for future survey work (Fig. 6.1b). Comparatively, the remaining, heavily fragmented, habitat of *T. v. nestor* is significantly under-represented in Sri Lanka's current PA network. Connecting corridors between some existing PAs could increase gene flow and total protected area (Fig. 6.1a).

(a)

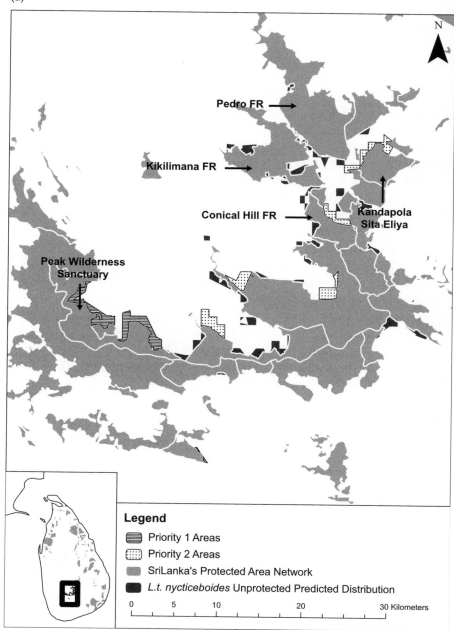

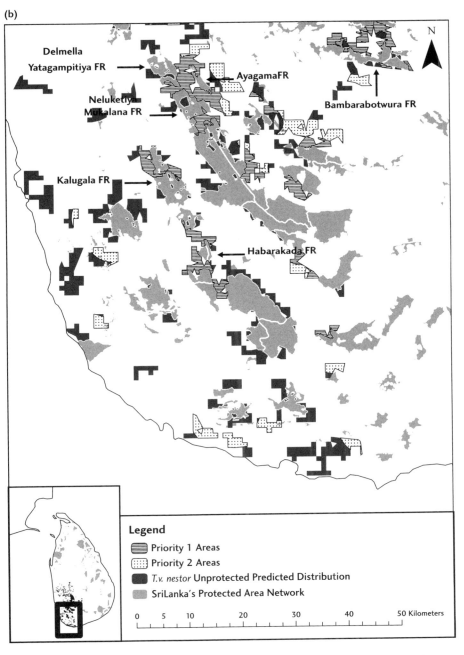

Fig. 6.1 (a) The Maxent predicted distributed for *L. t. nycticeboides*, including Peak Wilderness Sanctuary (suggested as a priority expansion area), and several sites suggested as key survey sites: Conical Hill Forest Reserve (FR), Kikilimana FR, Pedro FR, and Kandapola Sita Eliya. (b) The Maxent predicted distribution for *T. v. nestor*, including a recommended corridor from Delmella Yatagampitiya FR to Ayagama FR to Neluketiya Mukalana FR, and key survey sites: Bambarabotwura FR, Kalugala FR, and Habarakada FR.

6.3.6 Using presence only data

A majority of PNM use presence only data. Previous modeling methods have often depended on presence–absence data, as assessment of model accuracy was dependent on observed absence information. However negative data can be problematic and true absence data are rarely available, especially for rare and endangered species in regions that have been incompletely surveyed. A species can be easily and inaccurately classified as "absent" because (1) it was not detected even though it was present or (2) the species is absent for historical reasons even though the habitat is suitable (Hirzel *et al.* 2002). When a species' potential distribution is based on the fundamental niche, the use of absence data can adversely affect the model building process by excluding areas where the species may live (Anderson *et al.* 2003). Overall, it is beneficial to run models that use presence only data as that is the vast majority of information available to researchers.

6.3.7 Data collection for predictive niche modeling

6.3.7.1 Locality data

Good locality data for the study species is a critical first step in data collection. The most reliable sources of current data are field sightings, either from the published literature or from questionnaires of people working in the field. Reliable data can also come from natural history photographers and by searching online photographic and video databases that give reliable geographic coordinates (such as MaNIS, GBIF, and other online databases). Grey literature, such as annual reports from NGOs and funding bodies, provide a wealth of locality data, including information from recent surveys and fieldwork. Natural history museum collections are an additional resource (Loiselle *et al.* 2003; Gaubert *et al.* 2006). Data from museum collections can help to provide historic distribution information for species.

Once the locality data are assembled, it is a question of quantity (the amount of locality data) versus quality (geographic accuracy). As both GARP and Maxent predict relatively well with lower sample sizes (Stockwell and Peterson 2002; Hernandez *et al.* 2006), it is advisable to prioritize quality over quantity. For example, a researcher might remove localities that give only an approximation of coordinates, where the taxon is non-specific, or points that are repetitive within the data set. The species' localities should then be projected into GIS and examined for spatial autocorrelation. Spatial autocorrelation occurs when several points are grouped together because a species was sampled repetitively within a particular area (Phillips *et al.* 2006; Pearson *et al.* 2007). A suitable species-specific cut-off point should be selected, and one point from the cluster should be chosen to represent the area. This process is necessary because spatially autocorrelated data are not mutually independent and can negatively influence both model building and statistical analysis (Hampe 2004). It should be noted this may be a non-issue in newer versions of Maxent. Additionally, Maxent can be run using a "bias layer" to incorporate sample selection bias into the algorithm (Phillips *et al.* 2009).

6.3.7.2 Environmental data

Detailed and reliable environmental data are needed to define species' environmental relationships, which in turn are required to establish a species' predicted distribution outside of known localities (Rodriguez *et al.* 2007). Environmental variables must be current and high quality; the degree to which models accurately describe a species' range depends on how accurate the environmental variables are. Many environmental variables can be sourced from open source online databases (such as Worldclim). Vegetation layers can also be useful environmental layers, and are often created using methods described in the first part of this chapter. Applying such methods over very large geographic areas has the disadvantage of making these layers both categorical and subjective. Their use in modeling presents a special challenge due to the increased margin for human error and the large numbers of categories typically defined within such layers (Pearce *et al.* 2001). If used in model building, vegetation layers should represent a reconstruction of original vegetation types within the region as they would therefore have a temporal correspondence with species' localities (Phillips *et al.* 2006; Thorn *et al.* 2009).

6.4 Concluding remarks

The environments in which primates have evolved and live have influenced many things: body shape, dietary choices, social structures, and cognitive evolution. Measuring the complexity of that environment is no easy task, and seeing it from the perspective of an animal that is often 30 m high in a tree can be even more daunting. Applying field data to computer models so data from site-specific surveys can be used to predict broader trends can be a cost-effective and time-saving addition to field studies. We advocate the marriage of these two techniques. No computer model can be considered truly valid without ground-truthing, but at the same time, computer modeling allows for the use of data across a much broader scale, especially at a time when habitat loss is escalating at an alarming rate. We hope we have given you some tools to measure primates and their habitats, and that in the near future trends of habitat loss will reverse and future primatologists can focus on measuring forest regeneration rather than forest fragmentation and loss.

Acknowledgments

We thank the editors for asking us to contribute to this volume. We thank various students on the MSc in Primate Conservation whose data inspired various aspects of this paper: A. Arnell, A. Morales-Jimenez, F. Ruperti, I. Russell-Jones and J. Thorn. An Oxford Brookes Central Research Strategy Fund grant to Nekaris facilitated the writing of this paper.

7

Characterization of primate environments through assessment of plant phenology

Andrew J. Marshall and Serge Wich

7.1 Introduction

Plant allocation to reproduction and growth is not evenly spread over time. In most plant species, production of flowers, fruits, and leaves is confined to a limited portion of the year (Leith 1974; van Schaik and Pfannes 2005). In some taxa, bouts of flowering, fruiting, and leaf flushing may be separated by periods of many years (e.g., Southeast Asian mast flowering and fruiting: Wich and van Schaik 2000). Plant phenological patterns create substantial temporal and spatial variation in the availability of plant foods and consequently primate diets, with a number of important implications for primate reproduction, grouping, ranging, and sociality (van Schaik *et al.* 1993; van Schaik and Brockman 2005). For example, in many primate species, reproduction is timed to coincide with periods when plant food availability maximizes the chances of offspring survival (Myers and Wright 1993; Knott 1998). The size and cohesiveness of primate groups often fluctuates substantially over time in response to the availability of plant foods (Chapman *et al.* 1995; Wrangham *et al.* 1996) and primate ranging patterns are often influenced by plant phenological patterns (Buij *et al.* 2002; Hemingway and Bynum 2005). Finally, fundamental differences in sociality among primate taxa are hypothesized to depend on the nature and temporal stability of a species' plant food supply (Wrangham 1980; Sterck *et al.* 1997; Marshall and Wrangham 2007). Accurate characterization of plant phenology is therefore centrally important to many facets of field primatology.

In this chapter we provide an overview of the methods primatologists employ to assess plant phenology. Our focus is on practical issues of most relevance to field primatologists who seek to characterize the environments that their study subjects inhabit. We discuss the benefits and limitations of various sampling methods,

7

Primate Ecology and Conservation: A Handbook of Techniques. First Edition. Edited by Eleanor J. Sterling, Nora Bynum, and Mary E. Blair. © Oxford University Press 2013. Published 2013 by Oxford University Press.

selection of a sample to monitor, and the scale and scope of sampling. We end with a brief discussion of ways in which phenological data can be described, analyzed, and presented.

7.2 Field methods

Primatologists employ a range of methods to assess plant phenology. As with all decisions regarding sampling strategies, one must carefully consider the strengths and limitations of different methods, weigh various trade-offs (e.g., sampling intensively vs. extensively, financial and opportunity costs associated with alloca-tion of time towards collection of phenology data), and select a method that is appropriate for the study to be conducted. For example, assessment of primate feeding selectivity and preference requires data on the availability of all potential foods in the environment (Leighton 1993; McConkey *et al.* 2002); research examining the behavioral or physiological response of a primate species to variation in its food supply can restrict sampling to known food plants for the primate species in question (e.g., Chapman and Chapman 1996; Knott 1998; Newton-Fischer *et al.* 2000); and studies of habitat use require simultaneous monitoring of plants across the full mosaic of habitats that the study population inhabits (Leighton and Leighton 1983).

Regardless of the method employed, a phenological sampling strategy should produce a description of a primate's environment that is accurate, precise, and representative. In this context, accuracy refers to the degree of conformity between the sample (e.g., the percentage of trees in plots bearing flowers in a particular month) and the actual value (e.g., the percentage of all trees within a focal primate group's range that are flowering that month). Phenological accuracy will increase with larger sample sizes of stems, greater numbers of independent sample replicates (e.g., more plots or phenology transects), and when sampling units are placed in an unbiased way (e.g., plots are randomly placed).

Precision refers to the amount of sampling error in measurements. Highly precise phenological sampling regimes will have low sampling error: repeat meas-urements of plant productivity at a given time, either by the same individual or multiple ones, will produce very similar results. Precision can be increased through adequate initial training of all who collect data, periodic cross-checks to ensure all involved researchers obtain concordant results, and use of clear protocols and unambiguous decision rules for field data collection.

Finally, phenological measures should be representative of the full range of environments that the target primate individual, group, or species encounters. Even when there is substantial spatial heterogeneity in a primate's environment, representative results can be obtained via monitoring a large number of samples that are randomly located throughout the primate's habitat or by collection of data in a stratified manner that explicitly incorporates habitat variation into the sampling strategy (e.g., stratification of phenology plots based on forest type: Cannon *et al.* 2007b; Marshall *et al.* 2009a; Wich *et al.* 2011).

Here we focus on phenological methods appropriate for closed-canopy forests, as the majority of primates inhabit tropical or subtropical forests. Although we do not explicitly consider the assessment of plant phenology in woodlands, savannahs, or other environments, many of the principles discussed in this chapter apply to any primate habitat.

7.2.1 Selecting the sample to be monitored

All sampling methods have strengths and limitations, and some are better suited to particular studies than others (Table 7.1). Below we consider several methods and,

Table 7.1 *Comparison of phenological sampling methods discussed in the text.*

Sampling method	Description	Strengths	Limitations	Best suited for
Area based	Monitoring all plant stems that meet some basic criteria (e.g., size) within a delineated area (i.e., plot), usually multiple replicates are placed throughout the focal species' range	Unbiased, assuming randomly placed; stems sampled in proportion to their abundance; data can be used for other purposes, providing opportunities for collaboration	Time consuming; expensive; can provide too few stems of important, but rare, foods	Studies of unknown taxa or populations, where foreknowledge of diet is limited; studies that have a heavy ecological focus, where detailed description of plant phenology at the community level is necessary (e.g., preference studies)
Focal plant taxa	Monitoring a predetermined set of important food plants, often located along set phenology routes for easy access	Generally less work intensive; more efficient for some purposes as data are only gathered on species known to be food resources for the focal primate taxon	Requires foreknowledge of the species' diet in order to select target plant taxa; provides no information of habitat-wide productivity so can't assess	Studies of taxa for which long-term data exist, so it is relatively easy to determine which plant taxa are important to monitor; studies focusing on specific ecological

(continued)

Table 7.1 *Continued*

Sampling method	Description	Strengths	Limitations	Best suited for
			selectivity; stems are not monitored in proportion to their abundance, so difficult to place results in appropriate ecological context	interactions between primates and a particular set of plants (e.g., seed dispersal or predation studies)
Hybrid method	Use a representative number of plots to assess densities of all taxa, then monitor an additional set of rare but important taxa outside the plots. Then, in the analysis phase, be sure to scale the estimate of food availability by the stem densities	Provides unbiased sample of forest wide phenology and a more focused description of the food availability for the focal primate taxon	Difficult to know, *a priori*, how effort should be allocated among area-based versus focal taxon-based sampling	Collaborative studies with multiple researchers interested in different questions, so work can be divided up; forests that are not too spatially heterogeneous, so extrapolation from a limited sample is defensible
Fruit trails or traps	Walk along a predetermined route, counting fruits on the ground or placing traps of a standard size and periodically counting fruit fall into traps	Simple, replicable across sites; especially with traps, easily quantified and counted	"Only fallen fruits counted, fruits still in canopy not included; traps and fallen fruit can be disturbed by animals"	Simple cross-site comparisons (provided differences among sites that could bias comparisons are considered)
Remote sensing	Use spectral signatures of leaf flushing/fruit production	Broad spatial and temporal sampling; can sometimes be done retroactively, depending on imagery	Often imagery expensive, especially at high resolution; requires specialized training, skills; little taxonomic resolution; little information about understory	Very broad scale comparisons across space (and potentially time)

where applicable, illustrate our discussion with examples drawn from our own phenological studies in Indonesia.

7.2.1.1 Area-based sampling (e.g., plots)

Assessing plant phenology via area-based methods entails monitoring all plant stems that meet some basic criteria (e.g., diameter at breast height (DBH), height) within a delineated area (e.g., plot, transect). Plant stems in plots are normally permanently marked (e.g., with aluminum tree tags) and identified to the highest level of accuracy possible. Although desirable, perfect taxonomic identification is not required provided the confidence in botanical identifications is clearly reported and, for studies on primate diet, as long as the names used for plant stems in the plots match those used when gathering feeding data—even if the identifications are to morphospecies only. Usually, multiple sampling replicates are placed throughout the focal species' range, either randomly or by using a stratified random design that accounts for gross-level spatial heterogeneity (See Section 7.2.2). Provided they are accurately laid out and unbiased decision rules are made about inclusion of stems that fall on the edge of a plot, any shape can be used. We suggest, however, that the use of squares or rectangles is easiest and least prone to mistakes. In general, as mistaken determinations about whether or not to include stems that fall on the edge of a plot may bias results, we suggest minimizing the amount of plot edge for a given plot area (e.g., a 20 m × 50 m plot is preferable to a 5 m × 200 m plot as the former has roughly a third of the latter's perimeter).

When stems do fall on the edge of plots (e.g., when part of a tree's bole is inside the plot and part is outside), a clear decision rule must be created and consistently applied. In order to accurately sample the density of stems, the midpoint of the stem should be used to determine whether the plant falls inside or outside the plot (Fig. 7.1a). For example, if the plot edge passes through the bole of a tree with a 100 cm DBH in such a way that only 30 cm of the diameter is inside the plot, the stem should be excluded from the sample. Inclusion of trees that are less than halfway inside the plot would result in an overestimation of stem density, a bias that would be greater for larger sized trees (as large trees with midpoints that lie outside a plot are more likely than small trees to have part of their boles inside plots; Fig. 7.1b).

Lianas (woody climbers) present particular problems when establishing phenology plots. In addition to the challenge of determining the taxonomic identity of lianas, it is often difficult to determine which individual stems should be included and excluded from plots, and it is frequently difficult to determine the number of individual lianas present when lianas span multiple tree crowns. Although it is tempting to include in the sample any liana that enters the canopy of a tree inside a

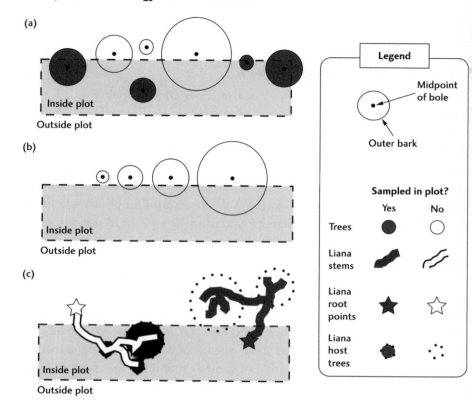

Fig. 7.1 Schematic depiction of phenology plots. The shaded region indicates the area inside plot, the heavy dashed line the edge of the plot. Circles depict stylized tree boles, with the midpoint of the stem indicated by a dot. (a) Trees whose midpoints are located inside the plot (shaded circles) are included in the sample; trees whose midpoints are outside the plot (open circles) are not sampled, even if a portion of their bole is located inside the plot. (b) Cartoon demonstration showing that the boles of larger trees are more likely to enter a plot than boles of smaller trees, although their midpoints are equally close to the plot border. This demonstrates why inclusion of trees based on whether any portion of the stem is found inside the plot biases estimates of stem density, a bias that is increasingly pronounced at large tree sizes. (c) Lianas should be sampled and monitored for phenology only when they are rooted (star) inside the plot (dark stem), regardless of which tree crowns they enter. Conversely, lianas rooted outside the plot (white stem) should not be included in the sample, irrespective of whether their host trees are sampled. Note that the size of the plot is not shown to scale relative to the size of the plant stems.

phenology plot, this method can lead to substantial biases since the inclusion of lianas in plots is then not directly related to their stem density. In order to obtain an accurate, unbiased measure of liana abundance and the phenology of lianas in a forest, we recommend including in the sample to be monitored only and all lianas that are rooted inside the perimeter of the plot. Thus, a stem rooted in the plot but which inhabits the canopy of a tree well outside the plot would be included, whereas a liana rooted outside the plot but which occupies a tree inside the plot would not (Fig. 7.1c). When a liana is rooted in multiple places, the last rooting point prior to ascending into the canopy should be used (Gerwing *et al.* 2006). This system requires knowledge of and clear marking of liana stems within the plot to ensure mistakes are not made in monthly observations. As epiphytes and hemi-epiphytes generally inhabit only a single tree, inclusion of individuals exhibiting one of these growth forms can be determined based on the location of its host tree.

Area-based methods provide a number of benefits. First, provided plots are placed randomly, they provide an unbiased sample of habitat-wide phenological patterns because stems are sampled in proportion to their abundance. Second, if one is beginning a primate study on an unstudied species or in a new environment, area-based methods are practical since they do not require foreknowledge of which plants are foods for the focal primate taxon. Third, these methods provide accurate data on the phenology of all potential foods in the forest, both things eaten and not eaten, permitting analysis of feeding selectivity. Finally, data from phenology plots are useful for a range of other questions. For example, subsampling of data permits assessment of temporal changes in food availability for any taxon in the forest for which dietary information is available (Marshall *et al.* 2009c).

Although providing the most comprehensive description of habitat-wide phenological patterns, area-based methods have considerable limitations. They are relatively time-consuming and expensive to establish and monitor (Chapman *et al.* 1994). For example, ongoing phenological work at the Cabang Panti Research Station (CPRS) in Gunung Palung National Park, West Kalimantan, Indonesia (Marshall 2009, 2010), comprising monthly monitoring of ten plots (totaling 1.5 ha) in each of seven forest types (roughly 6000 stems total), requires two full-time employees. In addition, since in many forests the majority of the effort spent conducting area-based sampling will be expended sampling non-food stems, and because many foods (especially preferred foods) are rare (Marshall and Wrangham 2007; Fig. 7.2a,b), area-based methods often produce sample sizes per food species that are inadequate for many types of analysis (i.e., fewer than 15 individuals per plant species, see section below).

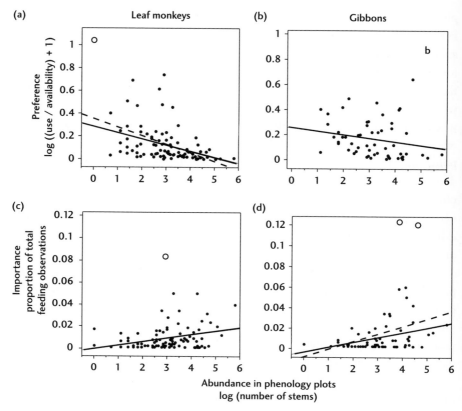

Fig. 7.2 Preferred foods are rare: (a) preference (use/availability) of leaf monkey food items (log (preference + 1)) is negatively correlated with abundance (log (# stems in phenology plots)). Spearman's rho = −0.52, p < 0.0001, n = 89 plant food taxa; when outlier in top left (the very rare *Adenanthera* (Fabaceae), open circle) is removed, results are consistent: rho = −0.50, p < 0.0001, n = 88 taxa. Regression lines provided for visual clarity only (solid and dashed lines fit with the outlier excluded and included, respectively). (b) Same as panel a, but for gibbon foods. Spearman's rho = −0.32, p = 0.015, n = 58 plant food taxa. Important foods are common: (c) importance (proportion of total independent feeding observations) is positively correlated with abundance (as in panel a). Spearman's rho = 0.35, p = 0.0005, n = 98 taxa. Inclusion or exclusion of the highly important *Dialium* (Fabaceae, open circle) does not alter the results. (d) Same as panel c, but for gibbon foods. Including two highly important foods (both *Ficus* taxa (Moraceae) open circles) Spearman's rho = 0.41, p = 0.0005, n = 66 taxa; excluding outliers, Spearman's rho = 0.38, p = 0.002, n = 64 taxa. Regression lines provided for visual clarity only (solid and dashed lines fit with the outliers excluded and included, respectively). Preference, importance, and plant abundance data from Marshall (2004), Marshall and Leighton (2006), and Marshall et al. (2009b). Analyses include only the portion of the diet comprising fruit pulp and seeds. Samples sizes for leaf monkeys: n = 734 independent feeding observations; gibbons: n = 481 independent feeding observations. Plant data from 126 0.1 ha plots placed using a stratified random design across the seven forest types at CPRS (n = 7288 plant stems; Cannon et al. 2007b).

7.2.1.2 Sampling focal plant taxa

Primatologists frequently collect phenology data on only a limited, predetermined set of important food plants (Chapman *et al.* 1994). These stems are often located along set phenology routes and near existing trails for easy access. Generally, monitoring at least 5–10 stems of each species has been recommended (National Research Council 1981), although recent simulation analyses suggest that samples of 15 stems per taxon are necessary to adequately describe phenological patterns (Morellato *et al.* 2010). We note, however, that the target number may need to be larger, sometimes substantially so, to adequately address certain research questions (e.g., assessing the degree of phenological synchrony within a particular plant taxon) or where there is substantial spatial heterogeneity in the study system, when an adequate sample should be collected in each forest type. When using focal plant taxa to assess phenology, it is important that plants are correctly identified by experts to ensure the appropriate taxa are being monitored.

Plant species exhibit a broad range of complex sexual systems, including dioecy, where individual stems are either male or female (Croat 1979). When monitoring the phenology of dioecious species for the purposes of assessing fruit abundance, care must be taken to ensure that female individuals are selected because male stems never produce fruit (National Research Council 1981). Roughly 4–5% of the world's plant species are dioecious (Richards 1996), although the proportion of dioecious plants can be considerably higher in tropical forests. For example, it has been estimated that 26% of the tree species in Bukit Raya, Sarawak (Ashton 1969) and 21% of the trees and shrubs on Barro Colorado Island, Panama (Croat 1979) are dioecious; the proportion of stems that are dioecious seems to increase at larger tree sizes. As large trees are relatively rare in plots, ignoring the sex of sampled stems could introduce substantial error. Determining the sex of trees is usually impossible unless they are flowering, so for dioecious species, a larger number of stems should be monitored (e.g., ~30 stems per species) to ensure that an adequate number of female individuals are included.

Assessing phenology via study of focal plant taxa is popular among primatologists (e.g., Chapman and Chapman 1996; Newton-Fischer *et al.* 2000). In large part, this is because it is less time consuming than conducting comprehensive, area-based methods that require the establishment of plots. In addition, gathering data on solely focal food plants is an efficient way to measure temporal changes in the abundance of food, as time is not spent monitoring stems that primates do not consume. Although commonly used, the method does have substantial limitations (Chapman *et al.* 1994). First, this method requires foreknowledge of which plant taxa are important foods for the focal primate taxon, something that cannot be known for an unstudied species or population. Even after years of long-term study, new food items are regularly added to the list of plants that a study group consumes (e.g., Rogers *et al.* 2004) making it difficult to be certain that all key foods are being monitored (Hemingway and Overdorff 1999). Second, as only known food plants are monitored, no information is available regarding availability of plants that are not food, meaning that assessment of feeding selectivity—which requires assessment of relative use and relative availability—is impossible (Leighton 1993; Marshall and Wrangham 2007). Third, as monitored trees are often clumped together along set phenology routes that are usually situated in convenient locations (e.g., near camp), they may not be representative of phenological patterns across the site, potentially biasing results, especially at large study sites. Fourth, as the number of monitored stems of a given plant food is not necessarily proportional to that food's importance in the diet (cf. Fig. 7.2c,d) or its relative abundance in the forest, measures of food availability derived from lists of focal plant taxa cannot be directly interpreted as indices of either the amount of food available for the target primate taxon or broader patterns of forest-wide fruit availability. Finally, these data cannot be used in most simple optimal foraging models, as these methods require habitat-wide measures of average food availability (Pyke *et al.* 1977).

7.2.1.3 A hybrid method

The two methods discussed above each have associated strengths and limitations. A hybrid of the two methods may be used that provides some of the benefits of each while limiting their attendant shortcomings. One form of this hybrid would be to establish a limited number of area-based plots to provide accurate, unbiased estimates of the stem densities of plants in the environment and to monitor their phenology. An additional set of relatively rare, but important food resources could then be tagged and monitored outside of the area-based plots to increase the sample sizes for these key resources. In the analysis phase, estimates of food availability from a particular plant taxon would be scaled by that plant's stem densities in the area-based plots. A second form would monitor only focal plant taxa (ideally across the full range of habitats that the study group or population occupies), as described in the previous section, but in the analysis phase data on plant productivity would be scaled by an objective index of habitat-specific abundance derived from botanical plots. These botanical plots would be surveyed once to determine floristic composition, but plant phenology would not be monitored in these plots. For an example of this approach see Willems *et al.* (2009). Assuming botanical plots provided an accurate and adequate sample of the entire range of the focal primate group or population (e.g., there are a sufficient number of replicates, the sampling strategy is adequate to characterize spatial variation), both of these forms of hybrid method result in a phenological measure that provides an unbiased assessment of productivity while reducing (or eliminating) the amount of time spent sampling non-food plants.

7.2.2 Scale and scope of sampling

7.2.2.1 Spatial scale

Careful consideration of the spatial scale of the study is necessary to ensure that the phenology data collected are adequate to characterize the system. In general, as substantial spatial heterogeneity is common even within a forest type, multiple, randomly-placed replicates of smaller plots are generally preferable to a single, large plot encompassing the same sampling area (Hayek and Buzas 1997). If differences among habitat types that are relevant to the focal primate taxon or specific research question can be predetermined, stratified random samples are likely to be an efficient sampling strategy. In some cases, however, the delineations among habitat types of most relevance to the focal taxon may not be readily distinguishable at the start of a study; in these cases, random transect placement across the study area is most appropriate. The number of replicates placed will clearly be affected by the type of research question being posed and the availability of resources. We recommend that a minimum of 10 reasonably sized (e.g., ~ 0.1 ha) plots be used to characterize an area, be it a group's home range, a study area, or a forest type within a study area. In this context it is worth remembering that in some systems, contiguous forest types or home ranges can exhibit phenological patterns that are highly distinct, even to the point of being uncorrelated or negatively correlated with each other (Marshall and Leighton 2006; Cannon *et al.* 2007b; Marshall *et al.* 2009a; Wich *et al.* 2011). Sampling regimes that do not account for this source of variability or, at a minimum, permit the assessment of its potential magnitude, are not recommended.

7.2.2.2 Temporal scope

Determination of the temporal frequency at which to sample should be based on the goals of the study and knowledge of the phenological patterns of the species and plant parts in question. Although goals of primate field studies vary widely, and despite substantial variation in the duration of different phenophases (e.g., flowers are generally more ephemeral than fruits), most researchers default to sampling at monthly intervals (van Schaik and Pfannes 2005). As early as 1975, some argued that bi-weekly sampling was more appropriate for accurate characterization of tropical phenology because it would capture ephemeral events more accurately

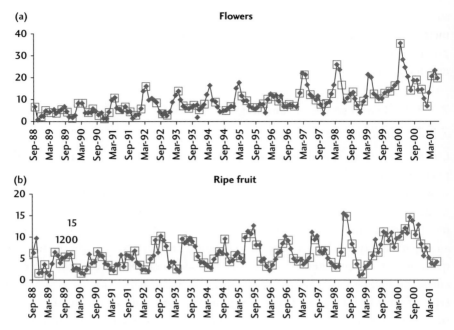

Fig. 7.3 Monthly (closed diamonds) and bi-monthly (open squares) percentage of trees carrying flowers (a) or ripe fruit (b) in the Ketambe study area. Data are from September 1988 until May 2001. Details on methods can be found in (Wich and van Schaik 2000).

(Fournier and Charpantier 1975). A recent study of specific plant species showed that sampling once a month or even at bi-weekly intervals missed flowering peaks and that therefore weekly measures better characterize the number of flowering events and the duration of flowering (Morellato *et al.* 2010). Analyses conducted at the community level show similar patterns: higher sampling frequency provides a more accurate picture of temporal variation (e.g., bi-weekly vs. monthly fruit trail counts: Chapman *et al.* 1994). There is, however, an obvious trade-off between sample size and sampling frequency. Assessing phenology on a weekly basis, as Morellato *et al.* (2010) advocate, would likely result in sampling of far fewer stems than is required to accurately characterize most primate foods (See Section 7.2.1.1), except in sites with extremely low floristic diversity. Sampling less frequently than once a month would permit monitoring of a larger number of stems, but could fail to record small, but potentially ecologically important, peaks in flowering and production of ripe fruit (e.g., Fig. 7.3a,b). On balance, monthly sampling is probably adequate for most primatological field studies, although researchers studying species for whom highly ephemeral resources (e.g., flowers) are important should consider sampling more frequently. A final consideration is the frequency of primate sampling. For example, there is fairly limited utility in sampling plant phenology every other week if primates are only observed every other month.

Sampling duration is another consideration. Most researchers monitor phenology for the same period as they gather data on their focal primate species, which permits examination of the interactions between phenology and primate behavior for the duration of the study. In many areas, it is probably reasonable to interpret the results of a 2–3 year field study as indicative of the ecology of the system more generally—although this is unwise in regions where important phenological patterns occur on much longer timescales. For example, a field study of the duration of a typical Ph.D. project is unlikely to capture the full range of meaningful ecological variation in many Southeast Asian forests. At Gunung Palung, fruit production during mast fruiting events and during periods of low fruit availability can differ by more than an order of magnitude (Marshall and Leighton 2006; Cannon *et al.* 2007b). Periods of extreme food abundance or scarcity can last for extended periods, with important effects on primate physiology and behavior (Knott 1998; Marshall *et al.* 2009b). Short field studies of 1–3 years may well miss masts or the following periods of extreme fruit scarcity, meaning that only a subset of the full range of ecological variation experienced by most of the primate study subjects is sampled. Studies of limited duration also are inadequate to understand the reproductive biology of the plants themselves: in an intensive 5.5 year phenological study of more than 7000 plants at Gunung Palung, a large percentage of stems never reproduced during the sampling period (Cannon *et al.* 2007a).

7.2.3 Inclusion of various plant growth forms and sizes

7.2.3.1 Plant growth forms

Another consideration before establishing plots is which plant growth forms to include. Often, existing information about the basic biology of the study species will be helpful in making certain determinations. When studying primate taxa that never or very rarely go to the ground (e.g., Sumatran orangutans, gibbons), monitoring the phenology of ground vegetation is unnecessary. In contrast, some primates (e.g., gorillas, bonobos) feed extensively on terrestrial herbaceous vegetation, and phenological monitoring must therefore include adequate samples of these plants (Ganas *et al.* 2009). For arboreal species, often only trees are monitored in phenology plots. Lianas are very frequently excluded as they are difficult to identify and monitor and because determination of whether individual stems should be sampled is often time consuming (See Section 7.2.1.1). Nevertheless, lianas are important food resources for some primate species, especially during periods when trees are not fruiting (Leighton and Leighton 1983; Moscovice *et al.* 2007;

Marshall *et al.* 2009b). In such cases, exclusion of lianas from phenological samples results in an incomplete, and quite possibly inaccurate, characterization of food availability for the target primate species. For some primates, epiphytes can be key food resources (e.g., Irwin 2007)—in such instances they should be systematically sampled and monitored. Finally, as figs are extremely important resources for a variety of primate taxa, we strongly recommend sampling all growth forms of figs in phenology plots, including epiphytes, hemiepiphytes, trees, and lianas.

7.2.3.2 Size cut-offs

Ideally, samples used to monitor changes in food availability for a target primate species would include only stems of a size that the target species actually feeds upon. This would be desirable both because it would reduce time spent assessing stems that the target species never uses and because it would produce an index of food availability that was more accurately tailored to the primate species being studied. In practice, however, such information is not available at the beginning of a study and researchers instead must select an arbitrary cut-off below which plant stems are not sampled. Setting a size cut-off can be advisable for several reasons. First, there is an inverse relationship between stem size and stem abundance, so excluding small stems substantially reduces sampling effort. Second as small trees are more likely to be below the size at which trees start producing flowers and fruits, if your target primate species are frugivores, excluding them may not substantially influence the estimates of fruit availability you may be seeking. In the same vein, as there is a steep positive relationship between tree size and crop size (Leighton 1993; Chapman *et al.* 1994), exclusion of small trees that do produce fruit is unlikely to dramatically influence estimates of food availability in most cases. Of course, if your target primate species are folivores, it may be wise to retain even small stems in order to monitor leaf phenology.

A widely-used cut-off for trees is $\geq$ 10 cm DBH, which is a reasonable choice given the considerations listed above. (Note: breast height is generally taken to be 130 or 137 cm above the ground, although conventions are not universally followed.) Furthermore, use of the standard 10 cm DBH tree cut-off facilitates comparisons between sites (Marshall *et al.* 2009a; Wich *et al.* 2011). Lianas are less often measured by primatologists and size cut-offs used vary among studies. A published protocol for liana censuses advocates inclusion of all lianas with stems that are $\geq$ 1 cm in diameter 130 cm from the ground (Gerwing *et al.* 2006; Schnitzer *et al.* 2008); as with trees, the smaller the cut-off selected, the greater the sampling effort required. Liana crop size correlates positively with liana diameter (Marshall, unpublished data), but the relationship is fairly weak— meaning that it is less easy to be confident that exclusion of small lianas will not have important effects on estimates of primate food abundance (cf. trees, previous paragraph). Finally, as hemiepiphytes can be extremely important food resources for primates in some forests (e.g., *Ficus* spp.), we suggest including in phenology samples all hemiepiphytes that are rooted in the ground, with at least one root with a diameter $\geq$ 1 cm, 130 cm from the ground.

Although most size cut-offs are reported as whole numbers and most DBH measurements are also recorded in the field as whole numbers (both in centimeters), care must be taken to ensure that the same rules for rounding are used in all plots. A generally accepted rule is that any measurement for which the tenths place is $\geq$ 0.5 is rounded up to the next whole number; measurements of < 0.5 are rounded down (e.g. 11.6 is recorded as 12, 11.4 is recorded as 11). Therefore, for example, if a DBH cutoff of $\geq$ 10 is used, the true cut-off should be $\geq$ 9.5 cm. If the cut-off used were actually $\geq$ 10 cm, then the first whole number category for DBH measurements would include only half the number of stems as the subsequent ones (e.g., stems measuring 10 to 10.49 cm would be included in the 10 cm DBH grouping, whereas for 11 cm DBH and groupings thereafter trees measuring 10.50 to 11.49 cm, etc. would be included).

As noted above, the use of a small size cut-off for trees is time consuming because small trees are very common. Therefore, for a given amount of sampling effort, the smaller the size cut-off used the smaller the area that can be sampled. Small sampling areas can result in phenology samples that may not be representative of the area to be sampled—particularly in the case of rare figs, lianas, and large trees that are often key food resources. One effective way to partially offset the trade-off between sample size and tree size is to use nested plots, in which small stems are monitored in only a subset of the plot. For example, Marshall and Leighton (2006) used nested plots to assess the stem density of gibbon and leaf monkey foods in each of seven forest types at CPRS, Gunung Palung. All fig roots and liana stems within 10 m on either side of the plot midline were included. The same rule was used for trees with DBH $\geq$ 34.5 cm. Trees with boles $\geq$ 14.5 cm and < 34.5 within 5 m of the plot midline were included. Thus, the sampling area for lianas, figs, and large trees that are often rare, important food sources was increased while limiting the time spent monitoring small, common trees. If data are collected using nested plots, care must be taken when calculating total stem densities because different sampling areas are used for different size classes.

An alternative to using a standard size cut-off for all stems is to tailor size cut-offs based on the plant taxon-specific minimum size at adulthood (i.e., the smallest size at which a plant bears flowers and fruit). Exclusion of non-reproductive individuals can produce a more accurate phenological estimate (e.g., if data are characterized as the percentage of stems with fruit) and reduce the amount of time spent sampling immature stems. It is rare, however, that such data will be available, unless previous work at the same site was conducted prior to the beginning of a primate study. In addition, use of taxon-specific cut-offs requires a high degree of certainty about botanical identifications in phenology plots. Finally, its use severely limits the comparability of plot-level variables among studies. For these reasons, use of taxon-specific cut-offs will rarely be useful for primatologists establishing new phenology plots. In situations where monitoring of additional stems of rare but important plants is desirable (e.g., the hybrid method, Section 7.2.1.3), however, knowledge of the minimum size at adulthood can be very useful to ensure that the additional stems to be monitored have the potential to be reproductively active.

Regardless of the tree size cut-off used, care should be taken to establish and consistently apply systematic rules for situations where standard protocols are not applicable. Examples of such circumstances include when tree stems branch below breast height, where trees grow on slopes or have stilt roots or high buttresses, or when hemiepiphytes or lianas have multiple roots or stems. Discussion of every such eventuality is beyond the scope of this chapter, but forestry manuals and inventory protocols (e.g., Gerwing *et al.* 2006; Schnitzer *et al.* 2008) should be consulted in such circumstances and consistent rules applied.

Surprisingly few studies have, with hindsight after the study was completed, tested whether the size cut-off used for trees in the phenology plots accurately reflected the sizes of stems that the study species utilized. Wich *et al.* (2002) examined this for three species: Thomas langurs (*Presbytis thomasi*), long-tailed macaques (*Macaca fascicularis*), and Sumatra orangutans (*Pongo abelii*). For all three primate species the average DBH of trees fed upon was much larger than the size cut-offs in the phenology plots, but the minimum size of foraging trees was at or only slightly higher than the cut-off size (Wich *et al.* 2002). Thus, for these species, the size cut-off used ($\geq$ 12.5 cm for trees) was appropriate, as it succeeded in capturing the phenology of all tree sizes fed upon while not overestimating food availability by using too low a size cut-off. We recommend that more such studies be conducted.

7.2.4 Characterizing the phenological phase

While patterns of plant growth and reproduction are, of course, continuous processes, generally primatologists use a limited number of discrete, visually-perceptible categories to classify the phenological phase of a plant stem ("Phenological development stages" *sensu* Brügger *et al.* 2003). Typically, the phenophases of flowers and fruits are assessed when present; often, new leaf flushing is also recorded. The precise categories used may vary between studies, but all methods should be operationally defined, consistently applied, and clearly explained when results are reported. For example, Marshall and Leighton (2006) classified plants as reproductively inactive, or containing flower buds (i.e., visibly developing, but not at anthesis); mature flowers (i.e., at anthesis); immature fruits (i.e., fruits in which the seed is undeveloped); mature fruits (i.e., full-sized fruits that are unripe but have seeds that are fully developed and hardened); or ripe fruits (i.e., the final development stage prior to fruit fall, usually signaled by a change in color or softness). Other classification systems have simply recorded the presence or absence of fruit (e.g., Chapman *et al.* 1994; Newton-Fisher *et al.* 2000) or used a reduced number of categories to score phenophases (e.g., noted the presence of young leaves, flowers, or ripe fruit; van Schaik 1986). Frequently, a single plant stem may possess attributes of several phenological classifications concurrently (e.g., bearing a mixture of mature and ripe fruits). In such cases, consistent decision rules must be applied and incorporated into analyses so as not to "double-count" stems. One simple rule is to score the entire stem as the most advanced phenological stage present when the plant is observed (e.g., a plant bearing both flower buds and flowers would be scored as having flowers, Marshall and Leighton 2006).

Often, some specialized knowledge of the plant taxon is required to accurately assign a plant to a particular phenological phase, as, for example, it can be difficult to distinguish flower and leaf buds through binoculars or external cues may not reliably indicate ripeness (e.g., some *Dialium* spp. (Fabaceae), where immature fruits rapidly grow to full size and the seeds mature slowly inside the fruits over a period of many months with no visible external changes). When possible, it is therefore advisable to consult with knowledgeable botanists or others with experience at the study site to develop this knowledge. When this is not an option, we suggest observers take detailed, dated notes and photographs of phenological categories to check previous classifications once deeper knowledge of a plant taxon has been gained.

In addition to classifying the phenological stage of a stem, it is often desirable to assess a plant's crop size. Crop size is measured in various ways: assessment of the proportion of the canopy bearing fruit, using a categorical scale, calculating crown volume, counting the number of fruits in the canopy, or by using DBH as a proxy (Chapman *et al.* 1992; Leighton 1993). Regardless of the method used, some assessment of the error and replicability of the estimate should be made, as the accuracy and precision of methods differ widely (Chapman *et al.* 1992). If such an assessment cannot be empirically made, then we suggest using an exponential scale that controls for the fact that for large crops estimates have disproportionately large associated errors (Leighton 1993). We note, however, that simple indices (e.g., percentage of stems bearing fruit) that do not incorporate crop size estimates are often adequate for many applications.

7.2.5 Alternative methods: fruit trails and litter traps

Two common alternative methods used to monitor phenology are fruit trails and litter traps. The fruit trail method entails an observer walking a predetermined route in the forest (often a clear trail so fruits can easily be observed) and counting the number of plant stems that have produced fruit that are encountered on the trail. Fruit availability is expressed as fruiting sources per km (Chapman *et al.* 1994; Buij *et al.* 2002). The litter trap method typically entails placing a number of square sheets of plastic or mesh of a standard size (e.g., 1 m^2) on four poles that are stuck into the ground (e.g., van Schaik 1986; Chapman *et al.* 1994; Wright and Calderón 2006; Chave *et al.* 2010). At predetermined intervals (weekly, bi-weekly, or monthly) the material caught in the trap is collected and separated into leaves, flowers, fruit, branches, and others. These components are then counted or weighed. When counted, production can measured as the raw or standardized number of items per unit time (e.g., seed production per year; Wright and Calderón 2006). When weighed, the wet weights of items of interest are recorded immediately; the dry weight is recorded after oven drying (e.g., Chapman *et al.* 1994). Litter trap data are then reported as weight/ha/year per item or in total (Chave *et al.* 2010).

Both fruit trail and litter trap methods are straightforward and simple. An added benefit of litter traps is that they provide quantitative estimates of forest productivity that can be compared easily across sites (Chave *et al.* 2010) and linked to factors such as soil fertility. Nevertheless, both methods have substantial disadvantages. A major shortcoming is that both methods assess what has fallen on the ground, not what is available in the canopy. This bias has several important consequences (Chapman *et al.* 1994). For example, fruit production as measured in litter traps is likely to be particularly underestimated during periods of fruit scarcity, when virtually all fruits are consumed in the canopy (Terborgh 1983). Similarly, litter traps are biased against preferred food items, as these generally are disproportionally targeted in the canopy. Small crops that are fully consumed in the canopy will be completely missed. In contrast, litter traps can be biased in favor of items that are aborted by trees or vines as a result of insect damage (Janzen 1983). An additional concern is that data could be biased due to items being directly removed from litter traps prior to measurement (e.g., by rodents or ungulates), although two studies suggest that the effects are not large (1–3% of fruits in the traps were removed: Chapman *et al.* 1994; Goldizen *et al.* 1988). Finally, a large number of traps are needed to obtain reliable values. Thus, the litter trap method has the added disadvantage of being very time consuming. A relatively large number of traps are required to sample the habitat in an unbiased way (Chapman *et al.* 1994), and the emptying of traps and separation, drying, and weighing of fallen items is time consuming.

Many of these criticisms can also be levied against the fruit trail method, although this method uses a presence/absence measure to assess fruit productivity and is therefore not as sensitive to resulting biases in data as litter traps. Although fruit trails can be rapidly sampled and therefore provide an efficient (although biased) index of fruit availability, they do not provide reliable data on the availability of flowers or leaves, both of which can be important components of some primate diets. However, if one is exclusively interested in a quick-and-dirty measure of fruit abundance for a primate species, for instance due to limited time and staff, this method is better than no sampling at all. If such a method is employed, the above sources of bias should be considered and care must be taken in the interpretation of results.

The extent to which various phenological methods yield similar results and which method provides the most accurate assessment of plant phenology are important research questions. This issue has received limited empirical investigation (Chapman *et al.* 1994; Hemingway and Overdorff 1999; van Schaik 1986) and certainly deserves additional study, as the most appropriate method will likely vary based on the research question, the primate species being studied, and the forests being sampled.

7.2.6 Remote Sensing

Phenological studies are increasingly incorporating data gathered via satellite (e.g., Pettorelli *et al.* 2011). The most frequently used index is the normalized difference vegetation index (NDVI, also see Chapters 5 and 9), which correlates with photosynthetic activity (Myneni *et al.* 1995) and therefore provides a useful index of leaf production and leaf cover. Specific reproductive phenophases, such as flowering and fruiting, cannot be discerned in these satellite images. Thus far, the use of remotely sensed data in primatology has therefore been mainly restricted to analyses at large geographical scales, such as continents or regions, where for instance the mean NDVI index values of primate home ranges are compared to that of a larger reference area (Zinner *et al.* 2002). A recent study compared measures of NDVI at monthly intervals to measures of food availability (flowers, fruits, and seeds) in focal trees on the ground and reported a correspondence between monthly NDVI indices and the day journey length and level of terrestriality in the vervet monkey (*Cercopithecus aethiops*; Willems *et al.* 2009). However, because the relationship was non-linear, the interpretation was not as straightforward as one might have hoped.

The benefit of using satellite imagery is that it permits broad spatial and temporal sampling and therefore provides researchers with the data necessary to examine periods prior to their own study or monitor areas beyond their own study site, provided the relationship between the NDVI index and food availability has been firmly established in the relevant forest types in the area. Another benefit is that NDVI data are freely available on various websites (e.g., <http://glcf.umiacs. umd.edu/data/gimms>), so once one learns the required analytical skills they can be readily applied. At present, NDVI data are not sufficiently accurate to permit monitoring of specific plant taxa (although other remote sensing techniques, such as high-resolution aerial photography, show promise in this domain; e.g., Jansen *et al.* 2008), therefore to date these data cannot be used to examine interactions between the phenology of certain food species and primate behavior. Another disadvantage is that such imagery does not provide information about the understory, thereby limiting its value for primate species that frequently forage beneath the main canopy. Nevertheless, remote sensing is a rapidly developing technology, and useful, novel applications are likely to be soon available that will extend our ability to monitor plant phenology and understand its effects on primate behavior and ecology.

7.2.7 Conservation drones

A promising new method for obtaining phenological data and NDVI data is the use of unmanned aerial vehicles (drones). Current technology and software allows for the development of low-cost drones (Koh and Wich 2012) that can be used to obtain high-resolution images that can potentially be very useful to monitor phenological activity and NDVI. Although the utility of this technology for phenological monitoring has yet to be tested, the technique could potentially allow for low-cost phenological monitoring over large areas.

7.3 Data considerations

7.3.1 Description and presentation of data

There are several ways to describe and present phenology data. The simplest is to report the mean and some measure of variance, although graphical presentation of data is especially helpful (Newstrom *et al.* 1994). Although more complex presentations may be required in some circumstances (e.g., Newstrom *et al.* 1994), often simple line or bar graphs that either present the monthly means or the raw values

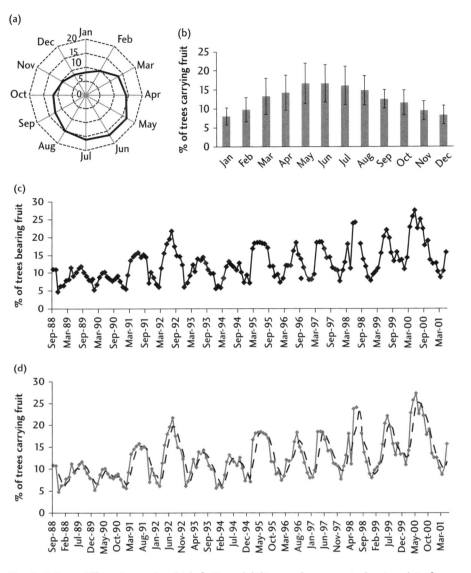

Fig. 7.4 Four different ways in which fruit availability can be presented, using data from Ketambe, Sumatra (Wich and van Schaik 2000). (a) As a circular figure where the mean percentage of fruit availability (black line) is depicted per month on a circle; (b) as a bar graph with the monthly means (gray bars) and standard deviations (black lines) across all years presented; (c) as monthly values connected with a line to show the time series; and (d) as monthly values as in 7.4c (solid gray line) with a three moving average plotted (black dashed line).

for all months (Fig. 7.4b,c) are adequate for primatological studies. Plots of moving averages are sometimes desirable, so that general cycles are clearly visible and short-term fluctuations (some of which may be the result of sampling error) are de-emphasized (Fig. 7.4d). Although not commonly used to date (but see Janson and Verdolin 2005), circular plots (Fig. 7.4a) and circular statistics may be particularly applicable to phenology data (Morellato *et al.* 2010); they have recently been applied to large-scale comparisons of phenology data based on litter traps in South America (Chave *et al.* 2010).

Although it is common practice, we do not advocate calculating or presenting composite phenological indices that collapse, for example, measures of both fruit crop size and the number of stems in fruit into a single index. When such indices are used, the same value can be achieved by changes in either the number of stems or the crop size per stem. As the number and size of food resources are likely to have important, and often independent, effects, summing the two obscures important ecological variation of great interest to primatologists.

7.3.2 Modeling and hypothesis testing

Comprehensive discussion of methods for modeling phenological data and testing hypotheses within this framework is beyond the scope of this chapter. An excellent treatment of many related issues is provided in Hudson and Keatley (2010). Here we merely note two important considerations. First, phenology data sets generally exhibit substantial temporal autocorrelation (i.e., non-independent error variance). This error structure must be explicitly incorporated into models to avoid falsely inflating the significance of results. A variety of tools for time series analyses that permit incorporation of temporal autocorrelation are available, many as freely available packages in R (R Core Development Team 2011). Second, the types of phenological analyses likely to be most useful for primatologists entail using measures of plant productivity to predict outcome counts (e.g., feeding group sizes, numbers of animals encountered along transects) or proportions (e.g., the importance of items in the diet, activity budget measures). In such cases, assumptions required for most commonly applied statistical tests (e.g., OLS regression, t-tests), such as that variables are normally distributed, are normally invalid because data are, for example, counts with an abundance of zeros or proportions bounded between zero and one. In such cases Gaussian approximations are not appropriate and alternative distributions (e.g., Poisson, binomial, negative binomial) should be used in models designed to test hypotheses.

7.3.3 Applicability to studies of global climate change

Changes in atmospheric concentrations of greenhouse gases are likely to have a variety of effects on patterns of plant reproduction. Phenological studies are indispensible tools in documenting these effects (Chapman *et al.* 2005a; IPCC 2007; Hudson and Keatley 2010). Field primatologists have been monitoring long-term patterns of plant phenology at sites across the tropics, making them well placed to both detect the effects of global climate change on tropical plant phenology and to assess the cascading effects of these changes on populations of primates and other tropical vertebrates. Consideration of the nature of these changes and their possible effects is beyond the scope of this chapter, but two examples suggest how fundamentally important these interactions may be.

First, a long-term study in the neotropics has demonstrated that production of flowers and fruits is elevated during El Niño Southern Oscillation (ENSO) events and that there have been greater increases in flower production among sampled lianas than trees over the past two decades (Wright and Calderón 2006). Wright and Calderón (2006) suggest that documented shifts to more severe ENSO events, changes in atmospheric CO_2 concentrations, and increased solar irradiance may further alter patterns of flower and seed production, which are likely to have major effects on the abundances of forest vertebrates and alter tropical forest dynamics. Second, mast fruiting events in Southeast Asian forests are correlated with ENSO events (Curran and Leighton 2000; Wich and van Schaik 2000). ENSO years have been getting progressively drier in Southeast Asia over the past half century (Salafsky 1998), suggesting that the climatic trigger of mast fruiting may be intensifying. As masts have enormously important effects on many forest vertebrates (Curran and Leighton 2000; Cannon *et al.* 2007a), potential alteration or disruption of this system could have important and far-reaching implications. In both examples, it is impossible to predict the precise effects of such changes, but without long-term phenology data we will not be able to detect climate-induced changes, document their impacts on vertebrate populations, or devise potential mitigation strategies.

7.4 Conclusion

Primate diets vary substantially as a result of temporal and spatial variation in the availability of plant foods, with important implications for primate reproduction, grouping, ranging, and sociality. Many primatological field studies therefore require accurate characterization of plant phenology. Sampling methods have different strengths and limitations, so care must be taken to select the measure most appropriate to the study species and the hypotheses to be tested. Regardless of the method employed, a phenological sampling strategy should produce a description of a primate's environment that is accurate, precise, and representative. In this review, we have aimed to summarize the major considerations related to the assessment of plant phenology by field primatologists and convey some simple rules of thumb that will facilitate the collection of high quality phenological data. We recognize that phenological data collection is often a relatively minor component of primate field studies, particularly in situations such as Ph.D. field studies when time and funds are severely limited. In such instances, while sampling intensity and methods may be less than ideal, we recommend that potential sources of error and bias be carefully considered. When suboptimal methods (e.g., fruit trails) are used, results must be interpreted with caution.

Acknowledgments

We thank the editors for the invitation to contribute to this volume. We also gratefully acknowledge the volume editors and Carel van Schaik for helpful comments that improved the quality of this chapter. Permission to conduct research at Gunung Palung National Park was kindly granted by the Indonesian Institute of Sciences (LIPI), the State Ministry of Research and Technology (RISTEK), the Directorate General for Nature Conservation (PHKA), and the Gunung Palung National Park Bureau (BTNGP). AJM gratefully acknowledges the financial support of the J. William Fulbright Foundation, the Louis Leakey Foundation, the Orangutan Conservancy, the Hellman Family Foundation, a Frederick Sheldon Traveling Fellowship, Harvard University, and the University of California at Davis. AJM also thanks Universitas Tanjungpura (UNTAN), his counterpart institution in Indonesia since 1996; R. Mangun, P. Moore, A. Harcourt, and especially B. Winterhalder for support and assistance; and M. Leighton for sharing his deep knowledge of the ecology and phenology of Bornean plants. Finally, AJM gratefully acknowledges the assistance and support of the many students, researchers, and field assistants who have worked at the Cabang Panti Research Station over the past two decades, particularly Albani, M. Ali A. K., Busran A. D., Edward Tang, Hanjoyo, J. R. Harting, Rhande, J. R. Sweeney, and L. G. Bell. SW gratefully acknowledges the cooperation and support of the Indonesian Institute of Science (LIPI, Jakarta), the State Ministry of Research and Technology (RISTEK), the Indonesian Nature Conservation Service (PHPA) in Jakarta, Medan, and Kutacane (Gunung Leuser National Park Office), Universitas National (UNAS, Jakarta), Universitas Syiah Kuala (UNSYIAH, Banda Aceh), and the Leuser Development Programme (LDP, Medan). SW also thanks a large number of assistants and in particular Usman for helping with data collection at Ketambe. Financial support for the period in which the phenology data were collected was generously provided by the Netherlands Foundation for the Advancement of Tropical Research (WOTRO), the Treub Foundation, the Dobberke Foundation, and the Lucie Burgers Foundation for Comparative Behaviour Research.

8

Methods in ethnoprimatology: exploring the human–non-human primate interface

Erin P. Riley and Amanda L. Ellwanger

8.1 Introduction

8.1.1 What is ethnoprimatology?

Ethnoprimatology, first coined by ecological anthropologist Leslie Sponsel (1997), is the study of the ecological and cultural interconnections between human and non-human primates (NHPs) and the implications these interconnections have for conservation (Fuentes and Wolfe 2002). This approach emerged in part due to the increased recognition in anthropology that, outside of Western culture, the boundary between humans and animals is more fluid than traditionally considered (Mullin 1999). This recognition spurred a call for a dynamic, interdisciplinary effort to understand human–animal relationships (Mullin 1999; Kirskey and Helmreich 2010). Another factor contributing to the emergence and embrace of the ethnoprimatological approach was the frustration with the increasing divisiveness between biological and cultural anthropology. In the face of a fractured discipline, ethnoprimatology offers a path toward reconciliation by integrating theory and methodology from both cultural and biological anthropology and a multitude of other disciplines (Riley 2006).

Emphasizing the potential for a fruitful collaboration between cultural and biological anthropology, Sponsel (1997) identified six realms of interactions: comparative ecology, predation ecology, symbiotic ecology, cultural ecology, ethnoecology, and conservation ecology. These broad categories served as a foundation for a subsequent wave of research using increasingly interdisciplinary techniques to explore a diverse array of topics including: the nature of human–non-human primate interactions (Fuentes *et al.* 2008; Sha *et al.* 2009); the role and place of NHPs in culture and religion (Wheatley 1999; Cormier 2003; Riley 2010); human–non-human primate disease transmission (Jones-Engel *et al.* 2005;

Primate Ecology and Conservation: A Handbook of Techniques. First Edition. Edited by Eleanor J. Sterling, Nora Bynum, and Mary E. Blair. © Oxford University Press 2013. Published 2013 by Oxford University Press.

Fuentes 2006), commensalism, conflict, and overlapping resource use (Sprague 2002; Paterson and Wallis 2005; Riley 2007; Hockings *et al.* 2009); and the incorporation of non-human primates in human economic systems through agricultural harvest (Sponsel *et al.* 2002), tourism (Fuentes and Gamerl 2005; Zhao 2005), and wildlife trade (Nekaris *et al.* 2010).

While the theoretical and practical significance of the ethnoprimatological approach has been well addressed in the literature (i.e., Riley 2006; Fuentes and Hockings 2010), scholarship on the methodological considerations of the approach is sparse. Our goal in this chapter is to provide a comprehensive piece that synthesizes and evaluates the appropriate methods to use for a number of ethnoprimatology's research foci, including: the impact of anthropogenic disturbance on a primate's ecology and behavior; overlapping resource use between humans and NHPs; and cultural perceptions of primates. Human–non-human disease transmission, while an important part of the ethnoprimatological endeavor, is not covered in this piece because methods in health assessment and epidemiology are covered in another chapter of this volume. We also direct readers to Jones-Engel *et al.* (2011); a book chapter explicitly devoted to methods in assessing human–non-human primate disease transmission from an ethnoprimatological perspective. Our chapter begins with a discussion of logistical issues associated with ethnoprimatological research, including how to navigate different cultural landscapes and deal with multifaceted ethical considerations. We then provide an overview and description of the key research foci of ethnoprimatology and the methods used to accomplish such research. We conclude with a discussion on the future trajectory of ethnoprimatological research.

8.2 Fieldwork logistics

Preparing for the field is a multifaceted endeavor. The more obvious facets include acquiring all the necessary supplies and equipment, taking appropriate coursework, and gaining appropriate methodological skills. But it does not end there; it also involves learning how to work with foreign collaborators and bureaucrats, which often means figuring out how to navigate different cultural landscapes and paying careful attention to ethical considerations. Here we discuss these logistical issues and make recommendations based on our experiences as North American scientists conducting ethnoprimatological research in East and Southeast Asia. Each research project will have a distinct set of challenges and constraints; we therefore encourage researchers to consider the discussion here as a broad framework from which to derive (and devise) guidelines appropriate to the specific site/situation.

8.2.1 Navigating different cultural landscapes

In most countries foreign researchers are required to work in collaboration with in-country scholars and universities in order to conduct research. The importance of this part of the overall research process cannot be overstated; a good in-country counterpart can make all the difference in terms of a researcher's access to sites and resources. If one is starting off fresh, it is crucial to devote sufficient time to the process of developing these relationships. Part of this process requires that one be cognizant of potentially different cultural norms about how research collaborations are devised and implemented. For example, in the USA, because straightforward negotiation is valued, you might be eager to immediately direct your correspondence toward the primary issue at hand (i.e., how to attain sponsorship). However, the cultural norms held by your potential scientific counterpart may differ drastically. Take time to understand how things work in the country where you aim to work.

In most, if not all, research projects that involve multiple investigators and institutions, there are a number of potentially touchy or difficult subjects, including: co-authorship, compensation, research responsibilities, and funding. These issues can be difficult to negotiate within one's own social-cultural community, let alone within a different cultural context. It is therefore crucial to devote time to understanding how negotiations are undertaken and how such uncomfortable issues are best discussed. If you are not from the region, one way of doing so is to communicate with other researchers or students who have experience working in the same country or region. In this case, you may find that your counterparts abroad are eager to explicitly discuss and resolve these issues; in other cases, perhaps not. Regardless, it is very important that these issues are addressed *before* the research begins so as to avoid any future conflicts that might ultimately result in the rejection of your research visa or permit applications. One useful tactic is to develop a Memorandum of Agreement (MOA) that explicitly outlines the expectations of all parties involved. Important components to include in these agreements include: co-authorship guidelines, statement of who is responsible for funding (and for which parts of the research), whether salaries or other compensation will paid to counterparts, and a data sharing plan. Your institution will likely want to review such documents prior to gaining the other parties' signatures; so again, be sure to allow sufficient time for administrators to provide their analysis and consent.

Without doubt, navigating different cultural landscapes is greatly facilitated by the ability to communicate in the languages spoken where one works. Investing time and energy early on in language acquisition is therefore an important first step in any ethnoprimatological project. Keep in mind that in many areas of the world this may mean learning a national language (e.g., Bahasa Indonesia) and the dialect, or even distinct language, used locally (e.g., Bahasa Lindu). In the USA, the Department of Education offers academic year and summer fellowships to assist students in acquiring foreign language skills. These FLAS (Foreign Language and Area Studies) fellowships are particularly useful for those seeking to learn less commonly taught languages (<www2.ed.gov/programs/iegpsflasf/index.html>). Complementary programs may be available elsewhere.

Many (ethno) primatologists work within protected areas, such as nature reserves and national parks, and as a result encounter another layer of negotiations. Based on our experience, there exists considerable diversity in the degree of control the protected area personnel exercise that varies both within and between countries. In some cases, you will be completely on your own in terms of working with local park staff; in others, your project will be heavily scrutinized by protected area management, sometimes to the point of appropriating its direction. Be prepared to work through this.

A final note: when abroad and dealing with potentially frustrating negotiations it *never* pays to become overtly angry. If all else fails, smile and return the next day; never underestimate the power and cross-cultural salience of a smile.

8.2.2 Ethical considerations

While the ethics of fieldwork has been discussed amongst field primatologists for decades, the last 5–10 years have witnessed a renewed discussion; one with a wider scope, particularly with regard to anthropogenic factors. At the 2009 American Society of Primatologists annual meeting, an entire symposium was devoted to current issues in the ethics of field primatology (see MacKinnon and Riley 2010). The motivation to jumpstart the conversation derives primarily from two factors, with the first being dissatisfaction with the irrelevancy of existing protocols on the ethical treatment of animals for field-based primatology. A second factor is growing concern about how to consider not only the well-being of non-human primate subjects but also the broader community that includes human livelihoods. The latter factor is largely the result of ethnoprimatology's emphasis on the human–non-human primate interface. Some key lessons learned (and issues that should be considered in future ethnoprimatological projects) include:

- How might your research design and methods exacerbate rates of human–non-human primate disease transmission?
- For those studying crop raiding behavior: does further habituation of primate groups increase rates of crop raiding? If crop raiding events are observed, what will be your reaction (i.e., intervene or ignore)? How might non-reaction

affect your relationships with local farmers? Will this lead to further farmer frustration that could result in retaliation against raiding primates?

- How might your research results inform conservation management plans? How might this in turn affect the livelihoods of local human communities?
- How might your research results inform management plans but negatively affect conservation efforts (e.g., culling a primate population rather than seeking ways to mitigate human–non-human primate conflict)?

8.2.2.1 Research permits

An important ethical responsibility of all field researchers, including ethnoprima-tologists, is to obtain all relevant permits required to conduct research. In fact, increasingly funding organizations require evidence of permits before they disburse funds. The process of acquiring research permits can be extremely taxing, so the best way to tackle it is to start early. Depending on the country or nature of your project, you may need to begin the process 9–12 months before you plan to begin your research. Another issue is that countries will vary in terms of the kinds of permits needed; some only require a general research permit while others require research visas as well as various levels of permits.

8.2.2.2 Research compliance: non-human primates

Academic and research institutions in the US have an Institutional Animal Care and Use Committee (IACUC) that is responsible for monitoring all research involving animals. Although many of the questions asked on these protocol forms are irrelevant for field-based behavioral research, there is evidence that this in changing, albeit in different directions. For example, the University of Wisconsin now has a Wildlife Waiver form for observational studies that if approved waives the requirement to complete a full Animal Care Use Protocol form (K. Strier, pers. comm.). On the other hand, other institutions are amending existing forms to include *greater* attention to the potential issues associated with field-based research, such as the costs and benefits of habituation and the potential impacts of trail formation on primate communities (see Fedigan 2010 for in-depth discussion). For example, San Diego State University requires a full Field Study Animal Protocol Form to be completed even if the research is strictly observational.

If your research involves the collection of biological samples, then expect another layer of protocol approval: Biological Use Authorization (BUA). Also, be aware that some primate species trigger considerable scrutiny from the overseeing committees (e.g., macaques generate a huge red flag because they have the potential to harbor the Herpes B virus which can be fatal in humans). Given these disparities in compliance regulation across institutions, be sure to check carefully with the requirement of your institution many months (at least 6 months) before you plan to begin your research (see Chapters 3 and 4 for more details on IACUC as well as permits for biological samples).

8.2.2.3 Research compliance: human primates

While most primatologists are familiar with animal care use protocols, ethnoprimatology's explicit attention to the human dimension requires another area of compliance: Human Subjects Use Approval by an Institutional Review Board (IRB). If you plan to interview people at your field site, whether informally or formally, you must first acquire IRB approval. The Use of Human Subjects form required by the IRB is a dense and detailed application. Completing this protocol will require the researcher to carefully reflect on the operational practicalities of conducting research, potential risks to the participants, and ways to mitigate risks. To gain IRB approval, applicants must also submit a written informed consent document, which should be read and signed by each respondent prior to their participation. Written consent may enable literate informants to better understand the relationship between the researcher, certain institutions, and research questions. However, obtaining written informed consent can be tricky for ethnoprimatologists and, more broadly, anthropologists. First, developing rapport with respondents takes time. Asking a respondent to sign a consent form may arouse suspicion or damage the potential to build a relationship built on mutual trust. Second, in some instances, rural human populations with whom ethnoprimatologists work may not be literate; hence, written consent forms are not feasible. As such, it may be necessary to ask the IRB to approve a verbal informed consent process. Verbal consent may also be most appropriate if interviews involve questions about potentially illegal activities, such as the hunting of primates. If the IRB approves the use of verbal consent in lieu of written consent, the researcher will submit a proposed script to the IRB that will be spoken out loud to potential respondents, detailing project goals, affiliations, risks, and benefits.

The first attempt to write the application may take as long as one month. If the research does not involve vulnerable populations, such as children or prisoners, it may be considered for exempt or expedited status. Exempt research is typically short term, no risk, and non-invasive. Expedited research is typically a long-term project that involves low risk, low to non-invasive collection of biological samples, behavioral observations, interviews, or collection of data using audio or visual recordings for research purposes. The overall timeframe varies widely by institution (so be certain to ask other colleagues at your institution), but in general be prepared for the approval process to take up to three months or more.

8.2.2.4 Capacity building

A final ethical consideration is the issue of capacity building of habitat country nationals. Many would argue that it is the ethical responsibility of scientists to actively engage in the training of university students, conservation managers, and protected area staff of the countries in which we work. It is therefore increasingly good practice to be prepared for this by designing projects with the potential roles of habitat country nationals in mind. Remember though that in some cases participation by habitat country nationals may be required. For example, when conducting research in a protected area in Indonesia, foreign researchers are required to be accompanied by protected area staff. Foreign researchers may also be required to train students in exchange for sponsorship needed to obtain research permits. Therefore, in addition to keeping in mind the importance of inclusion as an ethical responsibility, be prepared for these possibilities both logistically and financially (e.g., including travel costs and/or stipends for students in grant proposals).

8.3 Measuring the effect of anthropogenic disturbance on primate behavioral ecology

A critical component to investigating ecological interconnections between humans and other primates is to examine how human niche construction (i.e., the ways humans modify their environment; Odling-Smee et al. 2003) affects various aspects of a primate's behavior and ecology. This component typically involves assessing the nature and extent of anthropogenic disturbance and determining how such disturbance affects the activity budgets, ranging patterns, and habitat use of primate groups. Anthropogenic disturbance can include changing foraging conditions that emerge in the context of food provisioning (e.g., Zhao 2005) and human-induced habitat alteration (Fig. 8.1). In order to determine the effect of anthropogenic processes on primates one can either employ a longitudinal or a cross-sectional design. The latter is typically more common as one is more likely to encounter groups living in *already* disturbed areas rather than experience the disturbance during the course of the observations.

If employing a cross-sectional design to explore the effect of anthropogenic habitat alteration, one would first collect and compare basic habit composition data from habitats deemed disturbed and undisturbed (see Chapter 6 for techniques used in assessing habitat structure and composition). The extent of disturbance would be assessed by measuring specific indicators of anthropogenic disturbance (Fig. 8.2). For example, one technique is to establish vegetation plots (see Chapter 6

Fig. 8.1 Recently felled tree in home range of Tonkean macaque (*Macaca tonkeana*) group, Lore Lindu National Park, Sulawesi, Indonesia (photo by E.P. Riley).

for info on how to choose the appropriate plot size and shape) in which the following disturbance indicators are measured (e.g., Bynum 1999):

- The number of stumps.
- The "diameter at breast height" (DBH) of the stumps (a measure of total basal area cleared per habitat, since the percentage of area cleared might be as important or more important than the total number of trees cut).
- The number of exotic trees (an indicator of altered land use).
- The number of mid-canopy and canopy-sized trees suffering near complete loss of leaves and/or loss of the trunk out of the total number of those trees in the plot (this is assuming that near complete removal of leaves (50–100%) and cutting of the trunk affect the reproductive potential of a tree, and hence food availability for primates).
- The number of other indicators of disturbance of primary forest (e.g., rattan in Southeast Asia).

One would also assess activity budgets, ranging patterns, and habitat use (see Chapter 5 for details) of groups living in heavily disturbed and minimally disturbed

Fig. 8.2 Tree stumps, evidence of anthropogenic disturbance, in Lore Lindu National Park, Sulawesi, Indonesia (photo by E.P. Riley).

habitats. It is preferable that these groups exist in otherwise comparable ecological conditions (topography, altitude, soil types etc.), they are observed simultaneously (to control for potential seasonal effects on behavior), and that preferably two or more groups per habitat regime are observed. Another technique is to follow groups that regularly utilize different habitat regimes, and compare the relevant behavioral ecological parameters (i.e., activity, ranging etc.) across those regimes.

8.4 Exploring human–non-human primate overlap

A key focus of recent ethnoprimatological research is the extent and outcomes of overlapping patterns of spatial and resource use between human and NHPs. This focus includes investigations of the nature of human–non-human primate interactions in urban, urban/forest edge, and tourist sites (Zhao 2005; Fuentes *et al.* 2008; Sha *et al.* 2009) and examinations of overlapping patterns of food resource use by human and non-human primates (Wright *et al.* 2005; Riley 2007; Hockings *et al.* 2009). Food resources can mean both urban resources, such as human foods raided from kitchens and stores, and resources in more rural domains: forest resources and cultivated resources (i.e., subsistence and/or cash crops). Here we describe the techniques used to examine forest resource use, the use of anthropogenic resources (using crop raiding as the primary example), and the outcomes of human–non-human primate overlap.

8.4.1 Forest resource use

In order to determine the nature and extent of overlap in the use of forest resources one must examine the diet, ranging and habitat use patterns of the non-human primates (see Chapters 5 and 9 for description of techniques) as well as patterns of human resource use. The latter is best explored by asking people what resources are important to them and why, *and* by observing what resources people use, how often, and how much they use.

8.4.1.1 Ethnographic interviews: freelisting exercises

A simple yet effective interview technique to determine what forest resources people consider important is freelisting. In freelisting, a technique used in cognitive anthropology, one asks respondents to list everything they know about a particular category or cultural domain of knowledge (Bernard 2006). In this case, the category or cultural domain would be forest resources. Freelisting is widely used by ethnobotanists and other ecological anthropologists to explore domains of ecological knowledge (e.g., Casagrande 2004). See Riley (2007) for an example of the use of freelisting in ethnoprimatological research. Freelists are then analyzed using the program ANTHROPAC[1] (Borgatti 2004) to determine which items are the most salient. Salience is basically a measure of how much knowledge respondents share (i.e., how many people listed the same items) and the importance of that knowledge to respondents. It assesses this by taking into consideration not only the frequency of the item but also the order of the item on the freelist, as the most frequently cited items are not always the most salient. It is for this reason that we consider this technique and the salience scores it generates to be more informative than response percentages.

[1] ANTHROPAC is a DOS-based program for cultural domain analysis. It can be downloaded for free at <www.analytictech.com/anthropac/anthropac.htm>.

8.4.1.2 *Human behavioral observation: focal follows and spot observations*

Direct observations of human behavior have been used primarily in time allocation studies and by researchers interested in natural resource patterns. For example, to understand plant and animal use patterns of the Aché of eastern Paraguay, Hawkes and colleagues (1982) accompanied men and women on foraging trips during which they recorded the time spent searching, collecting, and processing resources. They also recorded the number of individuals involved in collecting, the species collected, the number of each species collected, and the weight of items.

To supplement data obtained in interviews, researchers can directly observe the behavior of respondents using methods derived from ethology. Following Altmann (1974), there are two main approaches to behavioral observations: spot observations and focal subject follows. Spot observations record data on the activity of multiple individuals at specific or random intervals. Focal subject follows involve observing a particular individual for up to a day at a time and recording the specific activity in which an individual is engaged (Borgerhoff Mulder and Caro 1985; Johnson and Sackett 1998). Focal subject follows can record data at fixed intervals or continuously (Johnson and Sackett 1998). While focal follows tend to be a preferred method in behavioral observation, they are not without limitations. In addition to observer fatigue, using continuous focal follows of respondents can be challenging. First, the presence of the observer may cause the focal respondent to alter behavior (Borgerhoff Mulder and Caro 1985; Johnson and Sackett 1998). Second, participants may find constant observation of their lives an irritating intrusion (Johnson and Sackett 1998). Monitoring and quantifying peoples' resource use can also be challenging. People may consume some plant foods while foraging, causing researchers to underestimate a harvest (Hawkes *et al.* 1982). In some cases, hunters may process animal matter prior to returning to a village, particularly during an extended hunting trip. Accordingly, if a researcher is relying on a check-point method to monitor resource use, it may not be possible to identify a species or the actual weight of the item (Cormier 2003).

As of yet, direct observation of resource use and behavior is fairly underutilized in ethnoprimatology, with some exceptions. Human focal follows have been used to characterize the nature of human–non-human primate interactions at temple and tourist sites (e.g., Fuentes 2006; see Section 8.4.3). In her study of Guajá interactions with, and perceptions of, non-human primates in Brazil, Cormier (2003) used random spot checks in addition to ethnographic interviews to collect information on the daily activities and diet of village residents. If an individual was observed eating during a spot check, the author identified the food being consumed. According to Cormier, this provided a more reliable estimate of diet than

Fig. 8.3 Bamboo collection in Fanjingshan National Nature Reserve, Guizhou, China (photo by A.L. Ellwanger).

weighing game, although food preparation and sharing complicated exact estimates. Recent work in Guizhou Province, China, has further developed the use of focal informant follows for ethnoprimatological research on resource overlap (Sheres 2010; Fig. 8.3). Following this approach, researchers select a focal informant to accompany during resource collection trips. During the follow, the researcher can ask detailed questions *in situ* about the types of resources collected, the frequency of collection, and the use for the resource. Global Positioning System (GPS) data points can also be recorded at sites of collection to create a visual representation of resource overlap. Despite the challenges of directly observing behavior or monitoring resource use, it is important to identify and quantify harvests in order to determine niche overlap between people and non-human primates.

8.4.1.3 Estimating niche overlap

In Hutchinsonian terms, niche space is a defined as a multidimensional hypervolume in which a species exists and persists. In order to maintain a viable population, a species must navigate through both a fundamental niche and a realized niche. The fundamental niche is defined by both biotic and abiotic factors affecting species survival including: environmental conditions, such as altitude or temperature; resource use; and physical traits of the organism that influence resource use. In addition to the factors that contribute to the formation of a fundamental niche, a realized niche accounts for the influence of competitors and conspecifics in shaping species survival (Hutchinson 1957). Although traditional ecological models predict niche separation due to competitive pressures, numerous studies in primatology provide evidence for niche overlap between both sympatric and parapatric non-human primate species. Niche overlap occurs when two or more species utilize the same resource or resources. Overlap between species' niches can be measured by collecting binary data (e.g., presence or absence of overlap), categorical data (e.g., high, medium, or low overlap), or quantitative data (e.g., vertical substrate use or percentage contribution to diet of a given resource; Pledger and Geange 2009).

Recognizing that realized niche space is dynamically constructed through complex interactions among organisms, ethnoprimatological studies explore the "multiple ecologies" (Fuentes 2010) and ecological niche overlap between humans and non-human primates. A common approach is to explore overlap by comparing observational data or published lists of non-human primate diets with ethnographic interviews or published lists of human resource use. For example, Riley (2007) used percentage feeding records and salience scores generated from free-listing data to identify the presence or absence of overlap in the use of forest resources between Tonkean macaques (*Macaca tonkeana*) and humans in Sulawesi, Indonesia. Using a categorical approach, Kinnaird (1992) examined competition between people and the Tana River crested mangabey (*Cercocebus galeritus galeritus*) for a forest palm, *Phoenix reclinata*, in Kenya by comparing the number of palms observed along transects with the number of palms harvested. In this study, palm harvesting was scored based on the degree of cutting: light, medium, heavy, or topped. One can also use quantitative data, for example, percentage contribution to diet of a given resource, to calculate an index of niche overlap. A common measure used by ecologists and primatologists is Morisita's index of similarity or the simplified Morisita index if one does not have data on the number of individuals of each species that use the resources in question (Krebs 1999). For example, Singh *et al.* (2011b) used data on the proportion of each food resource used to measure niche overlap between three species of sympatric monkeys (*Macaca silenus*, *Macaca radiata*, and *Semnopithecus entellus*). This technique could also be applied to determine the extent of overlap between human and non-human primates using data from behavioral observations (e.g., scan sampling on non-human primates and focal follows of humans).

8.4.2 Use of cultivated resources

In addition to relying on forest resources, many non-human primate species that live in close proximity to human settlements are able to incorporate anthropogenic resources into their dietary repertoire. Such anthropogenic resources include foods from kitchens, stores, garbage dumps as well as cultivated resources including both subsistence based crops (e.g., mixed vegetable gardens, fruit trees such as bananas and papayas) and cash crops (e.g., cacao, maize, rice). While initially viewed as aberrant behavior, most scholars now consider crop raiding behavior as highly adaptive, particularly in light of increasing alteration and destruction of non-human primate habitats. An interest in crop raiding is certainly not unique to ethnoprimatology, as there exists a rich body of work by social science scholars and conservation managers on crop raiding by wildlife and its impact on human livelihoods (Naughton-Treves *et al.* 1998; Hill 2000; Woodroffe *et al.* 2005). Nonetheless, ethnoprimatology's contribution to this body of work lies in its use of multiple methodologies to study human–non-human primate overlap and conflict. Drawing from cultural anthropology, ethology, and conservation ecology, ethno-primatological investigations explore *both* the human and non-human primate perspective by collecting and integrating quantitative data (e.g., crop loss measurements, the frequency and duration of raiding events) and qualitative data (e.g., why farmers act the way they do). Without measuring actual damage done by wildlife, a huge disparity may exist between reported and observed damage due to farmers overestimating the amount of crops lost. Moreover, people's perceptions of the extent of damage may be influenced by factors associated with the animals themselves (e.g., large body size, conspicuous behavior of most crop raiding primates). For example, in Lore Lindu National Park, while farmers believed the Tonkean macaque to be the most destructive raider, Riley (2007) found that forest rats were responsible for significantly more damage than macaques. Similarly, but in the opposite direction, in Kerinci Seblat National Park, Sumatra Linkie *et al.* (2007) found that the pig-tailed macaque was responsible for significantly more damage than wild boar; the animal that farmers claimed was the most destructive. Therefore, ethnoprimatological investigations involve a combination of (if not all) of the following: (1) assessing people's perceptions toward crop raiding; (2) measuring the amount and extent of the damage; (3) measuring the behavior of non-human primates during raids; (4) relating crop raiding behavior to patterns of forest fruit availability; (5) exploring ways to manage crop raiding.

8.4.2.1 Assessing people's perceptions toward crop raiding

This objective can be accomplished by conducting structured or semi-structured interviews with farmers. Structured interviews involve using a preset list of questions that is asked of each respondent. The questionnaire is the most common form of structured interviewing (Bernard 2006). Semi-structured interviews are similar to structured ones in that the interviewer may have a preset list of questions, except that there is more flexibility in the way the interview is conducted (e.g., questions may be added or deleted as deemed appropriate during the interview). It is critical that questions are reviewed at the site by a native speaker to ensure that they are culturally appropriate and will access the information being sought.

A common approach is to conduct interviews with the owners of the farms that one is simultaneously surveying for crop damage. This approach enables one to compare farmers' perceptions about crop loss with actual quantitative measurements of the damage inflicted by different raiding taxa. Key information to cover in the interview guides or questionnaires includes (Priston 2005; Riley 2007):

1. *Socio-demographic data*: age, sex, ethnicity, religion, length of residency, household size, highest level of formal education, socioeconomic indicators. For example, Priston (2005) noted the presence or absence of various household objects (e.g., TV, generators, DVD/player, refrigerator) that are indicators of socioeconomic status. The perceived importance of such items can be established prior to structured interviews by conducting informal interviews with village heads or focus groups.
2. *Farm data*: farm size, types of crops grown, distance of the farm to the village and to the forest.
3. *Attitudinal and behavioral data*: farmers' estimates of crop loss, farmers' guarding methods (e.g., what types, fences, weapons, dogs, how often,) farmers' opinions about which animals are responsible for what and how much, farmers' reactions to crop raiding.

It is important to remember that farmers may engage in retribution killings of raiding primates, which may, depending on the location and species, be an illegal activity. The interviewer must therefore take care to convey a neutral stance on such behavior. Another important concern with such work involves the issue of compensation for crop raiding animals (Priston 2005; Linkie *et al.* 2007). When I (Riley) first began my work in Sulawesi, a meeting was held for village officials during which the goals of the research project were explained. One villager present asked: "Since the monkeys will be your research subjects, will you compensate villagers for the damage that is done by them?" Although I was a bit put off by the question, in the end I was very glad the question was asked because it gave me the opportunity to explicitly state that I would not be responsible as well as to explain why. After this meeting, I never had a problem with farmers expecting compensation. Therefore, it is very important that local people have realistic expectations about what your research will and will not provide at the outset of the project.

8.4.2.2 Measuring the amount of the crop raiding damage

The techniques used to measure the amount of crop damage caused by primates (and other wildlife, for comparative purposes) will largely depend on the types of crops grown, the size of the farms, and heterogeneity of the farms. The following represent some of the approaches that have been used to measure crop damage.

1. *Measuring all damage within a farm*: This technique works well for small farms and for farms where only one crop is planted. For example, Riley (2007) used this method to quantify (bi-monthly) the number of cacao pods damaged/consumed by Tonkean macaques and two other wildlife taxa (Fig. 8.4). These data were then converted to percentages and to the number of kilograms lost (as a measure of the loss of potential harvest). For gardens where multiple crops are planted, numbers of damaged crops can be converted to an area using average planting densities for each

Fig. 8.4 Remains of cacao pod consumed by Tonkean macaques, Lore Lindu National Park, Sulawesi, Indonesia (photo by E.P. Riley).

crop (e.g., Naughton-Treves *et al.* 1998). Alternatively, one can simply measure the extent of the area damaged within each farm (e.g., Linkie *et al.* 2007).

2. *Setting up transect belts or plots within farms:* This technique is useful when full surveying of farms is not possible due to time, funding, or labor constraints. Hill (2000) established 2 × 10 m plots within crop stands from which the proportion of crops damaged was calculated. Priston (2005) established transects in the farms surveyed, in which each crop plant was assigned a score on the basis of the severity of damage. These data can then be converted to the percentage of plants damaged and/or the amount of crop area damaged (e.g., damage per m^2).

3. *Exclosure plots:* This technique involves establishing plots of agricultural land from which potentially raiding animals are excluded (Priston 2009). One then compares crop yields from open plots to the "excluded" plots to estimate crop damage. For example, Priston (2009) used this technique to measure damage to sweet potato caused by Buton macaques and wild pigs in Buton, South Sulawesi. While this technique may underestimate damage due to possibility of the plots restricting plant growth, it is a relatively cost-effective and low-labor intensive way to quantify crop damage (Priston 2009).

8.4.2.3 Measuring the behavior of non-human primates during raids

In addition to measuring the amount of damage caused by raiding primates, it is sometimes possible to record the frequency of crop raiding events. This is typically possible only when study groups are well-habituated, and when the groups are observed frequently enough relative to raiding frequency. All-occurrences sampling (Altmann 1974) during all-day focal farm surveys is the appropriate behavioral observation technique to record crop raiding behavior. As the name implies, this technique involves recording all occurrences of certain classes of behavior (see Chapter 5 for more information). At each crop raiding occurrence, the following information should be recorded: time of day, age/sex class of first individual to enter the farm, age/sex class or ID of all individuals involved (if possible), crop type damaged, location of farm, presence of farmers, reaction of farmers to raiding, deterrence methods used at farm, time last raiding individual leaves the farm (see Priston 2005; Hockings *et al.* 2009). If the primates remain within the farm for a considerable duration, one can then conduct group scans where the activity and position of every individual is recorded at predetermined intervals (e.g., every 2 minutes, see Priston 2005; see also Chapter 5 for more information on this technique).

8.4.2.4 *Relating crop raiding behavior to patterns of forest fruit availability*

In order to fully understand crop raiding behavior as an ecological strategy, it is critical that patterns of forest fruit availability are measured alongside measures of crop raiding. Patterns of forest fruit availability can be measured by setting up plots (e.g., Riley 2007) or transects (e.g., Hockings *et al.* 2009) within the primate group's home range in which the phenological phases of trees known to be food items are monitored on a regular basis (usually 1–2 times per month) over the study period. See Chapter 7 for more specific details on methods used to estimate patterns of forest fruit availability.

8.4.2.5 *Exploring ways to manage crop raiding*

Crop raiding by primates is without doubt one of the most difficult, if not *the* most difficult, challenges to primate conservation. An important dimension of many ethnoprimatological investigations is the application of research findings to mitigate and manage human–non-human primate conflict such as crop raiding. The ultimate goal is to devise strategies that minimize crop destruction, and thereby reduce the perceived need or desire held by farmers to trap, poison, and kill raiding primates. What is clear from the myriad of studies devoted to human–animal conflict research over the last 30+ years is that there is no one solution. What works and what does not work depends on the raiding species, the types of crops grown, seasonality, location of the crops grown, the spatial and temporal availability of "natural" forest foods, the behavior of the farmers, and the social-cultural context. One area of agreement, however, is that passive techniques to minimize crop raiding, such as fences, scarecrows, and the like are simply not effective against highly intelligent and flexible animals such as primates. As such, more active techniques to prevent crop raiding are more widely encouraged. Such techniques include active guarding (which for some primate species may be the only successful technique), taste aversion conditioning (Forthman *et al.* 2005), the planting of preferred natural foods along forest–farm edges as a buffer (Hockings and Humle 2009; Riley and Fuentes 2011), and strategic planting of crops (Linkie *et al.* 2007; Nijman and Nekaris 2010). The latter can be informed by a simple technique proposed by Priston and Underdown (2009) and modified by Nijman and Nekaris (2010) to predict the likely frequency of crop raiding damage based on the type of crop grown, and in relation to the farm's location (distance to forest edge). Emphasizing its simplicity and potential to be employed by local farmers and conservation managers alike, Priston and Underdown (2009) propose the following model to predict crop susceptibility:

$$IR(\text{risk of raiding}) = a/a + b$$

where, a = the total number of farms on which each crop is damaged and b = the total number of farms on which each crop was present and available to crop raiding primates,

such that IR values that approach 1.0 denote higher risk crops.

8.4.3 Outcomes of overlap: characterizing human–non-human primate interactions

An increasingly significant component of the human–non-human primate interface is interaction between human and non-human primates at sites of anthropogenic creation, such as temples and tourist sites. The provisioning of food at both of these locales, but more significantly at the latter, is the primary causal factor of direct interactions between humans and NHPs (Fuentes *et al.* 2008). An important first step in research characterizing this aspect of the human–non-human primate interface is to define what it meant by "interaction." A commonly used definition is a behavioral change in at least one as a result of the presence or behavior of the other (see Hsu *et al.* 2009). Because interactions can be initiated by people or by the non-human primates, researchers typically distinguish between these during data collection. Interactions are also typically defined as contact (e.g., grab, jump on, scratch, bite, hand food) and non-contact (e.g., visual monitor, threat, chase, lunge, avoid, toss food). The following techniques can be used to examine the nature of interactions and to determine which groups of people and primates (e.g., age/sex classes) tend to more likely be involved:

- *All-occurrence sampling* (Altmann 1974). This sampling technique can be used to record the following aspects of an interaction: number of humans and NHPs present, number of interactions, age/sex class of those involved, nature of the interaction (contact or non-contact, or both), context of the interaction (e.g., during feeding, provocation, or retaliation), and reactions to the interaction by those involved (e.g., did humans respond submissively, or in a neutral or dominant way?). In many contexts, it would also be appropriate to record whether the humans involved were tourists, locals, villagers, temple staff, and so on. See Zhao (2005), Fuentes *et al.* (2008), Hsu *et al.* (2009), and Sha *et al.* (2009) for more details on the use of this technique in ethnoprimatological research.
- *Focal follows*: This technique can be used in conjunction with the all-occurrences technique and focused on either the humans or NHPs, or both. When the focal subject is a NHP, one typically follows the standard technique of

focal animal sampling (Altmann 1974) where one individual is observed for a specific period of time, recording details of every interaction between the focal NHP and humans (e.g., Fuentes 2006; also see Chapter 5 for more details on behavior sampling techniques). As described in Section 8.4.1, one could also use this technique to observe humans. For example, Fuentes (2006) conducted 10-min focal follows on human tourist groups ranging between 1–7 individuals, noting the time of day, weather, age/sex of the NHPs and humans involved in the interaction, the presence, absence, and timing of food provisioning, and the type of food given.

- *Structured interviews*: In addition to the behavioral observation techniques described above, structured interviews conducted with tourists/temple staff/local people, may be useful in elucidating aspects of the human–non-human primate interface that strict observation cannot. For example, Loudon *et al.* (2006) found that while observations at the Mekori temple site in Bali, Indonesia suggested that long-tailed macaques were regarded indifferently or as mere pests by temple staff, interviews revealed that in fact the temple staff was planning to habituate the macaques as a means to transform the site into a tourist destination like other temple sites in Bali.

8.5 Exploring the cultural interconnections: knowledge and perceptions of nature

8.5.1 Who is the community?

In the past, human communities have been conceptualized as small, homogenous, and territorially fixed entities. This perspective obscures heterogeneity within communities in terms of differences in gender, age, status, ethnicity, education, length of residence, or economic income. A more appropriate understanding of community acknowledges multiple actors with multiple interests, how these individuals interact and influence each other, and the institutions that facilitate and dictate interpersonal interactions (Agrawal and Gibson 1999). Within-community diversity is important to consider during research design and analysis in that it may account for differing knowledge of and perceptions toward the environment. Riley (2010) found that indigenous *To Lindu* residents are tolerant towards crop raiding macaques because of folklore that emphasizes the interconnectedness of people and macaques. While these attitudes have resulted in a taboo against harming the monkeys, recent immigrants to the region do not strongly share the same attitude and may forcefully defend their crops from the macaques. In some cases, cultural norms may not permit a researcher to access certain groups.

For example, to explore Malagasy attitudes towards ring-tailed lemurs (*Lemur catta*) and Verreaux's sifaka (*Propithecus verreauxi*) Loudon and colleagues (2006) were only able to interview elder men due to local norms for public discussions.

8.5.2 Getting started

Prior to conducting any research, it is important to meet with local community leaders and obtain approval. After doing so the researcher should begin preliminary investigations using informal interviewing and household censuses to collect socio-demographic information. Informal surveys enable the researcher to identify demographic trends and cultural attitudes in the sample population. Informal interviews are an unstructured method in which the researcher jots notes throughout the day on conversations and interactions. At the end of the day, the researcher can use these abbreviated, short notes to construct more formal field notes (Bernard 2006).

Language barriers may complicate conducting surveys and interviews. Interviews should be conducted in a language in which respondents are fluent and, accordingly, survey instruments must be translated into the appropriate language or languages. In order to ensure that questions are phrased properly and utilize correct word choice to convey the intended meaning, survey instruments should be back- translated into the language spoken by the principal investigator. The first step in the back-translation process is to get either a native speaker or a fluent, non-native speaker to translate the survey instrument into the language spoken by respondents. Next, a different bilingual speaker should translate the documents back into the language spoken by the principle investigator. The original research questions can then be compared to the back-translated questions to look for differences in meaning (Bernard 2006).

8.5.3 Ethnographic interviews: accessing knowledge, folklore, and social taboos

Freelisting, the structured interview technique described in Section 8.4.1, is an excellent technique to determine people's knowledge and perceptions about primates and nature. For example, one could ask respondents to list items specific to the following domains: forest products/resources, known forest flora and fauna, objectives of a protected area, benefits of a protected area, negatives of a protected area, the role of non-human primates in forest preservation, and benefits of conservation. Furthermore, if one is interested in assessing the extent to which people agree then one could analyze the data using a reliability model (Weller 2007). Take for example, the interview question: "What role do primates play in forest preservation?" Assuming respondents would give multiple answers, one would first analyze the responses for the presence or absence of particular themes.

Then for each interview, the presence or absence of each theme is noted. The presence or absence of themes across all interviews can then be analyzed to determine the main themes and the agreement among informants (see Weller 2007).

Freelisting exercises can also generate preliminary data that can be used to construct additional structured interview techniques, such as multiple choice and true/false questions and ranking exercises. The data generated from these techniques are amenable to analysis by another approach that measures agreement among individuals: cultural consensus analysis (Weller 2007). Cultural consensus analysis estimates the extent to which cultural knowledge (e.g., knowledge about primate abundance) is shared among people. While this approach has been widely used by ecological anthropologists to explore, for example, local perceptions of wildlife abundance (van Holt *et al.* 2010), it remains underutilized in ethnoprimatological investigations.

Examining the role that non-human primates play in folklore, mythology, and social taboos can help elucidate the complex relationships they share with humans (Riley *et al.* 2011). Semi-structured interview techniques are a useful tool to develop an understanding of how non-human primates fit into peoples' worldview. Unlike structured interview techniques, the researcher is not strictly bound to a particular format and is able explore topics more fully at their discretion. Semi-structured interviews typically rely on open-ended questions, which elicit more organic dialogue and responses from respondents. For example, in their study of Malagasy folklore of the ring-tailed lemur and Verreauxi's sifaka, Loudon and colleagues (2006) used semi-structured questionnaires to ask questions like, "are the ring tailed-lemurs (*maky*) and sifaka sacred or special" and "what is the origin of the *maky* and sifaka?" It is important to record interviews (if respondents are comfortable with this) to document not only what was said, but how it was said, and in what context. Researchers should not rely completely on their memory to obtain interview data (Bernard 2006). Recordings of interviews should be translated into the researcher's native language and transcribed for analysis.

8.6 Conclusions

As the human population continues to grow, the contexts in which human and non-human primates interact will continue to expand. Primatological studies that focus on these interfaces are therefore timely, and will be so indefinitely. To complement existing literature on the theoretical and practical significance of the ethnoprimatological approach, our goal in this chapter was to provide guidance on how to *do* ethnoprimatology. A major theme that emerges in our review is the robustness and importance of multiple methodologies. Many of the facets of the human–non-human primate interface require the use of methods from multiple disciplines including cultural anthropology, ecology, conservation biology, and primatology. As ethnoprimatology continues to branch out to explore new questions, researchers will need to continue to diversify their methodological toolkits in order to meet the theoretical and applied challenges of this interdisciplinary study. We envision direct observations of human behavior becoming increasingly important in investigations of overlapping resource use. Ethnoprimatologists are well-positioned to include this dimension due to their training (as primatologists) in ethological methods of behavioral observation. Given ethnoprimatology's interest in the effect of human niche construction on non-human primate behavior and biology (Riley and Fuentes 2011), we also envision future ethnoprimatological research drawing from principles and techniques in evolutionary biology and anthropological genetics. Another likely trend will be research conducted at broader scales, particularly those projects that can inform conservation management policy. For example, Nekaris *et al.* (2010), in their study of the cultural context of the primate trade in three Asian countries, demonstrate the significance of a regional approach for informing trade mitigation programs. Lastly, we encourage a continued engagement with methodologies in cultural anthropology, including both method-rich traditions from cognitive anthropology (e.g., cultural consensus analysis; Weller 2007) and humanistic traditions (i.e., to understand why people do what they do). By continuing to integrate quantitative and qualitative analyses ethnoprimatologists can expand the present boundaries of disciplinary research methodology, providing a context for interdisciplinary publications and rich dialogue across disciplines on the human–non-human primate interface.

Acknowledgments

We are indebted to the editors for the invitation to contribute to this important volume and for the constructive critiques the editors and Agustin Fuentes provided on an earlier draft of this manuscript. Riley is grateful to LIPI, RISTEK, and PHKA for permission to conduct research, to Manto, James, Papa Denis, Pak Asdin, Pias, and Jaima Smith for their outstanding field assistance over the years, and to the College of Arts and Letters at San Diego State University for the Critical Thinking Grant that supported this project. Ellwanger would like to thank Fanjingshan National Nature Reserve and Yeqin Yang for logistical support and research permission, the San Diego Zoo Institute for Conservation Research and the American Society of Primatologists for funding, Dr. Chia L. Tan for research support, collaboration, and guidance, Kefeng Niu for research assistance, and the University of Texas San Antonio for institutional support.

9

Social and spatial relationships between primate groups

Michelle Brown and Margaret Crofoot

9.1 Ramifications of inter group interactions

9.1.1 Why study inter group interactions?

Primate groups are unusually (though not uniquely) stable, often persisting for generations. Though poorly understood, relationships between social groups have potentially far-reaching ramifications and can impact individual survival and reproduction via their effect on space use, resource access, vulnerability to predators, and exposure to disease. Group-level interactions also influence population structure by influencing the spatial distribution of groups within the habitat and patterns of dispersal, and are hypothesized to play an important role in shaping social relationships among group-mates (Sterck *et al.* 1997).

Compared to other aspects of primate socioecology, inter group relationships have received relatively little empirical attention. In part, this is because the data needed to systematically investigate inter group competition are difficult and expensive to obtain using traditional observation methods. Interactions between social groups are relatively infrequent, so a large number of observation hours are required to achieve the sample size needed to distinguish between competing hypotheses. Additionally, aggressive interactions between groups are usually chaotic affairs with many animals participating simultaneously; thus, to collect standardized, high-quality observational data, the event should be recorded from multiple perspectives to capture as much of the action as possible. Ideally, research on inter group interactions should consist of simultaneous follows of multiple, habituated social groups but such studies are labor intensive, time consuming, and logistically difficult. Despite these obstacles, the number of studies of primate inter group relationships is increasing, demonstrating a renewed interest in the subject.

Primate Ecology and Conservation: A Handbook of Techniques. First Edition. Edited by Eleanor J. Sterling, Nora Bynum, and Mary E. Blair. © Oxford University Press 2013. Published 2013 by Oxford University Press.

Our goal in this chapter is to encourage a more focused approach to the study of inter group interactions by detailing methods that researchers can use to facilitate inter-specific and inter-site comparisons. We begin with a clarification of some of the important terms underlying the study of inter group interactions, and follow with a discussion of the different types of relationships that exist among groups. In the third section, we parse the many components of naturally occurring interactions and identify methods for quantifying and analyzing these aspects. In the fourth section, we briefly mention experimental approaches to the study of inter group interactions, building upon the techniques outlined in Chapter 10. In the fifth section, we review methods for analyzing the impact of inter group interactions on ranging patterns and space us. Lastly, we conclude the chapter with a brief outline of some important avenues for future research.

9.1.2 Variation in inter group interactions: defining and clarifying terminology

Interactions between primate groups take many different forms. Previous studies have documented variation in terms of *who* participates (males or females, high- or low-ranking individuals, breeding or non-breeding individuals); *how* they participate (aggressively or affiliatively); *why* they participate (for access to food, mates, or other resources); *where* the interactions occur (at the edge or in the core of the range); and *when* they occur (in particular seasons or year-round). Unfortunately, the broad comparisons among species and populations that would allow us to understand this variation are hampered both by a lack of data and by muddled terminology. In this section, we identify and discuss the terms that have been used inconsistently. It is of the utmost importance that future discussions of inter group interactions involve explicit definitions and thorough descriptions of observed behaviors; without this clarity, it is impossible to identify the causes and consequences of different interaction patterns.

Primate species are typically labeled as either "territorial" or "non-territorial," but authors rarely define what they mean by these terms (Maher and Lott 1995). For some, "territoriality" refers to spatial exclusivity, while for others, it refers to aggressive interactions. However, there is no consensus on the degree of spatial exclusivity that is required, nor on the percentage of interactions that must be aggressive (as opposed to neutral or affiliative), for a group to be considered territorial. This ambiguity underscores the fact that "territoriality" is not a singular phenomenon, but instead takes many forms. The term territoriality should thus be clearly defined, and researchers should explicitly state what they assume its biological significance to be.

On a more basic level, it may appear self-evident that inter group interactions are occasions when the behavior of one group is affected by the behavior of another group. Some researchers, however, restrict the term to interactions that occur between two bisexual groups, whereas others include interactions between bisexual groups and solitary males, or between bisexual groups and all-male bachelor groups. As a bisexual group may have different motivations for interacting with another bisexual group than with a unisex bachelor male group, or a solitary male (Brown 2011), each type of interaction should be considered separately—at least until it has been demonstrated that they serve the same function. Confusion also stems from inconsistencies in the use of the terms "interaction," "encounter," and "relationship." For the purposes of this chapter, we define interactions as occasions when some or all members of a group react to the presence of another group; encounters as occasions when two groups are in visual, physical, auditory, or olfactory contact, regardless of whether either group reacts to the presence of the other in an observable manner; and relationships as the general quality of a series of such events. A relationship may be described as generally aggressive or neutral, or one group may consistently displace (dominate) another.

Inter- group interactions also vary in the types, frequency, and intensity of behaviors that individual animals exhibit toward an opposing group: these behaviors include aggressive agonism (chasing or physical contact), milder or non-aggressive forms of agonism (threat vocalizations and facial expressions, acrobatic displays, long-distance calls, submission, and/or fleeing), neutrality (no response to the opposing group), and affiliation (grooming, play, sitting in proximity, affiliative gestures, and mating). Several behavioral types can occur within a single interaction, and different interactions may consist of more or less of each behavior. While authors generally note whether their study population exhibits neutral or aggressive inter group interactions, quantification of the frequency and duration of each behavior is uncommon. Overall, we strongly encourage researchers to be explicit about the patterns of behavior they observe in the field, both to avoid arbitrary decisions as to what constitutes "territoriality" or an inter group interaction, and to acknowledge the diversity of these interactions.

9.2 Classes of inter group relationships

Relationships among primate groups are often competitive but, especially between groups of different species, they can also be mutualistic, commensal, or predatory. Potential functions of competitive interactions between groups include food defense, range maintenance or expansion, mate defense, and mate attraction. One common approach to differentiating between these types of competition relies on using differences in the participation rates of male and female group members to infer the function of the interaction. Male reproductive success is thought to be primarily limited by access to fertile females while female reproductive success depends more on access to food resources (Trivers 1972), and thus male participation in inter group conflicts is taken as evidence of mate defense while female aggression is assumed to be food defense. As we discuss in detail below, we feel that this approach is insufficient because it relies on a vastly oversimplified model of the factors that influence male and female reproductive success. In fact, in any given conflict, it is likely that multiple strategies are pursued, and thus alternate hypotheses are not necessarily mutually exclusive.

In addition to investigating why primate groups compete, studies have also explored the effect that this competition has on fitness-related measures. Both theoretical models (Stewart *et al.* 1997), and empirical studies (Peres 1989) indicate that patterns of resource exploitation by neighboring groups can have profound impacts on the ranging behavior and activity patterns of primate groups, and may serve to reinforce territorial boundaries via passive range exclusion. In contrast, direct resource competition between groups (contest competition; Nicholson 1954) has been hypothesized to promote tolerant social relationships among group-mates (Sterck *et al.* 1997), but the impact it has on individual reproductive success remains the subject of much debate. While some studies indicate that the energetic consequences of between-group contest competition are negligible compared to the effects of feeding competition among group mates (Janson 1985), others suggest that conflict among groups may play an important role in shaping patterns of population growth and demography (Robinson 1988).

Whereas within-species interactions are largely competitive, between-species interactions can be competitive, mutualistic, commensal, or even predatory. By associating with each other, groups may detect, avoid, and even deter predators more effectively than when not in association, but they may also be more detectable to predators. Similarly, groups may find and exploit food resources more efficiently when traveling together than when alone, but can also experience greater competition for those resources. Polyspecific associations are often simultaneously beneficial and costly to each group; for instance, groups may gain antipredation benefits but incur foraging costs during associations (Chapman and Chapman 2000b). Although within- and between-species interactions generally reflect different types of relationships, the same methods can be used to observe and evaluate them.

9.3 Field methods and analyses for studying interactions between groups: observations of naturally-occurring interactions

9.3.1 Multiple levels of complexity

Within-species inter group interactions are complex events that can be broken down into three levels of analysis: action phases, participant units, and motivating hypotheses (Fig. 9.1). Action phases are the chronological units of an interaction. For species that interact over long distances, the first action phase is *detection*; for example, after hearing a long-distance call, a group can determine the approximate location of a neighboring group and can ignore, avoid, or approach that neighbor. When groups are close enough to hear each other's short-range calls but cannot see each other, they can *initiate* visual contact by approaching. While groups are in short-range, visual contact, they may exhibit varying types and degrees of *agonism* or *affiliation*. Lastly, there is the *outcome* of the interaction, which can be measured in terms of wins, losses, or draws and the fitness consequences of these patterns.

The second level of analysis is participant units (individual animals, different sexes, or entire groups), each of which is associated with specific action phases.

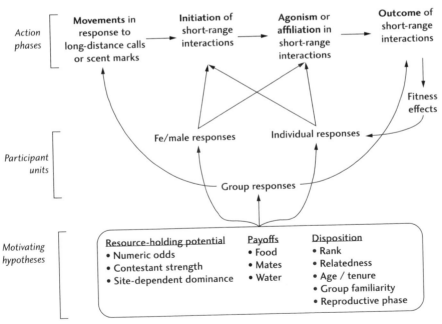

Fig. 9.1 Schematic overview of the relationships between action phases, participant units, and motivating hypotheses for within-species inter group interactions (parts of which are also applicable to between-species interactions).

For example, the *detection* and *outcome* action phases are evaluated at the group level, where the behavior under study is the direction or distance traveled by the group as a whole. In contrast, the participant units in the *initiation* and *agonism/affiliation* phases are less constrained and may be individual animals, a particular sex, or the group as a whole (for example, we might ask whether groups A and/or B, females and/or males, or specific individuals exhibit aggression during an interaction).

The third level of analysis, the motivating hypotheses, are potential explanations for the behavior of groups, sexes, and individuals, (Kitchen and Beehner 2007) and fall into three broad categories (Fig. 9.1). "Disposition" hypotheses are specific to individual animals, and reflect characteristics—like dominance rank, age, or reproductive phase—that may affect individual predispositions for participating in an interaction. "Payoff" hypotheses tend to be sex-specific (male mate defense hypothesis: van Schaik *et al.* 1992; female food defense hypothesis: Sterck *et al.* 1997; e.g., male food defense hypothesis: Fashing 2001) and identify the resources for which animals fight (Maynard Smith and Parker 1976). These hypotheses are not mutually exclusive and multiple factors related to individual disposition or payoffs can simultaneously drive the behavior of different group members. For instance, females may participate to defend access to food while males participate to defend their access to estrous female mates, and within each sex, high-ranking individuals may participate more often or more aggressively than low-ranking individuals. The third category of motivating hypotheses is related to the group's "resource-holding potential" (usually measured as numeric odds or physical strength of contestants), which is its ability to compete against neighboring groups. Whereas pay-off hypotheses serve to identify the underlying causes of aggression between groups, disposition and resource-holding potential factors modulate this aggression.

Many authors have evaluated the pay-off hypotheses by examining fluctuations in the rate at which interactions occur. However, a correlation between interaction rate and a payoff (e.g., food availability), or the occurrence of interactions at feeding sites, does not necessarily indicate food defense because groups are often independently attracted to the same resource (Harris 2007). Instead, the best means of testing the pay-off hypotheses is to evaluate agonistic behavior during interactions; for example, there is good evidence of food defense if participants exhibit significantly greater levels of agonism at feeding sites than at non-feeding sites. Simplistic analysis of inter group interaction rates has limited usefulness, and for this reason is not included in Fig. 9.1. To test specific motivation hypotheses, only the appropriate action phase(s) and participant unit(s) should be evaluated.

The preceding paragraphs have focused on the complexity of interactions within species, but a nearly identical framework can be used to understand inter-specific interactions. Between-species interactions consist of the same action phases and participant units as within-species interactions, but are driven by slightly different motivating hypotheses. In particular, groups may compete for food resources but one or both groups may simultaneously experience benefits of associating, such as a reduction in predation pressure or an increase in foraging efficiency.

9.3.2 Logistical considerations

Agonistic inter group interactions are generally fast-moving, noisy, and chaotic, making it difficult, if not impossible, for one observer to record detailed information about the behavior of all participants, particularly in densely foliated environments. For this reason, having multiple observers is desirable. In addition, a range of groups of different sizes should be studied because participant behavior often varies with the size of the focal and/or opposing group; however, only interactions between habituated groups should be evaluated because unhabituated groups generally flee from human observers. Because of these logistical challenges, studies of group-level interactions require more manpower, more study groups, and longer periods in the field than studies of individual-level interactions.

Ideally, multiple groups should be followed simultaneously, as this allows researchers to explore the diversity of relations between groups. Simultaneous follows of multiple groups are also necessary if long-range interactions are the focus of the study; by noting the time and location of a group's calls or scent markings, neighboring groups' subsequent reactions can be measured. The ability of human observers to hear the calls of neighboring groups is strongly affected by terrain, density of foliage, bird and insect calls, climatic conditions, and distance from the caller. Arboreal primates often seem able to hear calls missed by humans, so objective measures of call occurrence are necessary. Simultaneous follows also increase the quantity of data available within a sampling period, as multiple interactions can be observed simultaneously.

The density of groups in a habitat dramatically affects the rate at which interactions occur—not only because it affects the likelihood of chance meetings as groups travel through shared areas, but also because density is tied to the availability of resources, and thus the willingness of participants to contest access to those resources. Groups in low-density populations often have fewer and less agonistic interactions than groups in high-density populations (e.g., Sugiura *et al.* 2000). Consequently, it is important to consider the density of groups at prospective field sites; a site with many tightly packed groups will yield a larger number of interactions per unit of time, and potentially more intensely competitive interactions than a site with few groups.

9.3.3 Evolutionary game theory

In addition to the practicalities discussed above, simultaneous follows of multiple groups also allows for contests to be evaluated from an evolutionary game theory perspective. Though game theory was developed to model contests between individual animals (Maynard Smith and Price 1973), it also provides a basis for understanding group-level contests. Contestants appear to use strategies for determining when to escalate to outright aggression and when to retreat from a contest, which are based on each contestant's competitive ability (resource-holding potential) and the benefits gained by winning the contest (payoffs; Maynard Smith and Parker 1976). In a conflict between two opposing groups, the group with the lesser resource-holding potential and/or potential pay-off is expected to avoid escalation by retreating from the contest. If contestants possess roughly equal resource-holding potential or expected payoffs, they are expected to escalate to intense levels of aggression to determine the winner of the contest. Importantly, it is *asymmetries* in these values that are predicted to affect the initiation, escalation, and outcome of animal contests. Thus, observers need to quantify both payoffs and resource-holding potential, and to do so for both groups. Expected resource payoffs are sometimes more influential in determining interaction outcome than resource-holding potential (Crofoot *et al.* 2008), though the reverse may also be true (Wilson *et al.* 2001).

9.3.4 Measuring dependent variables

Different dependent variables are associated with each action phase. When examining long-range interactions, the variables of interest may be travel direction, distance, or speed after the long-distance call, latency to call in response, or the number or duration of response calls. The dependent variables relevant to the initiation of short-range interactions are the occurrence (or absence) of a close approach by specific individuals or a particular sex, or the latency or speed of approach. For agonistic/affiliative behavior during interactions, the responses of interest include the occurrence of a particular behavior (such as chasing), or the number or total duration of those occurrences. Lastly, an interaction can end in a decided outcome (win/loss) or a draw; determination of these outcomes is based on each group's pre- and post-interaction movements (e.g., Kitchen *et al.* 2004a).

Ultimately, it is important to evaluate the effects of winning, losing, or simply engaging in a competitive inter group interaction on proxies of individual fitness. These effects can be evaluated by exploring the energetic consequences of winning or losing an interaction, which are measured by determining post-interaction travel distance and velocity, the amount of foraging time lost (i.e., latency to resume feeding), or the quality of food items consumed after the conflict for members of winning and losing groups. At longer timescales, it is important to investigate how the competitive success of groups relates to home range size and quality (Harris 2006; Mitani *et al.* 2010), especially during periods of resource scarcity (Mertl-Millhollen *et al.* 2003), or to demographic variables like inter-birth intervals, sex ratios, and rates of infant, juvenile, and adult survival (Robinson 1988). Using genetic analyses, it may also be possible to determine whether competitively successful groups make larger-than-expected contributions to the gene pool than less-successful groups.

Similar measures can be used to assess the consequences of competitive interactions between species, but when relationships are mutualistic or commensal, the identity of the species responsible for maintaining the association is also important. This can be determined by calculating the frequencies of approaches and withdrawals by each group (calculated by comparing the travel speed of each group before and after the interaction), which are then combined using the Hinde index (Hinde and Atkinson 1970).

9.3.5 Measuring independent variables: payoffs

Among the payoff hypotheses, food-related variables are often the most difficult to quantify accurately, in part because it can be difficult to define "food." One approach is to consider all fruits, leaves, or other items in the habitat as potential food; however, primates rarely consume every available plant, so this approach is unlikely to accurately represent consumption patterns. The alternative is to measure only those plant species that are actually consumed, but it can be difficult to predict in advance which plant species should be monitored, as diets often vary dramatically over time (Chapman *et al.* 2002). Defining food is more than a semantic issue because the underlying objective is to quantify it in a way that is meaningful to the study subjects.

Several aspects of food resources may influence the likelihood that groups will defend access to them: the density of food items (Isbell 1991); the spatial distribution of feeding patches (Wrangham 1980); the size of feeding patches in relation to the size of the group (Koenig and Borries 2006); the rate at which food items renew within patches (Waser 1981); the quality of consumed foods, measured in nutrients or as a preference ranking (van Schaik 1989); and the frequency with which a group feeds in a specific location (Fashing 2001). In addition, food defense is often reflected in a spatial bias, such that interactions occurring near important feeding sites, deeper within the home range, or at defended territorial boundaries elicit more agonism than other interactions. Though authors typically choose just one food or spatial variable to evaluate, Brown (2011) found that this approach can lead to incorrect conclusions. For example, agonism by male and female redtail monkeys corresponded to site feeding value (the percentage of weekly feeding records occurring within a 50 m × 50 m quadrat), but not with food abundance, distribution, or patch size. Conversely, agonism by male grey-cheeked mangabeys corresponded to food abundance, distribution, and patch size, but not site feeding value. Had the study been based only on one food variable, there would have been evidence of food defense by only one of the two species. These examples highlight the importance of measuring multiple food variables. Though some might argue that this constitutes fishing for significant results, we contend that this approach is necessary to identify the aspects of food availability (if any) that are meaningful from the subject's perspective, as different primate species or populations will value different food qualities.

The food variables mentioned above should be evaluated separately for each home range, as they can vary significantly from one group to the next. These variables should also be measured in different seasons, because changes in food availability should drive changes in food defense behaviors. The traditional methods for doing this, described in Chapter 7, consist of phenological censuses, description of botanical plots, and measures of fruit- or seed-fall in traps.

New methods that rely on remote-sensing technology to estimate resource distribution and abundance over large spatial scales may also prove useful in studies of inter group interactions. The normalized difference vegetation index (NDVI), which provides information on primary productivity and plant phenology, has been used as a measure of habitat quality in a study of vervet home range use (Willems et al. 2009). This spectral index is calculated from satellite imagery of the Earth's surface reflectance patterns in the red and near-infrared regions of the electromagnetic spectrum, which are absorbed and reflected, respectively, by live plant cells. NDVI provides an index ranging from −1 to 1, where low values signify an absence of photosynthetically active vegetation and high numbers correspond to heavy vegetation cover (see Chapters 5 and 7 for additional discussion). It may thus be more useful as a measure of resource availability for folivores, or for primates living in relatively open habitat, than for frugivores or species in tropical forests. High resolution imagery from the QuickBird satellite can provide useful information on resource distribution for frugivorous primate species. For fruit tree species that have visually distinctive flowering events, these images can be used to map the location of reproductively mature individuals (Sanchez-Azofeifa et al. 2011). Aerial photography has also been used to generate distribution maps for tree species with distinctive canopy architecture (e.g., Caillaud et al. 2010). Unfortunately, only a relatively limited number of tree species are sufficiently distinctive to be mapped using satellite imagery or aerial photography. However, recent successes with hyper-spectral imaging (e.g., Feret and Asner 2011) suggest that high-resolution mapping of tropical forest canopies at landscape scales may soon be possible.

Another new technique is the direct measurement of participants' energetic status, via urinary c-peptide analysis. The advantage of this method is that it largely eliminates the need to define and measure food availability—a commonly used, but imprecise proxy for energetic status. Energetic status, or the relationship between energy inputs (food consumption) and outputs (basal metabolism, growth, activity, and reproduction), is thought to be the underlying motivation for food defense. The reproductive success of animals is limited by access to food resources when their energetic balance is neutral or negative, so they are expected to contest access to monopolizable foods at those times (Brown 1964). In contrast, the reproductive success of animals with positive energetic status is less limited by food, so in theory, they should be less willing to fight for access. C-peptide, a marker of energetic balance created during insulin production, is excreted in urine and can be easily collected and stored in field conditions because samples do not need to be frozen (Emery Thompson and Knott 2008).

As with food defense, mate defense can be expressed in different ways. The most obvious evidence of male mate defense is heightened levels of male agonism when estrous females are present within the group. Agonistic behavior by resident males is typically directed toward males in opposing groups and is exhibited primarily by high-ranking males and males in consortship (savanna baboons: Kitchen *et al.* 2004a). In many cases, resident males direct their aggression toward estrous female group-mates (referred to as "herding") as a means of keeping females away from extra-group males. In a few species, males attempt to attract new female mates during inter group interactions. In these cases, the male attacks dependent, unweaned offspring in the neighboring group; if he succeeds in killing an infant, its mother follows the infanticidal male and transfers into his group (mountain gorillas: Sicotte 1993; Thomas' langurs: Steenbeek 2000; however, not all inter group infanticide functions as mate attraction—chimpanzees: Watts *et al.* 2002). Males may also aggressively target same-sex potential migrants in neighboring groups as a means of dissuading them from immigrating. Conversely, potential migrants of both sexes may participate aggressively to assess the strength of neighbors, and non-aggressively (by grooming or copulating with neighbors) to establish positive relations with potential future group mates.

In the context of between-species interactions, the measurement of payoffs is slightly different from the methods described above. To determine whether one or both species experience feeding competition as a result of their association, the feeding patterns of each species should be compared when they are and when they are not in proximity to one another (e.g., a species may change its foraging niche by consuming different foods or foraging in different plant strata, or it may feed in the same trees as the other species, but only after the latter has vacated the tree). To determine whether either species experiences foraging-related benefits, individual feeding rates, duration of time spent in feeding sites, and time spent feeding per day should also be measured while in and out of association. To determine whether groups experience anti-predation benefits, predation attempts and positioning behavior should be evaluated. For instance, predation attempts may have a lower success rate and prey species should be more effective in deterring predators when groups are in association; in addition, individuals may be more willing to forage in exposed locations when in a mixed-species association.

9.3.6 Measuring independent variables: resource-holding potential

The size of a group, or its size relative to that of the opposing group, is an important measure of resource-holding potential (Sugiura *et al.* 2000; Kitchen and Beehner 2007). Group size should be tracked through frequent group counts, particularly in species where group composition changes regularly due to transfers, temporary absences or visitations, or frequent mortality events. However, the number of individuals present in the group may differ from the number of participants in an inter group interaction: participation is often limited to adults and subadults, and even within those age classes, not every individual participates in every interaction. For this reason, both the *actual* and *potential* number of participants should be counted in each group for each interaction (Zhao and Tan 2010). In species where only one or a few individuals participate in interactions, an alternate measure of resource-holding potential is the physical strength of the contestants. Strength is typically approximated by an individual's body mass, or by the frequency or duration of its long-distance calls (if these calls are correlated with body size, stamina, or competitive ability; Kitchen *et al.* 2003; Harris 2010).

Dominance relationships, where one group consistently displaces another, are sometimes evident among competing groups, but inter group relationships also vary in the extent to which interaction location influences the outcome of conflicts. At one end of the spectrum, location is irrelevant and outcomes are determined solely by the identity, size, or strength of the two groups (site-independent dominance). At the other end of the spectrum, location is the sole determinant of interaction outcome (site-dependent dominance) and the intruding group retreats from the resident group. (Note that site-dependent dominance may be considered a pay-off if resident groups gain access to resources by winning, and a form of resource-holding potential if resident groups win as a result of ownership conventions.) In between these two extremes, location is a partial determinant of interaction outcome (partial site-dependent dominance); larger groups tend to displace smaller groups, unless the interaction occurs within the core of the smaller group's range (Crofoot *et al.* 2008). Location can be measured in a number of different ways (e.g., as the distance from the center of the home range, a zone within the range, a particular side of an explicitly defended boundary, etc.); the most appropriate measure will depend on the behavior and ecology of the study groups. Importantly, different pairs of groups within a population can experience varying types of relationships (Sugiura *et al.* 2000), which points to the need for evaluating each pair of groups separately.

Resource-holding potential is less commonly measured for mixed species associations but is relevant for determining whether species compete for access to food resources. The ability of one species to displace another from food trees or canopy layers may be based on body size, group size, or aggressiveness.

9.3.7 Measuring independent variables: disposition

Different individuals experience varying costs and benefits of participating aggressively in (and of winning or losing) inter group interactions. Individual dominance rank is perhaps the most important of the disposition-related variables affecting individual participation during inter group interactions. Although high rank generally corresponds to high rates of participation in inter group conflicts (e.g., Cords 2007), in some populations, low-ranking animals are the most active participants (e.g., Sinha *et al.* 2005). Age is another relevant factor, as adults generally participate more often than subadults or juveniles. For individuals of the dispersing sex, tenure within the focal group may correspond with participation rates, where established individuals with offspring or strong social ties are more likely to participate than new residents. Among females, the reproductive phase is an important variable affecting participation, such that individuals with clinging infants are less likely to participate than cycling or gestating females. Individual participation may also be affected by familiarity with or relatedness to group-mates and neighboring groups. For instance, male vervets are more aggressive when interacting with their original natal groups than with non-natal groups (Cheney 1981), male Thomas langurs are less aggressive toward familiar groups with which they interact frequently than toward unfamiliar groups (the "dear enemy" effect: Wich and Sterck 2007), and subordinate male black howler monkeys participate more vigorously when they have close, long-term associations with high-ranking male group-mates (Kitchen *et al.* 2004b).

9.3.8 Statistical considerations

As with all research, statistical analyses must be planned in detail before beginning data collection. Because multiple factors simultaneously affect participant behavior, and because these effects may be undetectable with univariate statistics, it is crucial to use multivariate analyses (Harris 2007). Most or all of the research described in this chapter is appropriately tested using linear, logistic, or ordinal regression and many analyses will need to incorporate both fixed effects (independent variables) and random effects (repeated measures; e.g., group identity). Tests with both fixed and random effects are variously referred to as hierarchical, multi-level, or mixed models. To determine which combination of fixed effects best explains variance in the data, forward or backward stepwise regressions can no longer be recommended as these methods lead to high rates of Type I errors as well as unstable "best" models that do not generalize to other data sets (Mundry and Nunn 2009). Instead, the information-theoretic approach is generally recognized as the most appropriate solution to the dilemma of choosing the best-fit model, as it does not rely on arbitrary values for significance testing, allows a set of best models which may explain the data equally well, and can accommodate non-linear relationships among variables (Richards *et al.* 2011).

Statistical preparation consists of more than deciding which tests to conduct, but also a calculation of the sample size needed to test the hypotheses. G*Power (Faul *et al.* 2009) is a free power analysis program that allows users to quickly and easily determine minimum sample size requirements. Analyses of binary dependent variables (e.g., the presence/absence of aggression during an interaction) require larger samples than analyses of continuous data (e.g., number of chases during an interaction). For instance, a two-tailed test with power = 80%, $a = 0.05$, effect size = 0.30 (odds ratio = 2.19), three non-collinear fixed effects, and one random effect requires 38 interactions for a linear regression and 91 interactions for a logistic regression (in which either the 1 s or 0 s constitute 20% of the sample).

9.4 Field methods and analyses for studying interactions between groups: simulating (and stimulating) interactions

Adopting an experimental approach alleviates some of the difficulties associated with observational methods by allowing researchers to manipulate the social and ecological context of inter group interactions, and control their timing and location to facilitate observations and increase the total number of observed events. Chapter 10 of this volume focuses on experimental methods in field primatology, so here we limit our discussion to approaches that are useful specifically for investigating inter group interactions.

The most commonly used experimental approach involves broadcasting recordings of vocalizations to simulate the presence of an individual or group. Playback experiments can be used to simulate long-range interactions and the initial stages of inter group interactions (the first two action phases in Fig. 9.1), allowing researchers to investigate how variables like the location of a group within its home range, its size relative to its opponent, or its proximity to an important feeding site influence individual and group responses (McComb 1992). The behavioral responses that can be assessed in an experiment will vary by species and habitat type. Whereas the amount of time focal individuals spend looking toward the playback speaker or the distance they travel toward the source of the vocalizations can be measured for terrestrial savanna species, the limited visibility in forest environments can make it infeasible to collect these data for arboreal species (McGregor 2000). Instead, researchers might focus on the number of animals that move toward the speaker, the group's travel direction, or the number of calls produced after the stimulus. The type of vocalization used in such playback experiments (e.g., between- or within-group contact calls) will depend on both the species under study and the questions being addressed, but researchers should keep in mind that to simulate an inter group interaction—rather than an interaction with a solitary male or bachelor group—stimulus calls should include male and female vocalizations whenever possible.

Playbacks cannot fully simulate naturally-occurring interactions, because the level of agonism exhibited by members of the focal group, as well as their post-interaction behavior, depends on the actions of members of the opposing group. Attempts to experimentally investigate patterns of aggressive behavior during inter group contests or the factors that influence the outcome of such interactions have thus focused on stimulating, rather than simulating, interactions between groups. Baiting interactions by using feeding platforms or other provisioning sites to draw several groups to the same area (e.g., Zhao and Tan 2010) allows researchers to manipulate the location of interactions, as well as the abundance and distribution of food resources. It is also possible, although logistically quite difficult, to provoke interactions by using playbacks to attract groups to the same area (Crofoot, pers. ob.).

Experimental methods are especially useful for studying inter group relationships in populations where interactions between groups are rare. However, researchers should keep in mind that simulating the presence of a neighboring group increases the apparent competitor pressure for the focal group. For instance, if natural interactions occur no more than once a month but playback trials are conducted every week, the perceived competitor pressure has increased four-fold for the focal group. Populations that experience heightened competitive pressure tend to exhibit more intense aggression, particularly at feeding sites or in particularly important areas of the home range (Brown 2011). Experimenters should first determine the natural interaction rate and, from that, decide how often to conduct trials.

9.5 Field methods and analyses for studying interactions between groups: detecting the impact of inter group interactions on patterns of movement and space use

Whether it is mixed-species groups of monkeys traveling together to reduce their predation risk or subordinate social groups avoiding areas where they are likely to encounter dominant neighbors, primate groups have a profound impact on one another's patterns of movement and space use. Spatial data can thus provide insight into the nature of the relationships among groups, indicating whether they are attracted to, avoid, or are indifferent to each other. They can also reveal the effects that such relationships have on groups' access to resources.

9.5.1 Collecting spatial data

In studies of wild primates, it is common for researchers to record the location of their study group(s) at regular intervals, marking the group's position either by hand (based on landmarks like trails) or using a GPS unit. These spatial data provide a rich and often under-used source of information about how primates interact with both their social and ecological environments. However, to fully explore how groups influence one another's movements, and thus their patterns of space use and resource access, simultaneous data on the locations of two (but preferably more) social groups are needed. Collecting such data requires substantial manpower and/or the use of tracking technology.

Radio telemetry facilitates finding and following radio-tagged study animals and thus can be an important tool for projects where multiple groups must be followed simultaneously. Radio telemetry allows researchers to remotely monitor the movements of radio-tagged study animals. In addition, radio telemetry reduces the amount of time needed to encounter groups and minimizes bias in the locations where groups are found. This latter point may prove particularly important in studies of inter group relations. Sorties outside a group's normal range are probably important in shaping relationships between neighbors, but these types of unusual excursions may be missed in studies that do not use some kind of remote tracking because groups are less likely to be found where they are not expected to occur.

Although commonly used in the broader field of mammalian ecology, and a key tool in research on nocturnal primates, radio telemetry has not been widely adopted in studies of diurnal primates. In part, this is due to ethical concerns about animal safety related to both the risk of injury during capture and to possible side effects of collaring. Another commonly voiced concern is that capture will cause primates to lose their habituation to human observers. Most importantly, however, from the perspective of studying the interactions between primate social groups, radio telemetry fails to solve the manpower problem (but see Crofoot et al. 2008 for details about an automated radio tracking system). Whether using radio tracking to simply find and follow groups, or to estimate the location of radio-tagged animals, a researcher still needs multiple people on the ground to collect location data.

Satellite-based tracking promises to greatly increase our understanding of primate inter group relationships because it solves the aforementioned manpower problem, allowing the movements of numerous study animals to be tracked automatically and simultaneously. Although several satellite-based tracking systems exist (e.g., ARGOS, GlobalStar), the Navstar Global Positioning System (GPS) is most likely to be of use in studies of primate ranging and space use, due to its high degree of spatial resolution (i.e., meters rather than kilometers). The Navstar system consists of an array of 24 geosynchronous satellites owned and maintained by the US Department of Defense. Rather than emitting signals like radio tags, GPS tags act as receivers for the signals broadcast by these 24 satellites. Locations are calculated based on the time lag between the radio signals sent by the satellites and received by the tag, using information about the orbital path of each satellite (Rodgers 2001). Data are typically stored on the collars and either downloaded remotely (via radio or satellite link) or manually (when the collar is recovered).

Until recently, the weight and cost of satellite tags and their poor performance in forested environments limited the utility of these systems for tracking most primate species. Although the US Department of Defense discontinued its policy of degrading the accuracy of civilian GPS receivers in 2000, decreasing the error of GPS location estimates to ≤ 10 m (Rodgers 2001), GPS reception under forest canopy remained poor (Phillips et al. 1998; Jiang et al. 2008). However, new high-sensitivity GPS microcontrollers have improved the rate at which the newest satellite tags acquire location estimates (Markham and Altmann 2008), making them appropriate for use in forested habitats. Reductions in the energy requirements of these new GPS receivers have also allowed the development of significantly lighter tags (< 10 g; Brown and Brown 2005), which are small enough to use on a wide range of primate species. The lifespan of GPS collars is highly dependent on sampling frequency, but many of these smaller collars have lifespans ranging from several months to two years (Brown and Brown 2005; Crofoot, unpublished data).

The most effective method for collecting data on the movements of primate social groups will depend on a number of different factors, including the number of groups that are to be monitored, the area over which they range, the distance they travel per day, the length of the study, and the relative cost of labor. GPS collars remain expensive (~ $1000–$2000 per collar) and in some cases it may be more cost-effective to hire people to track groups on foot. GPS collars also have a short battery life compared to radio collars, and this may prove a limitation, especially if high temporal resolution is desired (i.e., six vs. one fix per hour). Finally, in tracking studies where the location of a collared individual will be used as a proxy for the location of the group as a whole, careful thought should be given to which individuals to fit with collars. While some age/sex classes may respond to interactions with neighboring groups by approaching them, others may move away to avoid interacting; thus, depending on the species, it may be important to collar individuals of more than one age/sex class.

9.5.2 Sharing space

Although patterns of primate space use are highly variable, the majority of species show some degree of site fidelity. This tendency to revisit areas means that most primates spend their lives in an area that is small compared to their movement capabilities. The degree to which these home ranges are shared or defended against intrusions by neighboring groups is treated as a salient trait that differentiates among primate species (i.e., territorial vs. non-territorial species). Theory predicts that primate groups will defend a territory when it is economically feasible to do so; that is, when the benefits gained through exclusive use of an area exceed the costs of excluding conspecific neighbors (Davies and Houston 1984). Mitani and Rodman (1979) showed that the ability to monitor range boundaries, indexed by the ratio of day range length to home range diameter (the D index), strongly predicted whether a species was classified as territorial. However, the significance of this relationship is not as straightforward as it might at first appear because use of the term "territorial" has been inconsistent in the primate literature. Thus, even among "territorial" species, there is a great deal of biologically important variation in the degree to which neighboring groups share space (Maher and Lott 1995).

The most intuitive, and thus most commonly used, method for assessing the extent to which primate social groups share space involves mapping the two-dimensional home ranges of adjacent groups and calculating the percentage of area that is shared by the ranges. This method has the benefit of being easy to calculate and, at least superficially, lends itself to straightforward interpretation: x% of Group i's range is also used by Group j. However, a number of methodological cautions and theoretical considerations should be kept in mind. First, for many methods of home range estimation, there is a strong correlation between sample size and home range area (Kernohan *et al.* 2001). Increasing the number of location estimates used will increase both the size of home range estimates and the degree of range overlap. Thus, when calculating measures of home range overlap, it is important that sampling intensity and duration be even among groups or, at a minimum, that the potential ramifications of uneven sampling be explicitly considered. Likewise, when comparing results from different periods or studies, it is critical to keep in mind that differences in the degree of home range overlap might reflect nothing more than differences in sampling. Second, the most commonly used home range estimator, the minimum convex polygon (MCP), does not perform well with irregularly shaped ranges and when used to measure joint space use, may reveal extensive home range overlap where, in fact, none exists (Kernohan *et al.* 2001). This second problem is ameliorated by using methods of home range analysis, like kernel home range estimators (Worton 1989), that take into account the interior anatomy of home ranges rather than focusing strictly on the outer boundary. However, the most critical limitation of 2D measures of range overlap, regardless of the type of home range estimator used, is that they ignore the wealth of information that location data provide about patterns of space use within the confines of the range boundary. They cannot distinguish between groups whose ranges overlap in infrequently used areas and those that share the focal center of their range. In fact, by focusing on the amount of area that is shared and disregarding the intensity with which those areas are used, 2D methods of assessing home range overlap equate these functionally disparate patterns of joint space use.

Incorporating the intensity of space use into measures of animal spatial interaction is not a new idea. Rasmussen (1980) proposed a grid cell-based method for calculating an index of space use similarity, based on the correlation between the utilization distribution (UD) of two animals' ranges. However, the development of kernel-based estimators which calculate UDs as probability density functions (Worton 1989) now make it far easier to represent the temporal dimension of animal space use and to treat the measurement of home range overlap as a 3D, rather than a 2D, problem.

9.5.3 Directional and non-directional measures of UD overlap

Utilization distributions calculated using kernel estimation are 3D surfaces that represent the probability of an animal being found in a given location. A number of different methods have been proposed for using UDs to characterize the degree of overlap between home ranges (Fieberg and Kochanny 2005). The biggest difference among these methods is that some provide *directional* measures of shared space use, representing the probability of finding Group A in Group B's range, while others yield a *non-directional* measure of overall similarity in the two groups' UDs. The most appropriate method to use will, of course, depend on the goals of the study.

Fieberg and Kochanny (2005) review several methods for estimating the degree of overlap between adjacent home ranges. The directional measure of UD overlap they propose can be envisioned as the volume under a 3D surface (Group A's UD) contained within the extent of a 2D polygon (Group B's home range). The volume of intersection (VI) index proposed by Seidel (1992) provides a non-directional measure of the overlap between two groups' home ranges based on the area of intersection between the full UDs for both groups. Unlike the directional method described above, the VI index yields a single value describing the range overlap for a pair of groups, ranging from 0 (when there is no overlap between two home ranges), to 1 (when the UD of the groups are identical). Fieberg and Kochanny (2005) compared the performance of the VI index to two related methods for quantifying home range overlap. Based on the ability of these indices to accurately rank pairs of home ranges in terms of their degree of overlap, they recommended the UD overlap index (UDOI) as the most appropriate index for quantifying space-sharing.

All indices of home range overlap discussed in this section are implemented in the R package adehabitatHR (Calenge 2006), making them easy to incorporate into studies of spatial interaction. However, they have comparatively high data requirements: more than 200 locations per group may be needed to accurately estimate kernel home ranges (Seaman *et al.* 1999; Garton *et al.* 2001). In addition, difficulty in determining confidence limits may limit the utility of these indices (Fieberg and Kochanny 2005). Although it is possible to use bootstrap techniques to develop confidence bounds for UD home range estimates (Worton 1995; Kernohan *et al.* 2001) or for the overlap statistics themselves (Seidel 1992), this approach may be problematic due to bias in kernel density estimators (Davison and Hinkley 1997). Finally, Fieberg and Kochanny (2005) stress that considerable caution should be taken when comparing measures of overlap across studies, because overlap estimates will almost always be biased, and the extent of this bias depends heavily on sample size.

9.5.4 Dynamic interaction analyses

Conceptually, analyses to quantify the degree of spatial interaction among groups of animals can be separated into two main categories. "Static" approaches (such as those mentioned above) measure spatial overlap without considering whether groups use the same space simultaneously or at different times. For example, computing the area of home range overlap between neighboring groups tells you how much space they share, but not whether they encounter one another in this area. "Dynamic" analyses, in contrast, inherently incorporate the temporal nature of these relationships.

When the movements of two or more primate groups are tracked simultaneously, it is possible to use these data to gain insight into inter group relationships by determining whether groups spend more or less time in close proximity to one another than would be expected by chance. The most straightforward approach to assessing spatial attraction or avoidance involves comparing the observed distribution of inter group distances to the expected distribution if groups were indifferent to one another (i.e., did not influence each other's space use). For example, Kenward and Hodder (1996) suggest that the observed median distance between two animals (D_o) could be compared to an expected median distance (D_e), calculated as the median of all distances between all recorded locations, using a sign test. Alternatively, resampling techniques can be used to generate a distribution of expected mean inter group distances by repeatedly randomizing each group's relocation data set with respect to time. The advantage of this latter method is that it could be applied to any metric describing the distributions of inter group distances, not just the mean. For example, if two groups avoid one another, we would expect to observe very few instances where they were close to one another—in which case, skew may do a better job describing the important characteristics of the distribution of inter group distances.

An alternative to analyzing the distances between primate groups is to explore the degree to which groups share the same area, at the same time. Minta (1992) proposed a method for assessing such spatio-temporal overlap by summarizing the space use of two groups as a set of binomial frequencies. At each sampling period, each group is classified as either present or absent from the shared area where their home ranges overlap. Observed patterns of co-occurrence in the shared area are compared to the rate of co-occurrence that would be expected if each group moved independently of the other. Minta's coefficients of interaction can then be used to describe the relationship between two groups as random, attraction, or avoidance.

Because many of the analytical methods for quantifying animal spatial interactions were developed for use with radio tracking data, they make a number of assumptions—most notably, that successive location estimates are statistically independent of one another—which are violated by most primate movement data sets. Whether collected directly by human observers, or remotely using GPS tags, data on primate movements generally consist of a relatively high-resolution time series of location estimates. To our knowledge, no one has yet investigated the effects that this serial autocorrelation has on the static and dynamic interaction analyses reviewed in the preceding section.

One possible solution to the statistical problems caused by autocorrelated movement data is to subsample movement data sets to the point where subsequent observations are no longer correlated (c.f., Swihart and Slade 1985). In theory, as the sampling interval increases, serial correlation should diminish until the lag time needed to achieve independence between successive observations is reached ("time to independence," Swihart and Slade 1985). However, animals with home ranges, by definition, have recurrent patterns of movement and thus such data independence may not be biologically feasible (de Solla *et al.* 1999; Powell 2000). In fact, de Solla and colleagues have argued that in the context of home range estimation, eliminating autocorrelation from location data sets reduces both statistical power and accuracy, and destroys biologically relevant information. They found that using more data increased both the accuracy and precision of home range estimators as long as the sampling effort was even among groups.

High-resolution location data contain a wealth of information, not just about where animals go, but about how they get there (Byrne *et al.* 2009). Thus, rather than seeing autocorrelated data as a statistical problem, we urge researchers to adopt modeling approaches that allow them to incorporate these details of animal movement patterns into their analyses.

9.5.5 Null models of inter group contacts

One of the most commonly used approaches to modeling rates of interaction among primate social groups derives from a classic physics equation for predicting collision rates in an ideal gas (Maxwell 1860). These "ideal gas" models treat animals (or groups of animals) as if they are molecules moving independently, and have generally been used as a null model to generate predictions about the rate at which individuals or groups would be expected to encounter one another if they did not influence each other's movements. For example, Waser (1976) found that observed encounter rates between grey-cheeked mangabey groups were substantially lower than the rates predicted by his ideal gas model, and based on this discrepancy, concluded that "cryptic" (i.e., long-distance) avoidance between social groups was occurring. Ideal gas models also yield predictions about the expected duration of interactions between primate social groups. A number of investigators have used these predictions to separate cases of mixed-species associations based on behavioral attraction from those that arise simply as a result of high population densities (reviewed in Hutchinson and Waser 2007).

The ideal gas approach makes numerous simplifying assumptions about how animals move: that travel velocity does not differ between individuals or across space; movement occurs in randomly oriented, straight lines; detection distance remains constant throughout the habitat; and individual or group density does not vary, to name a few. While these may be reasonable approximations of biological reality for some systems, they are violated in many others. Hutchinson and Waser (2007) have proposed a number of extensions to the ideal gas model that increase its realism and broaden its applicability. For example, they derive an adjustment factor that allows the model to be used to predict rates of interaction between groups whose home ranges do not fully overlap. Another major limitation of published ideal gas models is that while they can predict encounter rates, they do not provide a way of determining if observed values differ significantly from those predictions. Hutchinson and Waser suggest several different approaches for developing more rigorous statistical tests.

In our opinion, the most important limitation of the ideal gas approach to analyzing primate movement data is that it does not provide insight into the factors that drive patterns of interaction among groups. Knowing that groups encounter each other more (or less) frequently than would be expected if they were moving randomly and independently is merely a first step. To understand how these patterns arise, researchers need a means of directly examining assumptions about the dominant social and environmental factors that affect group movements.

9.5.6 Mechanistic models of inter group interactions

In contrast to the statistical and ideal gas models discussed above, mechanistic models of inter group interactions explicitly incorporate predictions about the ecological and social factors that influence animal movement decisions, and thus are responsible for creating observed patterns of space use. Rather than simply describing the size, shape, or overlap of adjacent home ranges, mechanistic models directly address the underlying causation of these patterns. A number of different modeling approaches, including agent-based modeling and mechanistic home range analysis, fall within this framework. Unlike the ideal gas models described above, these approaches can be used to generate testable hypotheses about the factors that drive patterns of interaction among groups. They also allow for prediction of the impact that environmental changes will have on patterns of interaction.

Although it has not yet been widely adopted as a tool for understanding patterns of interaction among primate groups, agent-based modeling presents an attractive, bottom-up approach to addressing such questions because it is flexible, can capture emergent phenomena, and provides straightforward, easily interpretable descriptions of biological systems. Agent-based models describe systems from the perspective of their constituent units, using simple behavioral rules to generate complex outcomes. Systems are modeled as a set of autonomous decision making entities, or "agents," that individually assess the situation in their programmed environment and make decisions about which of a predefined set of behaviors to adopt on the basis of a given series of rules. Models with different sets of decision rules serve as alternative hypotheses which can be tested by comparing observed patterns of movement or interaction to simulated data sets.

For example, Ramos-Fernandez and colleagues (2006) used spatially explicit, agent-based models to simulate spider monkey ranging and foraging behavior. In their model, the "agents" represent foraging spider monkeys that decide where to move in their habitat by selecting the target (i.e., a fruit tree) with the highest net benefit (target size/target distance). Their work demonstrates that in a habitat where resource distribution is heterogeneous and patch size distribution is exponential, and when individuals know the locations of available resources, the complex inter-individual association patterns that characterize spider monkey fission–fusion social organization can arise from this simple foraging rule without taking into account any social attraction among group members.

A major strength of agent-based modeling is its ability to incorporate aspects of a wide array of theoretical approaches. The spider monkey "agents" described above, for example, incorporate aspects of optimal foraging theory into their foraging choices (Ramos-Fernandez *et al.* 2006), and game theory is frequently used as the basis for decision rules in agent-based models (e.g., Hemelrijk 2000). Both deterministic and probabilistic movement rules can be implemented (Boyer and Walsh 2010). In addition, agent-based models can be easily adapted to the new movement ecology framework proposed by Nathan and colleagues (2008), which is built around four fundamental components of animal movement: internal state, external factors, motion capacities, and navigation capacities.

One weakness of agent-based models is that they are often computationally expensive; simulating the movements of many independent agents, especially when decision rules are contingent on the state of a changing environment, can be extremely time consuming. There is also a tendency with agent-based modeling to attempt to include too many aspects of a system. The most useful agent-based models are often simple, including the minimum amount of detail required to capture the phenomenon of interest. Finally, unfamiliarity with computer programming presents a barrier for many primatologists interested in adopting an agent-based modeling approach to data analysis. However, a number of software packages have now been developed for generic agent-based modeling which provide user-friendly graphical user interfaces (GUIs) and are often integrated with additional functionality like GIS, spatial statistics, and machine learning algorithms to facilitate the development of agent-based models by users with a range of experience levels. For a comparison of these software packages and helpful advice for using them, we refer readers to a review by Railsback and colleagues (2006).

Mechanistic home range models provide a well-developed alternative to agent-based modeling for exploring the impact of inter group relationships on primate space use (Moorcroft and Lewis 2006). These models represent animal movement behavior as a set of partial differential equations describing the random and directed components of movement (advection/diffusion equations), and yield spatially explicit predictions of space use patterns in the form of 2D probability density functions. They are mechanistic in the sense that each group's pattern of space use is calculated by an explicit mathematical scaling of its underlying movement rules. These movement rules serve as hypotheses for the causes of the observed pattern of space use, and competing hypotheses can be tested against one another to determine which one predicts a space use pattern that most closely matches the observed data. However, despite its promise to greatly increase our understanding of ranging decisions, mechanistic modeling has not yet been widely adopted in studies of primate ranging behavior or inter group relationships.

9.6 Directions for future research

Though many studies have focused on inter group interactions in primates and other animals, there remain many significant gaps in our understanding of the topic. Some voids persist as a result of inconsistent or inappropriate methods, which we have identified in the preceding sections. Others result from a lack of attention to particular topics, some of which we discuss here.

First, there is a need for greater cross-fertilization of ideas between primatologists and other behavioral ecologists. In this chapter, we have focused on studies of interactions between primate groups, but many of the methods we discuss, particularly those for analyzing movement data, were developed for other taxa. In addition, theoretical and empirical studies of other group-living species including social carnivores, ungulates, birds, and invertebrates are highly relevant; primatologists could learn much from their non-primate counterparts, and vice versa. One particularly important avenue of research is to determine if differences in the hypotheses used to explain patterns of inter group interactions in primates (e.g., Wrangham 1980; van Schaik 1989; Fashing 2001) and other social species (e.g., Emlen and Oring 1977; Ostfeld 1990) can be reconciled. It appears that the latter are not widely applicable to primates, because they predict that females will not defend resources of territories when living in social groups. This is patently untrue for many primate populations. Similarly, the primate-specific hypotheses stipulate fairly specific conditions under which food resources should be defended, yet these conditions may not be applicable to carnivorous species whose food sources are mobile. A key question which, to date, has not been satisfactorily addressed is whether, when certain social and ecological factors are held constant, the same set of general principles can explain the patterns of inter group interactions of both primates and other group-living animals.

Second, the role that the conflict between individual and collective interests plays in shaping inter group relationships deserves further theoretical and empirical attention. Inter group conflicts pose an inherent collective action problem: the benefits gained through cooperative defense (e.g., access to a food source) are generally shared among all group members, whether or not they contributed to its acquisition, creating an incentive for individuals to free-ride on their group-mates' efforts (Nunn 2000). How individuals make decisions about whether or not to participate, or how much effort to invest in an inter group conflict, and how these individuals' decisions then combine to shape group behaviors is thus quite complicated. We feel that the continued development of n-player game theory models and studies that focus on variation in individual behavior during inter group contests will be critical for advancing our understanding of primate inter group relationships.

Lastly, significant disagreement exists about the importance of group-level competition as a selective pressure. While inter group relationships appear to have significant, long-lasting effects on individual fitness in some primate populations (Cheney and Seyfarth 1987; Robinson 1988), in others, they do not (Janson 1985). Population density may account for some instances in which inter group relations have little or no effect on fitness, but for populations existing at a high density relative to habitat carrying capacity, it remains unclear what other factors affect the strength of inter group competition.

Acknowledgments

We thank the editors for inviting us to contribute to this volume, and to the editors, an anonymous reviewer, and Damien Caillaud for their input on an initial draft of this chapter. M. Brown was supported by a Postdoctoral Research Fellowship from the National Science Foundation (award # 1103444); M. Crofoot was supported by a post doctoral fellowship from the Smithsonian Tropical Research Institute, the Division of Migration and Immuno-ecology at the Max Planck Institute for Ornithology, and the Department of Ecology and Evolutionary Biology at Princeton University.

10

Experiments in primatology: from the lab to the field and back again

Charles H. Janson and Sarah F. Brosnan

10.1 Introduction

Behavioral primatology has a popular reputation as a non-experimental science (e.g., Tomasello and Call 2011). This reputation is only partly deserved, as experimental work on the behavior of captive primates has a solid tradition dating back nearly a century and covering many thousands of publications. Experiments on the behavior of wild primates have a shorter history of about 50 years and are far less common, yielding approximately 200 publications. This difference is not the result of disdain by field primatologists for experiments, but rather a likely outcome of different research traditions and constraints. Those who study primate behavior in the wild have often emerged from anthropology departments where detailed but non-interventionist observation used to be the reigning methodology, whereas it tends to be the case that researchers who study behavior in captivity have been trained in psychology, which has a long history of, and emphasis on, experimentation. These discipline-specific differences are likely related to the ease of doing experiments. Field experiments are difficult and require a subtle compromise between the needs of the researcher and conformance with the lifestyle of the wild study animals (Janson 2012); experiments in captivity are a ready extension of the control already exerted by humans in housing and breeding primates. Nonetheless, there is a middle ground: experiments on captive but free-ranging primates that live in large spaces combine some aspects of both captive and wild experiments. For instance, it is often possible to present a task to a single free-ranging individual (as in captivity), but the study subject may choose to ignore the choice (as in the wild). Likewise, recent laboratory experiments provide subjects with the option to refuse to participate, which is considered an experimental response.

Primate Ecology and Conservation: A Handbook of Techniques. First Edition. Edited by Eleanor J. Sterling, Nora Bynum, and Mary E. Blair. © Oxford University Press 2013. Published 2013 by Oxford University Press.

Experiments and observations can be arrayed along a continuum of when and how a researcher controls variables (Janson 2012). Experiments focus on *a priori* control of some variables. In the most extreme experiments, all variables are held constant except for one focal variable of interest that may vary freely or is fixed at specific levels. Observational studies rely on measurement of unmanipulated variables, and typically use statistical methods to "control" for the effects of extraneous variables in assessing how a focal variable affects the behavior under study. By careful choice of focal variables and study design, observational field studies can approach the rigor of field experiments, producing "quasi-experiments" (Janson 2012).

The rigor and flexibility that the use of experiments provides to researchers studying captive primates should make them a model to emulate by field researchers, but instead there tends to be a schism that divides these two groups of researchers. Studies of captive primates are cited mostly by researchers studying captive primates, and field studies are cited by researchers studying animals in the wild (but there are exceptions, e.g., Fragaszy *et al.* 2004). To some extent, this self-reinforcing referencing is symptomatic of the very different questions that are best addressed by studies in the two locales. Questions related to behavior are broadly divided into two types, based on Tinbergen's categorization (1963). First are those addressing the functional significance ("ultimate mechanisms") of a behavior, including the survival value of a behavior to the organism, and from what earlier behavior it evolved from. Second are questions addressing the mechanisms underlying the behavior ("proximate mechanisms"), including the physiological correlates of the behavior, such as hormonal, neural, and cognitive mechanisms, and how the behavior develops within the organism, including genes, but also learning and culture. The field is the only place where one can hope to understand the functional significance of primate behavior as it relates to natural components of an individual's environment other than its group mates. Research on captive primates can provide a window onto developmental and physiological mechanisms of behavior that are difficult or impossible to study in free-ranging animals. In addition, studies on captive primates can reveal responses to unusual "environmental" conditions that are rarely available to wild populations, thereby allowing a more complete description of a species' "norm of reaction" (Janson 1994). Ideally, any researcher hoping to understand a given behavior would strive to combine both functional and mechanistic perspectives, either within their own research program, or by actively referencing both literatures. However in practice, this does not occur frequently. Below we describe some aspects of experimental studies in captive and field situations; we tend to emphasize studies on capuchin monkeys, as we are most familiar with this group of primates, but this does not at all mean that comparable citations are absent from other taxa—indeed the number of studies on chimpanzees and macaques outnumbers that on capuchins by a substantial margin. We finish with a consideration of several research areas that would benefit from more direct collaboration among researchers using captive and field studies.

10.2 Contrasting benefits of field vs. captive experiments

10.2.1 Field experiments

The benefit of doing experiments in the wild is to replicate possible scenarios that could have led to selection for the behavior in a controlled way, while including a broad representation of natural variation in the environment. This replication of specific, predetermined variables within a natural context allows the researcher to disentangle possible confounding effects of variables that may be correlated with the behavior of interest. This is achieved by averaging the effect of a manipulated variable across the range of natural variation (e.g., Janson 1996). However, because there are a large number of natural variables that the experimenter cannot control, the researcher can never be fully confident that the studied behavior is (or is not) caused by the variable of interest, rather than some extraneous variable that happened to be correlated with it (Janson 2012).

Greater confidence in inferring the cause of a behavior can be obtained by manipulating more variables, as is possible in captivity, where it may be feasible to control all aspects of the primate's environment. Captive experiments vary considerably in the degree of control of variables imposed by the researcher, from the extreme simplicity of the isolation studies by Harry Harlow in the 1950s and 1960s to studies of free-ranging groups or populations of primates enclosed in areas from a fraction of a hectare to many square kilometers. For instance, the breeding population of macaques on the island of Cayo Santiago represents a group of monkeys not in their native habitat, yet it comes close to replicating a natural environment. Cayo Santiago has been a fertile ground for experiments in which researchers present free-ranging individuals with small-scale interventions (a play-back or a presentation of food choices, e.g., Santos et al. 2006). Studies on Cayo macaques can often amass considerable sample sizes quickly, as it is easy to find suitable study subjects and set up manipulations using these very habituated animals living at high densities (neither feature is common in wild populations).

10.2.2 Captive experiments

There are several distinct benefits of studying behavior in a captive setting. First, the degree of control over most variables is significantly greater than in the field.

Researchers can not only control the values of important variables, but also eliminate the confounding effects of extraneous variables altogether. Thus, "replicate" experiments in captive settings may truly replicate all the external conditions of the test, both for manipulated and unmanipulated variables. In some kinds of research (e.g., cognitive or perceptual abilities: Cramer and Gallistel 1997), it is essential to isolate subjects for testing, both to reduce social distractions and to eliminate inadvertent cueing from conspecifics; such temporary removal of individuals is readily done in captivity but is difficult in the wild for most primate species. Primates in captivity can be housed in specific social situations to monitor long-term patterns of development of social behaviors, or they can be placed together in specific pairs or small groups to test particular aspects of social cognition (e.g., reciprocity and exchange). Finally, more complex designs and complicated experiments (e.g., those requiring multiple repetitions or specific time delays) are possible in the laboratory, again due to the greater control available. These studies can be used to elucidate specific quantitative relationships between environmental variables and behavior (Herrnstein 1961; see also below).

Second, captive maintenance of study subjects allows the researcher considerable control over their ontogenetic trajectory. This includes not only their prior history of social housing, but more importantly the ability to train subjects to become familiar or proficient with certain kinds of tasks, equipment, and testing environments. A majority of psychological research on primates requires at minimum allowing them to become familiar with the test apparatus (e.g., tool use) or explicit training to work with a particular test situation (e.g., cooperative problem solving) or apparatus, from simple alternative-choice devices (e.g., "self control" studies) to complex sets of symbols or computers. Because it can take days to weeks or more to train primates to criterion performance with more complex tasks, it is not realistic to expect them to be able to learn such paradigms in the wild where time is limited and more immediate fitness-enhancing activities claim their attention. Additionally, the fact that the entire history of the primate is known, including his previous exposures and social history, allows for studies that address whether specific behaviors are innate or learned. For instance, rhesus monkeys exhibit extreme alarm to snakes. Careful captive studies demonstrated that this is not innate, but is learned through a single trial's exposure to an adult's reaction to a snake (Mineka and Cook 1988).

Third, having primates in captive conditions, whether in the lab or free-ranging, typically provides the researcher with observation abilities far better than those available to researchers on wild primate populations. Not only may the observer see more of any interaction or behavior of interest, but the proximity to the subjects allows clear video or audio recording of subtle behaviors that would be missed entirely in the field, including soft vocalizations, transient facial expressions, rapid shifts in gaze, and so on. Once such behaviors have been documented in captive conditions, they might become subjects for additional study in wild populations (e.g., gaze direction and duration are routinely used as measures of attention in captive studies, but have been used routinely only in field research on terrestrial primates, e.g., Cheney and Seyfarth 1988).

Fourth, the freedom from the constant "struggle for existence" of living in the wild allows captive primates to engage in exploratory, play, and social behaviors to a far greater degree than is typical in the wild. For instance it is not surprising that spontaneous tool use in capuchin monkeys was documented in captivity (Klüver 1933) at least 60 years before credible evidence of habitual tool use was found in wild populations (Fragaszy *et al.* 2004). Studies in captivity have always been fertile avenues to document and explore subtle or infrequent social behaviors that might be missed in the wild. For instance, post-conflict reconciliation has been documented in several species of capuchin monkeys in captivity (Verbeek and de Waal 1997), but evidence from wild populations is either weak (Manson *et al.* 2005) or absent (any population of *Cebus* "*apella*" in the wild).

Fifth, access to technology allows a given question to be approached from many perspectives. For instance, the study of cognitive maps in primates has been addressed using search for hidden food items by individual captive animals in a large enclosure (Menzel 1973), chimpanzees directing humans to look for specific long-hidden items in a space the chimpanzee cannot access (Menzel 2005), foraging tests by individual captive animals in small enclosures (Cramer and Gallistel 1997), and simulated foraging tasks on computer monitors (Sato *et al.* 2004). Use of such diverse methods allows greater confidence in any conclusions that coincide from the different approaches, and also helps to illuminate how different sensory modalities or ways of framing a task can affect a behavior.

10.3 Design constraints of field vs. captive experiments

10.3.1 Field experiments

In field experiments, a researcher typically leaves most parameters to vary naturally and manipulates only one to a few focal variables. Because the researcher can control only the few parameters of interest in the experiment, the notion of a "replicate" must be understood to mean "replicating only the experimental parameters," while the unmanipulated parameters may vary uniquely from trial to trial. This relatively weak level of researcher control is a consequence of the study situation and questions of interest. Frequently a field researcher is interested in the adaptive function of species-typical behaviors. Because adaptive function depends on the environmental features that selected for the behavior in the first place, field experiments should try to preserve the natural selective regime by manipulating the context of studied behaviors in a minimally-invasive way. Introducing too large a change in ecological or social context risks creating a behavioral response that is outside the adaptive range. However, if study animals can choose not to participate in a field experiment, then it must be designed to attract their involvement, treading between the need to conform to their natural lifestyle and being interesting or profitable enough to entice their participation (Janson 2012). In this situation, detailed observational studies are essential prior to designing meaningful experiments, as each new study population will experience different kinds and degrees of lifestyle constraints.

The difficulties of creating experiments that conform to a primate's lifestyle in the wild can be daunting. Studies of the functions of primate vocalizations in the wild have benefited greatly from the use of playback experiments (e.g., Cheney and Seyfarth 1988), but the constraints on the design of such experiments are stringent (e.g., for a general review of playback methods in bird song studies, see Catchpole and Slater 2008). First, because social primates can typically recognize which group-mates are producing a particular vocalization (e.g., Rendall *et al.* 1996), playback experiments must use only vocalizations derived from actual group members. Second, to avoid potential mismatches between visual and auditory cues, one should use the calls from a particular individual only when the focal animal being observed cannot see the actual group member whose vocalization is being broadcast. Third, because animals in general can rapidly learn the characteristics of a given recording and may habituate to it, ideally each playback should use a unique recording (Kroodsma *et al.* 2001), which is a problem for uncommon vocalization types. Finally, to avoid possible interpretation bias, the researcher involved in the design of the experiment should not score the resulting behavioral response, which requires recording the behavior on video for presentation to "naïve" observers. The end result of these and other constraints is that to perform even a modest number of valid field experiments may require many months or even years. For instance, Di Bitetti (2003) managed to complete only 39 experiments on food-associated calls in capuchin monkeys over a period of 16 months; it was difficult to find situations that matched all the required conditions in a dense forest environment, and about half of all experiments attempted had to be abandoned before completion as animals moved during the course of the setup process. Designing some kinds of foraging experiments in a wild monkey population can be equally challenging, although for different reasons (Janson 2012). The end result is that the number of replicates of field experiments is often modest and is under the control of the study animal or field conditions, rather than the researcher.

10.3.2 Captive experiments

Captive experiments have an entirely different suite of issues than do field experiments. Rather than concerns about lack of control, there are concerns about lack of validity. These stem from the much more restrictive living space, the increased "ease" of the subjects' lives (e.g., they are not required to forage for their own food or escape from predators, and have access to extremely high-quality medical care for those issues that do arise), and, in some cases, they live in a social organization that is not particularly naturalistic. Many of these are virtually unavoidable artifacts of captivity; work can be done to minimize their impact, but it is almost impossible to provide in captivity the full range of opportunities that exist in their natural environment. That being said, there are at least two issues that we have identified that are both extremely important and possible to address. These include what we refer to as the training bias and the context bias.

10.3.2.1 Training bias

Training bias is the issue of whether the behaviors seen in these animals are the result of their extensive training and experience. Primate subjects are tested repeatedly because they have very long lives, and ethical considerations prompt researchers to minimize the number of experimental subjects used. To what extent, then, are the behaviors seen in captivity the result of training of intelligent animals, which includes previous experiments or similar types of experiments, and to what extent are they reflections of "natural" propensities? There are ways to address this. First, of course, is a quick estimation of whether a result makes sense; if experimental tests are showing a behavior for which there are no obvious correlates in the wild, perhaps it's time to consider whether this is an artifact of captivity. A second way to explore the possibility of training bias is to look for similarities and differences between captive populations that differ on some of these dimensions. For instance, studies comparing chimpanzees with extensive cognitive training since infancy with those that lack such specialized experience have found both similarities and differences in task performance, indicating that training and experience may change behavior but do not necessarily do so. Both groups of chimpanzees show virtually identical patterns of trade (Brosnan *et al.* 2008), but differ in their responses to a coordination game derived from experimental economics (Brosnan *et al.* 2011). Thus, examining behaviors across multiple captive settings can be quite informative. If behavior varies, it is interesting to examine how a behavior depends on prior experiences; if it does not, the trait or behavior is likely relatively insensitive to training bias.

It is important to examine the effects of training on a behavior, as understanding what a study species typically does is very different from understanding what it can do. It may be interesting to see what a species can do, but what does it mean if only one individual is ever able to be trained up in this way, despite efforts in that direction (think of, for instance, Alex the parrot: Pepperberg 2002), or cases in which there is a large variation between individuals in their competence at a given task (such as navigating a maze: Fragaszy *et al.* 2003), a result that is not expected if the capacity were of high adaptive value to individuals of the species. We do not mean to imply that such feats are not impressive accomplishments. However, we do encourage lab researchers to consider what their results mean in the context of the primates' everyday lives. We also encourage field researchers to consider new findings from the lab and consider how they might be involved in behaviors they have witnessed in the wild. For instance, captive tufted capuchins are remarkably egalitarian compared to many well-studied female-bonded primates (De Waal 1986). A possible explanation of the more relaxed hierarchy in capuchins is the fairly recent discovery that, in the wild, high female rank is not well correlated with reproductive success in long-studied capuchin populations (e.g., Janson *et al.* 2012). Captive experiments frequently benefit from observational field studies because the latter may reveal the natural context in which a studied behavior in captivity makes sense. For instance, two species of callitrichids that differ in their feeding habits (one consuming widely-dispersed fruit, the other slowly-renewing plant gums) also differ in their use of spatial versus temporal memory of prior feeding episodes (Stevens *et al.* 2005). This difference makes sense only in light of knowing the natural foraging ecology of these species. Finally, lab studies may challenge field researchers to find the appropriate functional context of a behavior in the wild. For instance, many primates have demonstrated the ability to count objects in lab studies (Matsuzawa 1985). What pressures in the wild may have selected for such behavior, and how is this ability used in the natural environment? Although we can speculate (for instance, the need to identify clusters of food resources, or to avoid parties of competitors), such questions remain almost completely unexplored.

10.3.2.2 *Context bias*

The flip side of training bias is what we call context bias, or the issue that study subjects are often housed for convenience of researchers or the IACUC (Institutional Animal Care and Use Committee, the institutional committee that oversees all vertebrate research at US research institutions), not in ways that promote full development of social or manipulative behaviors. This may, for instance, result in smaller group sizes, fewer adult males, unnatural degrees of relatedness between group members, or restricting access to toys and other enrichment to those that can be sterilized. Thus, while training may lead to the expression of behaviors that are absent in the wild, the relatively more sterile captive environment may stifle the development of many very impressive species-typical behaviors. One example may be the inability to navigate in large spaces (Menzel and Beck 2000). Lack of practice with, or need to develop, large-scale navigation skills could account for the finding that captive capuchin monkeys appear to use an egocentric view of spatial relationships (Poti 2000), whereas evidence from the wild suggests that they possess some ability to understand spatial relationships allocentrically (Janson 2007). What is critical is to recognize situations in which captive subjects may not show us the real upper bounds for what these species can actually do. Experiments on spatial cognition in captivity typically use a small space, one in which the various goals are often all visible from any point within the test arena (e.g., MacDonald and Wilkie 1990). Under these conditions, primates may be able to mentally calculate the shortest possible foraging routes across multiple destinations (Cramer and Gallistel 1997). Experiments in the wild can offer situations in which feeding sites are widely dispersed so that no one feeding site can be perceived directly from other feeding sites (Janson 1998). In these larger spaces typical of wild foraging, primates may not plan minimum-distance routes, because constraints on memory, imprecise estimation of distances, encounters with predators and competitors, and a likely bias to prefer closer feeding trees may make complex multi-destination planning ability of little practical use (Noser and Byrne 2010).

Because of the restricted range of contexts available to captive primates, field primatologists often dismiss the results of captive experiments as "artificial." In captivity, test situations can be and often are deliberately quite different from any situation that a primate might encounter in the wild. Such deliberate artificiality may be appropriate for the many studies that seek to understand the mechanisms of behavior. Any behavior is the result of a developmental process of interaction between genes and the environment. Although natural environments are complex, they represent a subset of all possible environments, both ecologically and socially. Observing the expression of a behavior under the expanded range of environmental conditions possible in captivity can be very informative about the species "norm of reaction" (Schlichting and Pigliucci 1998) for that behavior. Particularly for primates, experiments in captivity can contribute important insights about the conditions required to develop certain behavioral capacities. There is evidence that some capacities, widely expressed under natural conditions, do not develop readily in restrictive captive settings (socialization: Anderson and Mason 1974;

navigational ability in large spaces: Menzel and Beck 2000). Conversely, captive conditions can afford situations in which latent behavioral abilities can be expressed readily, even when these are rare in nature (such as regular tool use in capuchin monkeys: Visalberghi *et al.* 2009). Finally, captive experiments may be the only way to document certain perceptual and cognitive limits or biases on their abilities to "solve" common adaptive challenges. Without a thorough understanding of these capabilities and limitations, it may appear that animals adopt suboptimal behaviors, when in fact they are making the best decisions that their cognitive machinery is capable of (e.g., impact of Weber's Law on "incorrect" calculation of average food intake rate when encounter rates vary: Kacelnik and Abreu 1998). Thus, precisely because they are artificial, experiments can reveal mechanisms of behavior that are important to a full understanding of its design features and evolution.

10.4 Experimental paradigms with primates in the field and captivity

We cannot in this chapter review every instance, or even kind, of field experiment carried out with primates. What is manipulated in field experiments generally falls into four broad categories: food, predators, vocalizations, and object/tool use. These can be used individually or in combination to examine a variety of topics (Table 10.1). A brief (and incomplete) survey of citations on PrimateLit (<http://primatelit.library.wisc.edu/>) revealed 124 papers using experiments to study behaviors of wild primates, of which 58 used playbacks to study the perception, structure, or functions of vocalizations 17 used playbacks to study social cognition among group members, 8 used playbacks or predator models to study recognition or perception of predators, 4 used playbacks or food manipulations to study food competition within or between groups, 14 used food manipulations to study foraging cognition, and 23 used introduction of novel objects to explore perception and understanding of object properties, usually in the context of making or using tools.

There is quite a large literature on captive experiments in capuchin monkeys, even though they are neither apes nor common captive primates. As in the wild, these experiments manipulate vocalizations, food, and tools/objects (Table 10.1), often with the same general research questions in mind. Capuchins in captivity have been exposed to far greater ranges and diversity of stimuli to study cognitive processes than in the wild. These cognitive studies generally cluster into two categories: ecological cognition (delay of gratification, numeracy, biased decisions, tool use, neophobia, tactical deception, and mental maps) and social cognition/behavior (food

Table 10.1 *Broad themes or questions addressed by experimental studies of primates in the wild, in free-ranging enclosed populations, and captive settings.*

Setting	What is manipulated	Research question(s) addressed	Representative publication(s)
Wild	Conspecific vocalizations	"Meaning" of vocalization, voluntary production, semantic or lexical use of calls, tactical deception	Cheney and Seyfarth 1988
Wild	Predator models, vocalizations	Predator recognition, detection, or reaction	Karpanty 2006
Wild	Food amount, distribution	Food competition, food-associated vocalizations, foraging cognition	Bicca-Marques and Garber 2005; Di Bitetti 2005; Janson 2007
Wild	Novel objects, potential tools	Neophobia, tool use and understanding, social transmission of manipulation behaviors	Fragaszy *et al.* 2004
Free-ranging	Food distribution	Food competition; number cognition	Belzung and Anderson 1986
Free-ranging	Vocalizations	Vocal cognition	Santos *et al.* 2006
Free-ranging	Food preference and delay	Self-control	Kralik and Sampson 2011
Free-ranging and captive	Competitor's knowledge of food	Perception of other's mental state	Flombaum and Santos 2005
Captive	Vocalizations	Perception, voluntary production	Pollick *et al.* 2005
Captive	Novel objects, potential tools	Neophobia, tool use and understanding, object/food transfer	Visalberghi and Limongelli 1994
Captive	Foraging complexity	Cooperative behavior, general cognitive ability	Brosnan 2010
Captive	Food distribution	Food sharing, number cognition	de Waal 1997; Matsuzawa 1985
Captive	Novel tasks	Social learning	Hopper and Whiten 2012
Captive	Food preference and delay	Self-control, loss aversion	Evans and Westergaard 2006

sharing, cooperation, social learning, and perception of social and mental states of others), with some very new work looking at neural correlates of behavior and cognition (Platt and Ghazanfar 2010). We have provided an (incomplete) summary of the major areas of research for both wild and captive experiments in Table 10.1.

10.5 Prospects for future collaboration between field and captive researchers

Implicit (and explicit) throughout this chapter has been the theme that we learn the most when field and captive experiments interact. We here highlight a few of the most important features. The over-arching theme, of course, is that field and lab research are complementary ways of viewing the world, not competitors. Captive research needs the insights about what the primates actually do in the wild to inform research questions, predictions about which variables are of interest to control, and interpretation of findings. Field experiments can likewise learn from lab experiments. An interesting result generated in the lab should pique the interest of field researchers, who may think of a context in which the behavior could be relevant. Moreover, of course, an experimental component is essential for providing data on causality and mechanisms, as opposed to correlating behaviors. While experiments are certainly done in the field, in some cases the control required may only be available in the lab. Ideally, individuals who are interested in the same questions or species could work together to help inform what the most interesting synergies are, and then work in that direction.

We provide here several potentially fruitful collaborations between field and lab researchers. In some of these cases, some work has already been done to pave the way, but we see connections that have not been made. In others, we recommend projects that to our knowledge have not been done. In all cases, we assume that the work will be experimental, although we also mention obvious non-experimental designs. We try to run the gamut between experimental studies from the field that should gain a captive-based approach and experimental studies from the lab that should gain a field-based approach; however due to the relatively larger number of lab than field experiments, we end up focused more on the latter than the former. In particular, we focus on tool use and cultural traditions, food sharing and cooperation, foraging cognition, and food associated calls.

10.5.1 Tool use and cultural transmission

There has been a recent explosion of research on tool use in capuchin monkeys (e.g., Ottoni and Izar 2008; Visalberghi *et al.* 2009). This has focused on the new discovery that capuchin monkeys spontaneously nut-crack in the wild, and the sophisticated understanding shown by the decision-making going on in this process. These experiments have been completed entirely in the field, and rely on the provisioning of appropriate (and not so appropriate) tools and observations of which tools they ultimately employ. Capuchins show a very high level of understanding of which characteristics of different hammer stones are functional, and which are superfluous and even potentially misleading. However, what is unknown is how the monkeys acquire this behavior. Presumably, it is socially learned, as are so many other behaviors in primates, including capuchins (for an excellent recent review of social learning, see Hopper and Whiten 2012). Nonetheless, to our knowledge, such a study on learning of tool use has not been done in capuchins.

The best studies of social learning involve long-term observational studies of the same individuals, to see which individuals spend time together and who acquires behaviors from whom. In particular, juveniles are expected to learn vertically, or from their mothers. Of course, this will hopefully remain possible in the field groups currently under study, but another possibility would be to introduce the behavior to a captive capuchin group and document the spread of the behavior through the group (e.g., Dindo *et al.* 2009). Such transmission studies have had great success in the lab, albeit with far, far simpler tasks. On the other hand, control groups could simply be given appropriate tools and nuts to see if the behavior emerges through independent learning. Although complex behaviors are potentially very difficult to acquire, there are anecdotal reports of capuchins using what is available in their group to crack nuts opportunistically (e.g., Brosnan, pers. ob.), so at least in principle learning should be possible, even if it is not due to social transmission.

Another interesting avenue, assuming that there is a social component to learning, would be to explore which social learning mechanism is responsible for the behavior. Which parts of the operation are most critical for learning? Must subjects watch the actual action, or is it sufficient to see which tools are most effective? Is stimulus enhancement sufficient, that is, will individuals seeing others eating nuts and the remains of the session (e.g., shells and tools) be more likely to explore the tools? A rather far out possibility (and one that we admit is rather unlikely, although see Poss and Rochat 2003) is to see whether captive primates can learn from watching videos of natural nut-cracking behaviors by wild capuchins. Video models provide a remarkable degree of control, both due to the consistent nature of the stimuli and the use of video manipulation or creative editing, both of which can be used to create controls which are not particularly feasible when dealing with real interacting animals. Finally, this sort of natural example could be used to study new areas of interest, such as conformity and prestige, through the manipulation of the model identity.

10.5.2 Cooperation

A very fruitful line of research in the lab is the investigation of cooperation in capuchins. Of course, some of the experiments would be rather hard to run in the field. Many rely on the ability to restrict the monkeys, thereby assuring that the first monkey works for one reward and the second for another. While this is in principle feasible in the wild, by providing separated areas for the individuals to work, we think it is in practice impractical. Monkeys who are separated by too great a distance may not realize that they are working together, and putting up barriers of some sort would likely only serve to create fear in the monkeys. On the other hand, some such studies have not required separation, and newer approaches are emerging that are more group-based. Despite the messy data and lack of control, appropriate statistical modeling can provide an idea of what factors are affecting observed behavior (Janson 2012).

Creating a joint task is challenging but feasible (for a recent review of the experimental cooperation literature in capuchins, see Brosnan 2010). Although the level of control expected in the laboratory—where each individual's test is carefully calibrated to their strength and ability—is probably impossible, it is not unreasonable to think that individuals might work together. One such task might be to place food in a clear container on a feeding station, covered by a heavy lid. Ropes or handles could be attached to the cover such that individuals working together would be able to uncover the food. In the experimental lab literature, the most common apparatus for joint tasks is the barpull task, in which monkeys pull two ropes to bring in a counterweighted tray. While obviously the standard design wouldn't work (a smart monkey would simply walk to the food!) perhaps the paradigm could be reversed in the field. In other words, instead of putting the monkey in a cage, we could put the food in a cage and make the monkeys pull a tray to bring the food close enough to reach (in captivity the monkeys also reach through the mesh to pick the food off of the tray). Unfortunately the "easier" design of having an automated feeder that dispensed when levers were pressed simultaneously has thus far proven to be untenable in experimental studies. One real challenge to captive studies has been finding tasks that are comprehensible by the monkeys; comparative work has shown that monkeys who happily pull in trays to get food refrain from pushing levers to achieve the same end, possibly due to the lack of kinesthetic feedback regarding their partner's behavior (Brosnan and de Waal 2002). A second issue is designing a task that the monkeys want to use; if the task is too hard, or the food not sufficiently motivating, they are likely to ignore the apparatus in the wild.

The abovementioned studies represent the most basic form of cooperation; individuals coordinate their activity in time for a joint reward. Nonetheless, interesting questions can be asked. For instance, lab studies have shown that capuchins are more likely to pull for rewards that are physically spread out than for those that are clumped (e.g., one pile of food), and do so from the first trial, indicating that they understand the contingencies of the task (de Waal and Davis 2002). Moreover, kin were more likely to cooperate for the clumped rewards than were non-kin. An interesting question is whether the monkeys in this experiment learned that working with a dominant for a single large reward is likely to be unfulfilling based on previous experiments, or if they understand this from general interactions in their social group. Field data suggest that reciprocity in wild capuchins is based on long-term perceptions of partner value (Tiddi *et al.* 2011), and the same might be true of cooperation both in the lab and the field. At a more complex level, cooperation may require complementary actions, rather than the same action; in principle, this is possible in the field as well.

10.5.3 Food-associated calls

Food-associated calls are given by primates in the context of food sources. While we understand well the ecological and social contexts in which these calls are used in the wild, the adaptive advantage to such calls is still not clear. They have been proposed to function in spacing, attraction, punishment avoidance, and statements of possession, amongst others (Di Bitetti 2001). In part, this knowledge gap is due to the dearth of relevant experiments. For instance, we do not know whether a capuchin that "discovers" a novel food source not known to the rest of the group would give a food-associated call even if it knew from the experimental context that the rest of the group could not interfere with its feeding on the source. The fact that food-associated calls in the wild are subject to a clear audience effect (Di Bitetti 2005) shows that they can choose when to give or not to give a call, but in this study it was not possible to control for the identities of potential audience members. This is an area in which lab studies might help to clarify the function in a way that field studies cannot. In the lab, primates can be separated from their partners into familiar (and therefore not stressful) testing areas. Experimenters can control whether or not these test areas are visually isolated from their group, and, if not, which individuals of the group can see the subject. This allows us to ask questions about the effects of bystanders' (and food possessors') rank, sex, and kinship on calling behavior. Moreover, experimenters can control what type and quantity of food is found, as well as whether others have access to food. These studies might also be combined with those examining food sharing or other behaviors to see whether food associated calling affects sharing, or the reverse.

10.5.4 Foraging cognition

One area for which field experiments have provided an extraordinary wealth of information is in foraging cognition (See Section 7.2.1.1). Unfortunately, there have not been a great number of captive experiments to complement this information. In part, this is due to the fact that most lab studies are not sufficiently ambitious in scope. This is understandable, as the monkeys' captive habitats are usually only a (very small) fraction of the size of a home range in the wild, and are comparatively less complex as well. As a result, it is difficult to design studies to test foraging cognition, and moreover it is possible that the lack of complexity leads to less advanced cognition in this respect. There are several potential ways to solve this.

First, if it is possible to give subjects exposure to a more complex environment, it may be possible to emulate the foraging cognition required in the wild. This is feasible either in very large captive enclosures, such as at zoos or sanctuaries, or for individuals who have had alternative experiences. For instance, Charles Menzel (2005) works with chimpanzees who, as juveniles, were allowed to roam a forest that they can still see from their current enclosures. His research examines their search strategies for food hidden in the forest, investigating the order in which they request hidden foods to be returned by an experimenter. There can be long time delays between the chimpanzees witnessing the hiding event and sending an experimenter to retrieve the food (> 1 day), yet the subjects typically collect the food in order of preference, rather than in order of the number of items, item type, or proximity to themselves or the experimenter. These results indicate that chimpanzees are very aware of the contingencies of the foods they are after and, while these subjects are certainly not trying to minimize foraging time, they are nonetheless making consistent decisions based on other aspects of the foods in question.

Another issue with studying foraging cognition in the lab is that the animals are usually under no pressure to maximize their intake, or minimize the time to acquisition. One study, also on chimpanzees, demonstrates this very nicely. Chimpanzees were trained to observe a food object being hidden in a scale model of their enclosure. Following this, subjects were allowed to go look for the treat. The majority of subjects failed to use an optimal foraging strategy (e.g., they consistently searched the hiding places in the same order, regardless of where the object was placed) until the task was constrained so that the subjects were only rewarded if they found the hidden reward in the first hiding place that they checked (Kuhlmeier and Boysen 2001). These results emphasize an important issue with respect to captive work; these animals do not exhibit the same constraints and daily activity patterns of animals in the wild, and so tasks that might be easy for an individual trying to optimize its food intake per unit search time will result in "failure" for a captive animal who experiences very few such constraints.

Finally, it may be possible to utilize the field data to form hypotheses about what mechanisms are relevant, and then design complementary tests in the laboratory. For instance, computerized designs in which subjects have to solve tasks to receive food could examine the trade-off between reward quantity and foraging difficulty or time investment (e.g., Stevens *et al.* 2005). Alternatively, researchers could introduce computer navigation programs or other opportunities to increase subjects' experience of spatial complexity (Fragaszy *et al.* 2003).

10.6 Conclusions

Combining data and efforts derived from field and lab experiments promises to provide a much more comprehensive understanding of primate behavior. Field research provides a better understanding of what primates do in their natural environment and allows for an understanding of adaptive function that is challenging to uncover in captivity. On the other hand, lab research provides opportunities for the experimental rigor necessary to tease apart causality and provide an understanding of mechanism that cannot easily be done in the field. Although most topics are primarily pursued in one or the other setting, the examples provided show the power of combining the approaches. In the few instances in which such synergy exists, for instance in social learning and tool use, it is clear that applying knowledge and techniques from the lab to the field, and the reverse, have allowed both "camps" of researchers to forward their science in notable ways. In the long term, ideally, the idea of camps will be broken down as researchers come to rely more and more on each other's expertise and methodologies. Although some researchers have successfully combined both approaches within the same working group (e.g., Visalberghi, Fragaszy, Matsuzawa), in most cases it is challenging in practice for a single research program to undertake both field and captive studies. Nonetheless, all of us can increase our interactions with each other, read each other's work, and draw examples from each other with the goal of forwarding our knowledge of primate cognition and social behavior.

Acknowledgments

CHJ is grateful to the Argentine National Parks Administration for permission to carry out field experiments on capuchin monkeys in the Iguazú National Park. CHJ's work has been funded by a succession of NSF grants (BNS- 8007381, BNS-9009023, IBN-9511642, BNS-9870909, BCS-0515007), a Fulbright fellowship, and grants from Sigma Xi and National Geographic Society's Committee on Research and Exploration. SFB's work has been funded by NSF SES 0729244, CAREER SES 0847351, SES 1123897, and NIH P01HD060563.

11

Diet and nutrition

Jessica M. Rothman, Erin R. Vogel, and Scott A. Blumenthal

11.1 Introduction

Across the Order, primates consume a wide variety of foodstuffs, including primarily fruits, leaves, and invertebrates, but also seeds, gums, lichens, bark, roots, and in some cases, other vertebrates (e.g., mammals, birds, and lizards), as well as invertebrates (e.g., crabs and insects). Assessing dietary properties is important to a number of areas relevant to primatologists, including life history, ecology, and behavior. Knowledge of the dietary requirements of primates can also help conservation managers better protect their habitats, and provide insight to captive care managers. Here, we suggest methods to examine the mechanical and nutritional properties of primate diets. We also discuss means to examine the diets of elusive primates through stable isotope analysis.

11.2 Observing the animals

There are a variety of methods available to quantify the behaviors of primates (Altmann 1974; Martin and Bateson 2007; see also Chapter 5). For examining diet, the choice of behavioral methods depends on the question that is being asked. For instance, to estimate nutrient intake, it is important to determine the amount of food (i.e., in grams) rather than the time it takes to consume a particular food, which is an indicator of foraging effort. The best way to gain an accurate estimate of a primate's nutrient intake is to complete full-day continuous follows because this measure will allow the researcher to record each item consumed, intake rate, and the time spent feeding on it. This measurement will provide the total amount of nutrients consumed per day. However, this method is dependent on the ability to obtain a large sample size of complete days for each animal, because if the day is not typical of the study period, it can introduce large errors into the analysis. In other instances, the researcher is interested in the diet diversity in a study group.

Primate Ecology and Conservation: A Handbook of Techniques. First Edition. Edited by Eleanor J. Sterling, Nora Bynum, and Mary E. Blair. © Oxford University Press 2013. Published 2013 by Oxford University Press.

Here, it is often necessary to obtain a sample of the majority of the individuals feeding, and record precisely the food species, part, and maturation stage of the food item. The best method in this case would be to conduct a scan sample of individuals of various age and sex classes feeding in a group at predefined times to capture the variety of food types. Thus, the best behavioral methods will be dependent on the questions asked (Rothman *et al.* 2012).

11.3 Sample collection

Assessing the nutritional quality of primate diets usually requires collection of plant and/or fecal samples. Importantly, the collection of samples requires permits from the host country to export samples, and to the researcher's home country for the import of samples (Rothman *et al.* 2012; Chapter 3). When collecting food items, it is essential that the food item be processed identically to how it was consumed. Thus, if the primate removes the outer peel of an herbaceous stem and consumes the inner pith of a young shoot, then this same processing should be employed by the researcher. Similarly, if leaf tips are eaten and the remainder of the food item is discarded, the leaf tips should be the only part collected. Researchers should pay careful attention to whether folivorous primates eat both the leaf lamina and petiole, or leave the petiole behind, as its nutritional composition may differ dramatically. Fruits present a challenge because if the fruit is swallowed whole by the primate, the seed usually passes undigested through the gut. However, seeds also fill the gut, preventing more food from being consumed. We suggest that three samples be collected for each fruit: the whole fruit, pulp, and seed (Rothman *et al.* 2012). In addition if primates selectively consume or avoid the skin, it should also be separated. In this way primatologists can use the information to answer a variety of questions depending on their goals. It is always important to make note of the species, part, and maturity/ripeness of the food item. Entomologists should be consulted in the case of insects. Like human-consumed fruits and vegetables, primate foods can also vary dramatically in their nutritional composition according to microhabitat, season, rainfall, and soil (Chapman *et al.* 2003), thus single species-specific nutritional profiles are inadequate. To characterize primate diets, multiple samples of each plant part–species combination are needed, which requires many samples to be analyzed. For mechanical analysis, we recommend collecting 4–5 representative food items to test for each feeding bout, and testing from several different trees or lianas where the primates are feeding when possible.

To obtain the food item, numerous methods are available. For terrestrial plants foods can be collected using a machete or plant pruners, while for tree fruits and leaves, tree saws are available that have extenders of several meters. Very tall trees can be climbed (Houle *et al.* 2004; Fig. 11.1). Fallen pieces of food items, even those with bite marks, which have not been masticated or damaged can also be collected and tested, as long as the tissues to be tested are not deformed. We suggest that researchers try to opportunistically collect insects. In other cases, researchers could take advantage of areas that are fogged by entomologists. For mechanical analysis, food samples should be transported back to the field station and tested as soon as possible, certainly within 12 hours of sample collection. If plant tissues that are prone to loss of rigidity once removed from the plant (e.g., young leaves) are collected, we recommend measuring such items immediately after they are collected. Samples can be placed in clean, sealed bags or plastic containers for transportation back to the field camp. For nutritional analysis, researchers should aim to obtain at least 30 grams of dry weight of material for analysis, which will provide 1–2 grams for each assay, with some to spare for replicates, and loss of

Fig. 11.1 Use of a tree saw to collect leaves eaten by monkeys in Kibale National Park, Uganda. Photograph © Amy Ryan, reproduced with permission.

sample during grinding and analysis. This may require collecting up to 500 grams of wet weight of a sample, as some plant parts are quite high in water content. Samples should be weighed to 0.01 grams immediately after collection, and then weighed again when they are at a constant dry weight in the field.

11.4 Drying samples

While samples for mechanical analysis need to be analyzed in the field, specialized instruments for nutritional analysis are usually not available in field stations, so the samples usually need to be exported for analysis. Most important in drying a sample is ensuring that plant enzymatic activities are halted and that the nutritional attributes of a sample are preserved. This can be achieved by preserving the sample in liquid nitrogen (Ortmann et al. 2006) or quickly drying the sample. We recommend the use of inexpensive temperature-controlled food dehydrators when electricity is available, or heating over a charcoal or propane stove where temperature is carefully controlled (Conklin-Brittain et al. 1998). Samples should be dried at or near 55 °C for macronutrient analysis and a bit lower if secondary compounds are of interest (45°C). Heating at higher temperatures will alter some aspects of carbohydrate chemistry (Van Soest 1994). It is critical to avoid mold.

When samples are dry, they should be placed in tightly sealed and labeled plastic bags, preferably with a desiccant package (silica gel). The silica gel needs to be placed in a permeable bag or sack otherwise it may adhere to the sample. We suggest that researchers place a label on the inside of the bag, as well as the outside. The samples should then be milled to a standard size; the most practical is to mill at 1 mm through a Wiley mill, which is a cutting mill. Various mills have different ways of grinding samples so it is important to take note of the type of mill used (Rothman et al. 2012). It is best to dry and mill samples in the field, as advance processing helps to prevent mold and pathogens (Rothman et al. 2012)

11.5 Mechanical analysis

Quantifying the physical attributes of food items that are accepted by primates is essential to better understanding variation in diet selection among our closest living relatives. Whether a primate ingests, chews, and swallows a food item will be partially dependent on the item's physical resistance to these processes. Indeed, the mechanical properties of primate diets have been closely linked to variation in craniodental morphology (Rosenberger and Kinzey 1976; Lucas and Pereira 1990; Wright 2005; Vogel et al. 2008), suggesting that primate food intake is limited by the items that can be physically processed, in combination with the ability to process them in the digestive tract. For example, a primate cannot obtain the energy-rich nutrients within a seed if the seed is too tough to bite into. Historically, food items were placed in categories based on the observer's perception of their properties (e.g., hard, soft, medium hard; Kay 1981; Boubli 1999). Recent methods derived from food material sciences have enabled primatologists to use standardized mechanical testers, facilitating the comparison of food mechanics data (Darvell *et al.* 1996; Lucas *et al.* 2001). We focus on these methods here but note that studies have incorporated different methods to test the physical properties of primate foods (Kinzey and Norconk 1990; Elgart-Berry 2004; Lambert *et al.* 2004).

All plant parts require mechanical defenses to resist environmental pressures, specifically those from the abiotic environment and herbivores. Defenses against herbivory are generally classified into two main groups: (1) the ability of the plant tissue to resist the initiation of a crack and (2) the ability of the plant tissue to resist crack growth once a crack has been initiated (Lucas *et al.* 2000). To examine these defenses, field biologists have become concerned with three main material properties of solid foods: Young's modulus, yield stress ("hardness"), and fracture toughness. To better understand these properties, a simple review of material sciences is warranted. When force is applied to any solid object, it will either deflect that force or change shape, known as deformation (Lucas 2004). If deformation occurs and the force continues to be applied, the object will crack. Thus, mechanical tests have been developed to record both the force that is applied to the object, and the extent of deformation or crack propagation that results. These data are of vital importance to better understand the link between tooth or jaw form and diet. The main components of these tests involve stress and strain. *Stress* is the force divided by the cross-sectional area over which it is applied, while *strain* is the amount of deformation divided by the length of the specimen (Lucas 2004). Stress and strain are used to quantify various mechanical properties of items consumed by primates. Below we describe the methods used to collect samples and quantify food mechanics in studies of primate dietary ecology.

11.5.1 Sample processing

To test plant samples, a portable tester is required. The Universal Portable Tester has become the standard for testing food material properties in the field (Darvell *et al.* 1996; Lucas *et al.* 2001), with a newer model recently developed (Fig. 11.2; <http://lucasscientific.com/>). The accuracy and consistency of data are standardized using the tester kit, facilitating cross-site and -species comparisons. This new model (FLS-1 mechanical tester) is a USB-powered machine, and thus

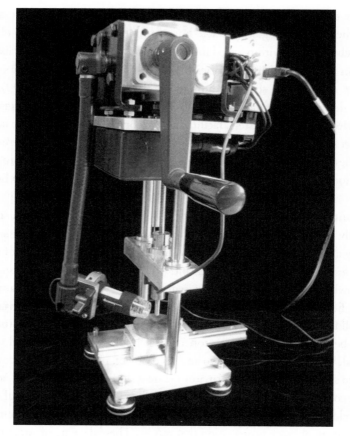

Fig. 11.2 The FLS-1 mechanical tester. The tester comprises a frame on which is mounted a handle attached to a sealed worm drive. The machine has three columns. The middle one is attached to the metal crosshead and has a ball screw threading giving very accurate positioning. The load cells, of which there are two (50 Newton and 500 Newton capacities), are mounted below the crosshead via screws. On the side of the worm-drive is the Linear Variable Displacement Transducer, a highly accurate method of measuring displacement. Photograph © Mark Wagner, reproduced with permission.

only requires an electric source to charge a laptop computer battery. It comes with five specialized software programs that generate force-displacement curves (Fig. 11.3) along with results, once the measurements (e.g., length, width, depth) of the specimen are entered into the program. Below, we briefly describe the most common tests conducted with the universal tester kit; however we refer the reader to Lucas (2004) for additional tests to quantify additional properties of food items.

11.5.2 Young's modulus

Young's modulus (E), also called elastic modulus, is technically defined as the ratio of stress to strain in the early phase of the object's deformation under a given load. More simply, it is the ability of an object to resist elastic deformation and is a measure of an object's stiffness or rigidity. Young's modulus is the initial slope from the force-displacement curve and is measured in mega- or gigapascals (MPa or GPa respectively; Fig. 11.3a). While Young's modulus is most commonly measured using a compression test by compressing cylinders of food, additional tests including bending or tension may be more appropriate for some food items (Lucas 2004). In the compression test, a cylindrical tissue is made with a core tool, measured, and then placed between two metal plates and compressed. A rule of thumb is that the height to diameter ratio of the specimen should be no more than 2:1 to prevent bending. This results in a force-displacement gradient at small deformations. Young's modulus is calculated as the force produced to deform the specimen, normalized to the relevant dimensions of the specimen. Young's modulus is often confused with hardness, but as you will read below it measures the stiffness of an object, not its hardness (Lucas *et al.* 2000). Young's modulus in orangutan foods has been shown to range from as low as 1.4 MPa for ripe fruit (Vogel *et al.* 2008) to as high as 7 GPa for seeds (Lucas *et al.* 1994).

With the introduction of the FLS-1 tester, a new test called the blunt indentation test has been introduced. This test implements a well-understood technique that is rapidly gaining ground in biology as a way of characterizing viscoelastic behavior in a succinct manner. Most moisture-laden materials will display time-dependent elasticity. This test gives you the possible range of the elastic moduli of a specimen accurately and quickly with a minimum of specimen preparation. For this test, you do not need to input any specimen dimensions. Either a block of material (BULK test) or a thin sheet (MEMBRANE test) is loaded slowly and evenly by a blunt probe of known radius for 10 seconds. The probe is then stopped and the load allowed to decay for 90 further seconds. The program then automatically fits a multi-coefficient exponential decay model to the data that allows the relevant calculations to be made (Begley and Mackin 2004; Chua and Oyen 2009).

11.5.3 Yield stress (hardness)

Although often used to describe primate food properties, hardness is not a material property in itself, but instead reflects yield stress (e.g., the stress at which a mark in the specimen becomes permanent). As Lucas (2004) emphasizes, "like many scientific terms, it is derived from everyday language but has no meaning in the food material sciences." In general terms, yield stress is the stress at which

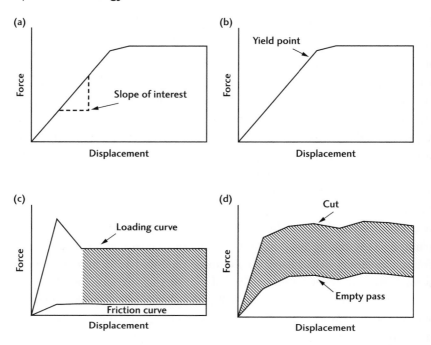

Fig. 11.3 Diagrams of the force-displacement curves for the mechanical tests described. (a) Young's modulus. The part of the curve that is of interest is the slope of the curve, preferably in the lower part of the curve. Young's modulus is then calculated based on the cross-sectional area of the cylinder, the cylinder height, the amount of displacement and the force. (b) Yield stress. Typical graph produced performing the indentation test. The yield point is the part of the curve that is of interest. The hardness is then calculated as the force divided by the area of indentation (in the specimen). (c) Fracture toughness-wedge test. In this graph, the top line is the resulting loading curve produced as the wedge is forced through the specimen. The initial increase represents the force building, with this force at its highest at the peak where a crack is initiated. At first this is unstable but it then levels out as the crack runs just ahead of the sharp wedge. The bottom curve is the friction curve; the area below this curve is subtracted from the loading curve. Thus, the area of interest is the shaded area. (d) Fracture toughness—scissor test. This graph is similar to the wedge test except that the entire shaded region between the actual cut and the empty pass is of interest. If a mid-vein or secondary vein is cut, this will be noticeable in the curve (not shown here). Figure redrawn from Lucas and colleagues (2003).

permanent deformation begins, and it is measured as hardness. Lucas (2004) provides a useful discussion on the differences between yield stress and hardness, which we will not elaborate on further here. Indentation tests, which involve a conical, pyramid, or spherical object being pressed into the specimen, are usually applied to quantify hardness. Typically, the hardness of a specimen is measured as the force divided by the area of indentation, calculated by measuring the diameter of the indented area and, similar to Young's modulus, it is reported in units of MPa or GPa (Fig. 11.3b). Lucas (2004) reports hardness values ranging from 175 MPa for insect cuticles to 267 MPa for seed coats. It is important to note that "hardness" can also be calculated as the square root of the product of Young's modulus (E) and fracture toughness (R). Objects with high values of $\sqrt{ER}$ are considered stress-limited and are often stiff and resistant to deformation, two characteristics that are often defined as "hardness" by researchers.

11.5.4 Fracture toughness

Fracture toughness (R) is energy consumed to propagate a crack of a given area through a substrate and is measured in joules per meter squared (Jm^{-2}). It is the most common and probably important material property measured in primate studies, as the majority of primate foods must be fractured to maximize energy intake prior to swallowing. There is often a negative relationship between fracture toughness and Young's modulus such that tough objects generally, but not always, have low modulus. The degree of toughness in plant tissues will depend on the structure and dimensions of the cell walls. For example, the lamina of immature leaves consumed by wild orangutans may range anywhere from 100–1400 Jm^{-2}, whereas mature leaves range from 220–2450 Jm^{-2} (Vogel et al. 2008).

   The two most common tests for quantifying toughness are the wedge and scissor tests. The wedge test involves forcing a sharp wedge through a pre-shaped and measured piece of plant tissue (Fig. 11.3c). Typically it is a block of material and the width is measured prior to testing. The wedge is forced through the specimen, resulting in the crack, which runs just ahead of the wedge tip. The depth of the crack is calculated by the software that comes with the kit. Once a crack is made, the wedge is then reversed back into starting position and an empty pass is run through the specimen again, which creates a friction curve. The friction is then subtracted from the force curve and the area between these two curves (shaded) provides the force to propagate the crack (Lucas 2004). The part of the curve that is of interest is typically the flat part of the curve (Fig. 11.3c).

   The scissor test for fracture toughness is typically used for foods that are difficult to shape (e.g., some fruit species' pulp) or that are in the form of flat sheets or rods (e.g., leaves, seed coats, vines). For leaves, it is typically recommended that several cuts (3–4) are made through each leaf, as different parts of the leaf may have different material properties. In addition, the mid-rib, veins, and lamina are measured separately in the software within the same cut. In this test, force is applied to the handles of very sharp specialized scissors, forcing them to propagate a crack through the material as the scissors are closed. Similar to the wedge test, an empty pass is conducted to account for the amount of friction created by the two scissor blades rubbing against each other. However, unlike the wedge test, this empty pass is generally done prior to the main test without the specimen in place. The force-displacement curve is similar to that of wedge tests (Fig. 11.3d). For materials that do not conform to the wedge or scissor tests for toughness, we refer the reader to Lucas (2004) for a description of additional tests.

11.6 Nutritional analysis

A basic nutritional analysis includes information about energy and protein in a diet. Non-structural carbohydrates and fats provide the main sources of energy, and, depending on digestive anatomy and physiology, a portion of energy may be provided by fiber (e.g., structural carbohydrates) in a primate food. For example, both howler monkeys (*Alouatta* spp.) and colobines possess digestive adaptations to gain energy from a leafy diet and have substantial energetic returns from fiber (Milton and McBee 1983; Edwards and Ullrey 1999). Protein may also provide energy, particularly when other sources are limited, but it is mainly used for provision of essential amino acids, which are used in tissue growth and repair. Minerals can also be limiting in primate diets, and should be measured whenever possible (Yeager *et al.* 1997; Rothman *et al.* 2006).

To estimate the nutritional chemistry of primate foods, a variety of analytical methods can be used. We suggest that readers take time investigating each method to determine its merits and limitations before analysis. The AOAC International regularly produces handbooks of nutritional protocols, and has a journal whereby methods that are rigorously tested and are standard in the agricultural and food sciences are presented. The most recent versions of these texts should be consulted to obtain protocols for various techniques. In addition, a variety of recent publications provide guides for analyzing primate foods (Conklin-Brittain *et al.* 2006; Ortmann *et al.* 2006; Rothman *et al.* 2012). Here we focus on energy and the macronutrients: protein, carbohydrates, fat, and fiber. We also briefly discuss the measurement of tannins, a secondary compound that is common in primate foods.

11.6.1 Dry matter

The first step in analyzing samples is determining dry weight. This step is critical because nutrients aside from water are only present in the dry portion, thus results of analysis should be expressed on a dry matter basis. To calculate a sample's moisture and dry matter, a two-step process is used, which includes field and laboratory drying (Rothman *et al.* 2012). As noted above, the sample should be weighed after collection, and then afterwards it is dried to constant weight. This represents the initial or field moisture content. In the laboratory on the same day when the sample is prepared for analysis, a portion of the dried sample should be placed in an oven for 16 hours to remove any additional adsorbed atmospheric water. This will provide a coefficient for determining the dry matter of a sample. This coefficient typically varies between 88 and 95% of the total sample weight and is used to correct the analytical result of each assay. Results should be provided on a dry matter basis. Nutritional results could also be expressed on an organic matter basis, whereby the ash, or inorganic matter, of the sample is accounted for. To obtain an estimate of the ash in a sample, a subsample should be burned at 500–550 °C and re-weighed to provide an estimate of the minerals, soil, and dust contamination. Some primate foods contain substantial amounts of ash; for example, some leaves eaten by colobus monkeys (*Colobus guereza*) in Kenya were 20–25% ash (Fashing *et al.* 2007a).

11.6.2 Protein

Protein is a limited nutrient in some primate diets, and it has been suggested as an important criterion for leaf selection (Milton 1979). There are various methods for examining the protein contents of primate foods. The most common methods used assess the total amount of nitrogen in a food because protein in typical agricultural feed ingredients is typically 16% nitrogen. Thus, multiplying 6.25 by the concentration of nitrogen provides an estimation of protein. The Kjeldahl procedure and Dumas combustion (AOAC methods 984.13 and 990.03; AOAC 1990) both measure the total amounts of nitrogen in a sample; however, it is also important to realize that there are various forms of non-protein nitrogen in plants and invertebrates, which should be accounted for. In addition, the Kjeldhal and Dumas procedures do not account for the digestibility or quality of protein. To further refine estimates of protein in primate diets, researchers should consider using these methods and additional methods. First, measurements of fiber-bound nitrogen should be taken and subtracted from the crude protein of a sample. This provides a good estimate of the protein available for digestion. Second, researchers should use an assay to account for the non-protein nitrogenous compounds. These methods are reviewed in Rothman *et al.* (2012). Alternatively, correction factors to account for unavailable protein may be used (Milton and Dintzis 1981; Conklin-Brittain *et al.* 1999), though these will not account for the variability in non-protein nitrogen among samples. However, the best way to estimate the protein available to primates is to quantify dietary amino acids, because proteins are composed of amino acids. Amino acids are analyzed using high performance liquid chromatography (HPLC) where each amino acid is separated according to its polarity.

11.6.3 Fats

Insects and some fruits are probably the most fatty parts of most primate diets. Most leaves are typically quite low in fat. For example, the mean amount of fat in 450 leaf samples eaten by primates in Kibale National Park, Uganda was 1.6 ± 0.9 with a range of 0–4% on a dry matter basis (Rothman, unpublished data). Some fruits, however, are high in fat, similar to avocados; for example, fruits eaten by Japanese macaques (*Macaca fuscata*) were 23–40% fat on a dry matter basis (Iwamoto 1982). To assess the fat composition in primate foods most studies use an ether extraction, which is a simple gravimetric procedure whereby food samples are placed in hot ether for a set amount of time, and the loss of material is recorded. Ether extract can be a good estimation for animal foods, but plants have non-fat components that are extracted by ether, such as wax, cutin, galactose, essential oils, chlorophyll, glycerol, and other compounds. Thus, the best measure of fats in a food sample is through fatty acid analysis whereby specific fatty acids are separated via gas chromatography (GC). If fatty acid analysis is not possible, we recommend that researchers use a correction factor to roughly account for the non-fat components that are extracted in the ether but are not fat. In the agricultural sciences, one is subtracted from the ether extract (Rothman *et al.* 2012), and we suggest this correction be applied when using ether extract to calculate fat's energetic contribution to primate foods as well (e.g., Rothman *et al.* 2011).

11.6.4 Non-structural carbohydrates

A large portion of energy is provided by the non-structural carbohydrates, which are part of the intracellular contents, or "cell sap." These non-structural carbohydrates include the simple sugars, such as glucose and fructose, and the storage reserve compounds, such as starch. Foods that are high in non-structural carbohydrates, particularly sugars, are highly sought by primates, particularly frugivores (Garber 1987; Reynolds *et al.* 1998; Vogel 2005). Even primates that can process a highly folivorous diet will eat sugary fruit when it is available to them (Milton 1980; Rogers *et al.* 2004), and many primates can apparently discern the tastes of different sugar solutions (e.g., Laska *et al.* 1996). Measuring sugars in primate foodstuffs can be accomplished by a variety of different spectrophotometric assays whereby a hot water or ethanol extract of the plant material is placed in an assay with acid, and any soluble carbohydrates produce a color. The most popular of these methods are the anthrone method and the phenol sulfuric acid assay (Rothman *et al.* 2012). These are very broad-spectrum assays and provide a rough estimate of the sugars in the sample based on a sugar standard solution like glucose or sucrose. The most precise way of estimating sugars in primate foods is via the separation methods of HPLC. Field methods have also been used to detect sugars in plants, but their efficacy is yet not well understood for tropical plants.

Methods to accurately determine starch may be difficult because of confounding compounds like phenolic antioxidants. A common method for estimating the total non-structural carbohydrates, including sugars and starches is to subtract the protein, fat, neutral detergent fiber, and ash from 100%. However, this measure is problematic because the errors associated with each of these analyses are confounded in the process (Rothman *et al.* 2012). We suggest that researchers use HPLC whenever possible to separate the specific sugars, and/or follow methods of Hall and colleagues for estimating nonstructural carbohydrates in primate diets (Hall *et al.* 1999; Hall 2009).

11.6.5 Structural carbohydrates

The structural carbohydrates compose dietary fiber, or the cell wall portion of plant cells that animal enzymes cannot digest. Primates have a variety of digestive adaptations for coping with fiber, and in particular many herbivorous primates host symbiotic bacteria in their foregut (e.g., colobines) or hindgut to convert fiber to usable energy (Van Soest 1994; Lambert 1998). Soluble fiber, such as pectin, is easily digested by gut microbes, but insoluble fiber, including hemicellulose and cellulose, is typically only partially digested. Lignin is indigestible. The Van Soest detergent analyses provide an excellent means to separate the different types of fiber in primate foods, and are widely used in primatology (Van Soest 1994). Researchers should note that each step of the detergent analysis recovers mostly the intended compounds (hemicellulose, cellulose, and lignin) but in addition other substances are also recovered (Van Soest 1994). Soluble fiber may also be of interest, particularly for frugivorous primates (Conklin-Brittain *et al.* 2006; Rothman *et al.* 2012), and assays are developed for its assessment (AOAC 991.43; AOAC 2005).

11.6.6 Energy

Energy acquisition is an important correlate of fitness, and when it is limited in primate diets, indices of reproductive output are lower (Bercovitch and Strum 1993; Thompson and Wrangham 2008). Accordingly, it is critical to measure in studies of primate nutritional ecology. Conklin-Brittain and colleagues provide a very useful step-by-step guide to estimating energy acquisition (Conklin-Brittain *et al.* 2006), and we reiterate some points here. Energy can be estimated in primate diets in two ways, through actual estimations of the energy provided by a food item via bomb calorimetry, and through estimation using equations whereby the contributions of energy (fat, non-structural carbohydrates, protein, and fiber) are assigned energetic values in calories or joules.

The choice of which method to use depends on available information. If samples are ignited in a bomb calorimeter, their total, or gross, energy is being gained. However, this measure does not take into account whether or not all of the nutrients in the food are actually digestible to the animal. For example, wood, which is mostly indigestible to primates, has a higher gross energy content than sucrose, which is completely digestible (NRC 2003). Thus, bomb calorimetry should not be used unless the digestible energy content is being measured, whereby the energetic content of feces is measured and subtracted from the diet's gross energy estimates in an appropriate time frame. This approach has been rarely used in field primatology, probably because it is difficult to estimate the exact quantities and nutrients of diet items consumed, know the transit times of the study subjects, and collect fecal samples. Without this information, we recommend that primatologists estimate metabolizable energy gains through equations; using an estimated 4 kcal/g for non-structural carbohydrates and protein, and 9 kcal/g for fats (NRC 2003). The energetic contributions from fiber should also be estimated for those animals that have adaptations for digesting fiber. To obtain digestibility coefficients, we suggest using lignin as a marker for digestibility (Rothman *et al.* 2008), or using fiber digestibility estimates arising from captive studies (Conklin-Brittain *et al.* 2006).

11.6.7 Plant secondary compounds

Primates encounter an array of potential plant secondary compounds, all of which have various costs, such as protein precipitation in the diet, and benefits, like antibacterial properties (Glander 1982). Despite their common occurrence in primate diets, these compounds are very difficult to accurately measure, so we know little about their effects on primate feeding patterns. Condensed and hydrolyzable tannins have been the focus of many such investigations because they are prevalent in primate diets and have the potential to negatively affect protein digestibility (Rothman *et al.* 2009a), but some primates may have adaptations for coping with these compounds, such as tannin-binding salivary proteins (Mau *et al.* 2011). Thus it is difficult to know how tannins impact primate feeding behavior, and whether they impact diet nutritional quality. We suggest primatologists employ three steps to estimating tannins in primate diets. For screening samples to see if tannins are present, we suggest extracting the food samples in 70% aqueous acetone (v/v), and then using the acid butanol assay to assess for condensed tannins, and a potassium-iodate assay to screen for hydrolyzable tannins. To quantify the amounts of tannins in primate diets, tannins should be purified via Sephadex. Lastly, estimations of the biological activity of tannins (protein-binding ability) could be obtained using a recently developed assay by DeGabriel *et al.* (2008). A very useful handbook of tannin methods and protocols is available on the Internet from Ann Hagerman (2011; <http://www.users.muohio.edu/hagermae>).

11.7 Stable isotope analysis

Stable isotope analysis is a flexible, quantitative technique for reconstructing diet in a wide variety of species and ecological settings beyond the scope of conventional observational methods. This technique is advantageous when examining the dietary behavior of primates that are difficult to observe or feed on items that are difficult to quantify (e.g., Dammhahn and Kappeler 2010). The stable carbon ($^{13}C/^{12}C$) and nitrogen ($^{15}N/^{14}N$) isotope ratios of an animal's tissues reflects the isotopic composition of its diet. Interpreting these values requires an understanding of how carbon and nitrogen isotopes are distributed in primate food webs and the identification of isotopically distinct dietary inputs.

11.7.1 Tissue choice

The time scale at which isotopes are incorporated into animal tissue determines the resolution at which ecological information is recorded. The bulk isotopic composition of feces, hair, enamel, and bone, for example, represent records of differing integrated time encompassing periods of days to years, and can be used to quantify both intra- and inter-specific dietary differences (Sponheimer *et al.* 2009). Short-term, intra-individual changes in feeding behavior and diet can be reconstructed by serially sampling tissues that form incrementally with rapid isotope turnover rates and are resistant to isotopic exchange, such as enamel and hair. Serial fecal sampling provides the most highly resolved isotope record of diet change, limited only by gut retention times. Comparable data can be generated when sampling tissue types from living, recently dead, and museum specimens that can elucidate dietary variability within and between individuals, populations, and species over a wide range of potential temporal and geographic scales. Different tissues are not enriched relative to each other or to diet in a uniform manner, and do not necessary reflect equivalent measures of diet (DeNiro and Epstein 1978; Crowley 2010). For example, most proteinaceous tissues in animals synthesize their carbon (and virtually all nitrogen) mainly from dietary protein, while carbon in bone and tooth apatite is derived equally from all dietary carbon sources, including carbohydrates and lipids (Ambrose and Norr 1993).

11.7.2 Sample collection and laboratory methods

For isotope analysis of organic matter in plants, tissues, and bones, samples should be collected in the same manner explained above to the extent possible. After decontamination, homogenization, and drying, samples are usually converted to purified CO_2 and N_2 in an elemental analyzer (EA), with the isotopic composition of resulting gases measured on an isotope ratio mass spectrometer (IRMS). Isotope ratios are conventionally expressed using the delta (δ) notation as difference in parts per thousand (permil, ‰) from a standard. For carbon, $\delta^{13}C = {}^{13}C/{}^{12}C_{sample}/{}^{13}C/{}^{12}C_{standard} - 1) \times 1000$, using the internationally accepted isotope standard Vienna Pee Dee Belemnite (V-PDB). $\delta^{15}N$ is calculated with reference to the $^{15}N/^{14}N$ ratio of air.

Collecting tissue samples from museum specimens is the easiest approach, but precludes the possibility of sampling locally available foods. In some cases, with proper permits and permissions, museums will permit sampling of enamel and bone, but non-destructive collection of hair samples may be preferred. Hair can be obtained from living populations by sampling at night nests, foraging sites, in feces, or from darted individuals. Standard treatment with 2:1 methanol and chloroform followed by water rinsing is sufficient to remove surface contaminants from hair derived from both museum specimens and living animals (O'Connell and Hedges 1999).

When sampling museum specimens, extensive care should be taken to avoid destruction of any morphologically significant surface of bone or tooth. Bone and tooth samples can be obtained from living populations by burying dead individuals and allowing for natural skeletonization. The isotopic composition of bone collagen and tooth enamel derived from either museum specimens or recently dead animals is expected to retain original values, as these tissues are generally resistant to diagenetic alteration for tens of thousands of years (Lee-Thorp and van der Merwe 1987). Preparation of bone and tooth collagen involves defatting and demineralizing specimens before isotopic analysis (Sealy et al. 1987; Ambrose 1990; Koch et al. 1997). Carbon and oxygen isotope analysis of apatite carbonate in bone, dentine, and tooth enamel involves purification by removing fats with chloroform and methanol and removing proteins with bleach or hydrogen peroxide, prior to reaction with phosphoric acid to release CO_2 for measurement on an IRMS (Ambrose 1993; Koch et al. 1997).

11.7.3 Stable carbon isotopes

Divergent photosynthetic pathways of C_3 and C_4 plants, so called because of the number of carbons in the initial photosynthetic product, result in a bimodal distribution of carbon isotope ratios among plants in tropical and subtropical regions (O'Leary 1981). Consequently, C_3 trees, shrubs, and grass can be isotopically separated from low-latitude C_4 tropical grasses. The CAM plants, which primarily include succulents such as euphorbias that are rare outside desert environments, utilize C_3 and C_4 photosynthetic pathways and can therefore exhibit $\delta^{13}C$ values encompassing the entire range characterizing both. Thus, among primate tissue, $\delta^{13}C$ values that are more enriched than the range exhibited by C_3 plants reflect feeding on a mix of C_3, C_4, and potentially CAM plants (Schoeninger *et al.* 1998; Codron *et al.* 2006). Primate tissues may exhibit enriched $\delta^{13}C$ values even where there are no locally abundant C_4 vegetation if there is human provisioning (Schurr *et al.* 2012).

Most primate taxa, however, subsist predominately or exclusively on C_3 resources even when C_4 plants are available. Baboons are a significant exception because they consistently consume C_4 grasses where available (Codron *et al.* 2006). Fortunately, there is substantial isotopic variability among C_3 plants. In closed canopy tropical forests, subcanopy and forest floor foliage exhibit more depleted $\delta^{13}C$ values than leaves in the upper canopy or near canopy gaps (van der Merwe and Medina 1991). This vertical isotope gradient of forest vegetation is passed on to consumer tissues, including primates, and can be used to distinguish terrestrial from arboreal folivorous feeders as well as separate primates consistently feeding at different heights within an individual canopy (Ambrose and DeNiro 1986; Cerling *et al.* 2004). In addition, the isotopic composition of forest vegetation can vary due to factors other than microhabitat. Non-leafy plant matter such as fruit may exhibit less depleted $\delta^{13}C$ values than foliage, which suggests it is possible to detect fruit feeding isotopically (van der Merwe and Medina 1991; Cerling *et al.* 2004; Blumenthal *et al.* In Press).

11.7.4 Stable nitrogen isotopes

The stable nitrogen isotope composition of vegetation integrates terrestrial nitrogen cycle processes, and is known to vary with temperature, precipitation, salinity, nutrient cycling, and resource availability (Ambrose 1991). Leguminous plants often have less enriched $\delta^{15}N$ values than non-nitrogen fixing plants, and primates feeding on leguminous plants have been shown to exhibit relatively depleted hair $\delta^{15}N$ values (Schoeninger and DeNiro 1984; Schoeninger *et al.* 1999). In addition to plant diet input, the nitrogen isotopic composition of animal tissue becomes more enriched by approximately 3–6% with stepwise increases in trophic level, and more omnivorous primates exhibit more enriched tissue $\delta^{15}N$ values (Schoeninger and DeNiro 1984; Schoeninger *et al.* 1999). Additional data are needed on how dietary and nutritional variability impacts the nitrogen isotopic composition of primates.

Primatologists must be cautious in their interpretations of isotopic data because other, more idiosyncratic factors may also sometimes play a role in primate tissue $\delta^{15}N$ variability. Among male bonobos, for example, $\delta^{15}N$ values were correlated with male rank, which may indicate the potential influence of sex and age (Oelze et al. 2011). Additionally, seasonal variability in mouse lemur hair $\delta^{15}N$ values, which might otherwise be interpreted as variation in arthropod feeding, may instead reflect sex and species differences in patterns of torpor (Dammhahn and Kappeler 2010). These findings highlight the importance of considering multiple potential sources of variability in the nitrogen isotope composition of primate tissues.

11.8 Conclusions

Field studies of primate diets are important in understanding patterns of behavior and life history strategies. While we have outlined the process for examining the physical and nutritional attributes of primate foods, there are additional types of methods that could be explored in primate field studies. For example, for researchers looking to examine a large number of samples of the nutritional attributes of primate foods in a particular habitat, near infrared reflectance spectroscopy can be a useful tool (Rothman et al. 2009b). Little is known about the digestive ecology of wild primates. Methods developed in the agricultural sciences demonstrate that we can use plant markers like alkanes to assess the digestibility of primate foods non-invasively (Mayes 2006). Field primatologists are also encouraged to use the results of captive digestive trials to better interpret aspects of primate diets in the wild.

Acknowledgments

We thank the editors for inviting us to contribute to this volume, and the reviewers for helpful suggestions. We thank Stan Ambrose, Margaret Bryer, Janine Chalk, Paul Constantino, Kendra Chritz, and Peter Lucas for comments on a previous draft. We also thank Colin Chapman, Debbie Cherney, Ellen Dierenfeld, Joanna Lambert, Alice Pell, James Robertson, Debbie Ross, Mike Van Amburgh, and Peter Van Soest for very helpful discussions about the selection of nutritional methods described here.

12

Physiology and energetics

Jutta Schmid

12.1 Introduction

Physiological and behavioral ecologists try to find out more details about the dynamics of how organisms operate in their natural environment. Everything organisms do, both physiologically and behaviorally, involves the use of energy. Thus, body temperature and consequently energy expenditure of animals have a profound influence on the ability of animals to perform effectively and energy is therefore one of the most significant currencies determining the genetic fitness of organisms (Tolkamp *et al.* 2002). Apart from the heat that is generated internally, animals must also cope with extremely variable thermal attributes of terrestrial environments. If not controlled, this variability in the external gradients would exert an enormous influence over body temperatures and thus energetic balances of animals (Schmidt-Nielsen 1997). Furthermore, it is fundamental to investigate how changes in ecological stressors such as predation, food availability, and habitat destruction may affect both physiological and behavioral components of fitness in natural populations. Thus, the measurement of energy turnover and, in particular, how it is allocated to specific activities, can provide quantitative insights into diverse phenomena such as resource competition, reproductive cycles, or foraging strategies.

Primates living in seasonal environments exhibit a wide range of physiological and behavioral adaptations that enable them to cope with fluctuations in climate and energy availability (Garcia *et al.* 2011; Pusey *et al.* 2005; Schmid and Ganzhorn 2009; Schmid and Speakman 2009). Knowing these mechanisms in response to both predictable and unpredictable cycles of food allocation is essential to further understand the nature and the limits of energy economy. Thus, understanding behavioral responses and physiological consequences of primates is a useful tool for conservation planning for endangered species regarding the suitability and conservation of their natural habitats (Ganzhorn and Schmid 1998).

Primate Ecology and Conservation: A Handbook of Techniques. First Edition. Edited by Eleanor J. Sterling, Nora Bynum, and Mary E. Blair. © Oxford University Press 2013. Published 2013 by Oxford University Press.

This chapter describes different methodologies of studying energy expenditure that you can apply to primates; shows how you can measure body temperature of primates using different types of sensory systems; and outlines practical details that need to be considered before using a particular technique. In this chapter I aim to explain how different methods basically work and the advantages and disadvantages that attend their use.

12.2 Methods of measuring energy expenditure

For the estimation of the rate of energy turnover of animals, four methodologies are presented in the following sections: (1) indirect calorimetry based on the fact that animals consume oxygen to oxidize organic compounds, (2) time budget analysis combined with energetic costs estimated for each component in the behavioral repertoire, (3) measurements of CO_2 production by the doubly-labeled water method (Lifson and McClintock 1966; Speakman 1997), and (4) heart-rate telemetry that depends on the physiological relationship between heart rate and oxygen consumption. Indirect calorimetry and time-energy budget analysis identify components of the total energy costs, whereas the doubly-labeled water method yields overall estimates but no insight into the contributions of different behaviors to the total energy expenditure. All four methods involve combinations of field observations with laboratory trials. There are considerably different methodological complexities and technical difficulties that need to be handled when applying a particular method.

The first practical aspect of using any particular technique to measure energy demands is to consider whether the chosen methodology is really the most appropriate technique for the problem at hand and whether the provided results will reach an adequate answer to the questions being posed. The choice between different methods depends on the ease of performing surgery in the field and obtaining calibration data and consideration of costs. Clearly, the most decisive aspect when choosing an appropriate technique is the size of the animal in question as well as its behavioral and physiological repertoire. In addition, if you want to use the doubly-labeled water method you have to consider if it is feasible to recapture your study animal within a certain time frame. Finally, when you have selected a particular methodology to measure energy demands of your study animal you need to take into consideration what training is required for its application, how and where you fulfill these training requirements, and how much time is needed.

12.2.1 Indirect calorimetry (respirometry)

Basal and maximal energy expenditures, and the amount of energy animals spend when engaged in specific activities, can routinely be measured using indirect calorimetry (respirometry). This technique to determine the metabolic rate measures the amount of oxygen (O_2) consumed or the amount of carbon dioxide (CO_2) expired. To convert O_2 consumption or CO_2 production to energy expenditure it is necessary to know the specific caloric equivalents of 1 ml of used O_2 or expired CO_2 which depends on the substrate that is utilized (for caloric equivalents see, for example, Schmidt-Nielsen 1997). Nowadays, open-flow systems with a continuous-flow gas analysis system are used for respirometry measurements. In an open-flow gas analysis system the animal is confined in a chamber through which fresh air is passed via a flow regulator. Within the chamber the animal takes in oxygen and expires CO_2. The amount of both oxygen and CO_2 is measured in the chamber as well as in the exhausted gases from the metabolic chamber using gas analyzers. The difference between the O_2 content in both gas channels is due to the respiratory activity of the animals in the metabolic chamber.

For at least 200 years respirometry measurements have been carried out only in the laboratory (McNab 1992). However, today portable metabolic O_2- and CO_2-analyzer assemblies are available to measure metabolic rates of free-living animals in their natural habitat. These portable analyzers are often self-constructed (see, for example, Fietz et al. 2010; Schmid 2000; Schmid et al. 2000), but can also be obtained from official manufacturers (e.g., FoxBox from Sable Systems International, USA; Nowack et al. 2010). The major advantage of a portable gas analysis system in comparison to the laboratory respirometry technique is that you can record physiological data of wild animals. Basically, there are two possibilities to apply this technique in the field. First, if the animal you study naturally uses tree holes or burrows for sleeping and/or resting, you can utilize these natural enclosures as metabolic chambers. This is ideal, because the animal can remain in its natural environment which significantly reduces disturbance to it. As an alternative, a second option is to keep your focal individuals in outdoor enclosures under natural conditions of temperature and photoperiod. Depending on the species you work with, the enclosures should be equipped with adequate nest-boxes that the animals can use as sleeping quarters, and at the same time they serve as metabolic chambers for the respirometry measurements. The latter setup has the advantage that continuous measurements of the animals under focus are possible without moving the experimental equipment.

Estimates of energy turnovers measured using field systems are, however, not as accurate as those recorded with laboratory systems. This is mainly because environmental conditions in the wild are subject to permanent variations that affect the metabolism of the measured animal. On the plus side, however, such outdoor systems are relatively cheap to set up, and they provide a reliable estimate of energy expenditure (Dausmann *et al.* 2004; Schmid 2000; Schmid *et al.* 2000). If you want to study energy expenditure of free-ranging primates it is obligatory to contact a physiological lab with experienced scientists and technicians. This is essential for the setup of the entire gas analysis equipment as well as for the final calculations of daily energy requirements.

Standard respirometry techniques have one major advantage over other methods for quantifying the energy requirements of animals (some further methods are presented in the following): the metabolic rate of animals is an accurate and precise estimation. One distinct disadvantage of using indirect respirometry, however, is that the animals cannot perform all activities of interest in the restricted area of a metabolic chamber (independent of whether you use a natural chamber in the wild or an artificial chamber in outdoor enclosures or the laboratory).

12.2.2 Time–energy budget estimates of daily energy expenditure

A standard method for estimating daily energy expenditures (DEE: the total amount of energy an animal spends during a 24 hour period) is to prepare time and energy budgets, in which the durations of an animal's activities are multiplied by the respective energy costs of the activities. This procedure involves observing an animal carefully throughout the day and monitoring each of its activities. Subsequently, the energetic costs associated with each of the behaviors recorded are required to convert the time budget into an energy budget. Bearing in mind that different levels of activity are associated with different levels of energy expenditure, and that at the same time these levels of metabolic rate are influenced by the thermal environment, it is, however, often difficult to allocate realistic values. There are various methodologies that have been employed to convert the time budget into an energy budget and to decide which the most appropriate approach is; it is indispensable to check with the relevant literature (see, for example, Weathers *et al.* 1984; Culik and Wilson 1991; Goldstein 1988).

A wide-ranging problem of time–energy budgets as a method for estimating energy turnovers in free-living primates is that it is often difficult to record the behavior of the focus animal in detail. This might be because the focal individuals cannot be kept continually in sight, the primates might not behave naturally when under observation, or because of time constraints of the observing scientist. To avoid disturbance of the primates by the presence of observers, field biologists often work with habituated animals. This option, however, must be carefully considered and anyone intending to habituate primates should think carefully about the pros and cons (Doran-Sheehy *et al.* 2007; Williamson and Feistner 2011). Alternatively, to record much more detail of the animal's activity without any effects of the human observer on the primates, one could use radio telemetry. The most modern recorders log information continuously while they are fixed on the animal and can be downloaded when the animal is recaptured. Usually the signal obtained from the transmitter reveals the animal's location, where exactly it went, how long it spent there, how fast it traveled, and a fluctuating signal may even indicate that the primate is moving. However, regardless of which technique is employed and even when automation is in use, to obtain detailed time and energy budgets is a time-consuming methodology.

12.2.3 Doubly-labeled water method

The doubly-labeled water method (DLW) for estimating total CO_2 production over a period of time is based on the fact that isotopes of oxygen in body water are in complete and rapid exchange equilibrium with the oxygen in respiratory CO_2 (Lifson and McClintock 1966; Speakman 1997; Speakman and Racey 1988). This technique uses stable isotopes, deuterium (^2H, hydrogen) and heavy oxygen (^{18}O), to label an animal's body water and determines the flow rate of water and CO_2 through the body over time. An isotopic label of ^2H will leave the body primarily only as water, whereas ^{18}O will be eliminated by the flow of water and CO_2 and so washes out faster than hydrogen. If both isotopic labels are injected at the same time (therefore doubly-labeled water) the difference between the washout rates will reflect the CO_2 production and thus indirectly the energy expenditure. For the conversion of CO_2 production into energy equivalence, the knowledge of the respiratory quotient (RQ), which is the ratio of CO_2/O_2, is required. To measure the RQ for an animal under study one can either estimate the RQ from the food consumption, or extrapolate from laboratory studies, or one can attempt to measure the RQ directly from the background abundance of the isotopes. Alternatively, RQ values can also be estimated using empirically derived equations and these can then be converted into energy demands (see, for example, Gessaman and Nagy 1988; Schmidt-Nielsen 1997). For a more detailed description of the biochemistry underlying the technique, as well as for the accuracy and precision of the method, it is therefore necessary to consider further relevant literature (Lifson and McClintock 1966; Nagy 1980, 1983; Speakman 1993, 1997; Speakman and Krol 2005).

The DLW can be applied both in the laboratory and in the field. In practice, the simplest form of the method requires that the focal animal is dosed with the isotopic label and that a sample of body water is taken once the injection has reached complete isotopic distribution in the body pool. The equilibration time in the bloodstream depends on the body size of the animal and is nowadays well quantified with 30 min to 1 h for animals weighing less than 100 g, around 3–4 h for animals weighing 30 kg, and around 6 h for humans and animals weighing more than 70 kg (Speakman 1997). The amount of the isotope required depends on the body mass of the study animal, the enrichment of the injectate (stable isotopes are in general supplied at varying levels of enrichment), and the duration of the experimental measurement. It is essential to know the exact quantity of stable isotopes injected and also the exact body mass of the animal. Weighing small animals should therefore be performed using a precision electronic balance. However, where mains electricity is unavailable (usually in the field), portable electronic balances that work off batteries should be used rather than spring balances which have only poor accuracy. When working with larger animals in the field some alternative weighing procedures must be applied (for example torsion or spring balance). The best way to achieve the exact amount of isotope that has been administered to the animal is to weigh the dose syringe using a four-figure balance immediately prior to injection and immediately afterwards and then calculate the difference in weights. Problems again arise in the field where there is no electricity. Given the fact that weighing the dose is not going to be feasible in the field, it is important to consult a specialized isotope laboratory to work out alternative methods with experienced technicians. There are different routes for injection (intravenous, intramuscular, intraperitoneal, and subcutaneous) and you must obtain training from a veterinary in the suitable dosing method before you perform it yourself.

Once the isotopes are completely equilibrated, the initial sample is taken, where the most desirable sample sources are blood and saliva since both sample sources are direct measures of the instantaneous state of the body water pool (Speakman 1997). Urine samples are not suitable since it is unlikely that you can obtain a urine sample at a specific time of day as you cannot control the timing of urination. The best method for taking a blood sample from primates both in the laboratory and in the field is some form of peripheral vein puncture, which is performed with a small needle. This must be carried out under anesthesia, and training is necessary to perform this procedure correctly. The least invasive approach for collecting blood samples, however, is using starved blood-sucking ectoparasites that withdraw blood meals from veins (Helversen *et al.* 1986; Helversen and Reyer 1984). For isotope analysis, the samples are best flame sealed into capillary tubes. The capillaries are best protected in sample tubes containing some cotton wool so that they don't break. Samples should be stored at room temperature or in a domestic refrigerator at about 4–8 °C. However, do not leave them in the cold because if they freeze they will burst and you will lose them.

Once the first sample is taken, the animal is released at its site of capture and it is allowed to continue its habitual level of activity and dietary intake for a variable time period until its recapture. This time span between capture and recapture once more depends on the body mass and can vary between 24 h for the smallest vertebrates to up to 30 days in humans. After recapture of the injected animal, you take a second sample to obtain a final isotope enrichment measure. The isotope enrichments need to be expressed relative to background enrichment and thus, some estimate of background enrichment is also needed. For large animals you take a blood sample before injecting the isotopes, and for smaller animals you usually sample some unlabeled individuals that are not part of the experimental measures.

The primary advantage of the DLW is that it provides a direct estimate of CO_2 production. In addition, the technique can be easily applied in the field on free-ranging animals since it only requires that an animal is captured twice, injected once and bled twice. Finally, a major advantage of using the DLW is that there is almost no practical limit on the size of animals. However, recapture of your released animal in the field within a certain time frame cannot be guaranteed. At present, the smallest animals on which energy estimates have been obtained through doubly-labeled water weighed less than 300 mg (bumblebees, Wolf et al. 1966), whereas the upper size limit relies only on the costs and availability of sufficient isotopes to label the animal.

One disadvantage of the DLW is, however, that although it will give accurate measures of daily energy expenditure, it cannot subdivide the total costs into its components. Another disadvantage of this method is that although it provides a direct estimate of CO_2 production, a significant error could be generated when converting to energy using an inappropriate RQ. The only thing you can do is to utilize a best-guess RQ based on the dietary information of your study animal. Furthermore, when studying free-living primates, there is also the potential difficulty of recapturing injected animals within the given time frame before all the isotopes have been eliminated and thus an estimate of CO_2 becomes impossible. To overcome this problem of losses of successful DLW analysis due to failed recaptures, scientists should be best educated in methods for darting and capturing non-human primates (see Chapter 3), but also plan with a large enough sample size. Finally, the major disadvantage of the DLW is its high cost both in materials and equipment required for the analyses of the isotopes. As of 2011, heavy oxygen costs US$55 per ml and hydrogen costs about US$75 per 100 ml. In addition, the costs of analysis are significant and at present a two-sample estimate of DEE comes to about US$150–200 per animal.

12.2.4 Heart rate method

A fourth technology as a means of estimating metabolic rate is the heart rate method that depends on the physiological relationship between heart rate and oxygen consumption (see, for example, Butler *et al.* 1992; Weimerskirch *et al.* 2001). This approach provides a direct field estimate of metabolic rate over prolonged periods of time and also enables subdivision of the total costs into specific activities. The standard procedure is that the recording system is surgically implanted and subsequently removed, which is a highly invasive method that should be carried out only by a veterinarian. Thus, it is obligatory to contact specialized laboratories before applying this technique. The modern implantable data-loggers usually also monitor body temperature and body position and thus provide detail about what animals are doing. Now even externally attached electrodes are available that produce signals of sufficient clarity for use, although often the animal does not tolerate the presence of external electrodes and they may become detached and lost.

A major advantage of the heart rate method is that there is no limited time period within which the focal animal has to be recaptured, because the data are stored in a lasting memory and can be downloaded at any later time (for capturing methods see Chapter 3). The most significant problem with the heart rate method is, however, the fact that in order to produce precise estimates of O_2 consumption, it is necessary to generate a relationship for each individual animal involved in your experiment. To derive such a calibration equation you need to hold the animal for some extra periods of time, which may be inappropriate if the animal is performing behavior in the field where such a long period in captivity would be distressing (for example when the animal is caring for young). The heart rate method cannot be used at lower body masses for technical reasons and at present implantable heart rate data-loggers are restricted to animals weighing approximately 1 kg (Butler *et al.* 2004). Smaller, non-logging transmitters that are attached externally are lighter and the lower size limit is 40 g. However, such transmitters require permanent access to the animal to record data.

12.3 Measuring body temperature

Information about body temperature (T_b) is elementary in a variety of physio-logical and behavioral investigations such as studies on thermoregulation, energy demands, reproduction, or endocrinology (see, for example, Dausmann *et al.* 2004; Kobbe and Dausmann 2009; Schmid and Ganzhorn 2009; Williams *et al.* 2011). Basically, temperature-sensitive sensors are either attached externally or implanted into the animal (these two methods are described in more detail below). Temperature loggers are now available that weigh less than 2 g (e.g., iBCollars from Alpha Mach Inc., Mont-St-Hilaire, Canada), making measuring body temperature of even the smallest primate species possible (Kobbe *et al.* 2010; Schmid and Ganzhorn 2009; Schmid *et al.* 2000). However, transmitters still differ in frequency, signal range and reception, battery lifespan, type of data recorded, and mass. The pros and cons of these practical details need to be weighed before embarking on a study using one of the described techniques to measure body temperature. A further practical aspect of applying a particular method is to think about the financial investment in equipment required. Finally, the size of the animal one wishes to study is an important factor to be considered and whether the method of choice will actually work for the species and situation under consideration.

12.3.1 Core body temperature

The simplest technique is to measure core body temperature ($T_b$core: the internal temperature of a human's and an animal's body) rectally by using a flexible or firm probe that is inserted 2–4 cm (depending on your study animal) into the rectum. Although the temperature reading is accurate, the problems are that you have to briefly immobilize the animal and you only get a single reading at time. A more invasive technique is to measure the $T_b$core of an animal by means of implanted temperature-sensitive transmitters. You can use temperature-sensitive transmitters where you have to record the signals from the radio transmitter by using a telemetry system consisting of receiver and antenna. Alternatively, you can use temperature-sensitive data-loggers with internal memories that store data automatically which you can then download to a computer. Custom-made computer programs are used for setting and resetting of the data-loggers as well as for retrieving the data and defining sampling intervals. The most significant advantages of data-loggers are that there is no need to follow the animal in the wild and that you can record data continuously for a long period of time. The disadvantage of using data-loggers is that you have to recapture your study individual to download the data because you otherwise lose the data. When using transmitters, it is necessary to calibrate each transmitter against a certified precision thermometer in a temperature-controlled water bath. In the case of data-loggers, calibration curves, usually provided by the manufacturer, can be used for improved precision of the loggers.

You can implant transmitters either into the peritoneal cavity or under the skin of the animal. To decide which of the two methods is best to apply depends on the size and anatomy of the animal under investigation. Before implantation into the animal, you should first seal both data-loggers and transmitters with a paraffin-wax coating, and then disinfect them to minimize risk of infections or allergic reactions. All surgical treatments should be conducted under anesthesia. Therefore, it is obligatory to weigh the animal to the nearest 0.5 g before surgery to calculate the correct justifiable amount of anesthetic (see a more detailed discussion of these methods in Chapter 3). In the case of a field study, allow the animals to recover from the invasive operation for several hours before release. When the experiment is finished, you have to repeat the surgery to remove the implanted temperature telemeters. Implanting transmitters or data-loggers is theoretically and practically feasible in the laboratory as well as in the field but appropriate veterinary guidance is essential.

A less invasive approach to measure $T_b$core is to use temperature-sensitive transponders. These passive integrated microchips (approximately 12 mm × 2 mm and about 0.06 g) are encased in glass or a bio-compatible coating and are suitable for use even in the smallest species. Temperature-sensitive transponders give you both a reading of the identification (ID) code of the animal and its body temperature. They are injected subcutaneously without anesthesia into the animal's back area by using a syringe or injector "pistol." Transponders function as radio-frequency transmitters and contain no batteries and are rather activated and read by using a special microchip reader. A practical difficulty is the need to get the reader sufficiently close to the chip (between 2 and 10 cm depending on the type of the reader).

12.3.2 Skin body temperature

To monitor skin body temperature ($T_b$skin: the external skin temperature of a human's or an animal's body) you can use radio-collars with integrated temperature-sensitive data-loggers or transmitters. Generally, $T_b$skin measured with radio-collars is a reasonable estimate of core T_b (Audet and Thomas 1996; Dausmann 2005). The radio-collar should be placed so that the temperature sensor is in contact with the animal's skin. You may mount solder on the temperature probe to enhance direct contact of the sensor with the skin of the animal (Kobbe *et al.* 2010). Depending on the sensor type, it might be necessary to calibrate each radio-collar against a precision mercury thermometer in a water bath before use. Transmitter signals are received telemetrically using a telemetry receiver and an antenna to pick up the signals from the radio-collared animal. Data can then be stored automatically in an onsite PC or manually in a normal data book.

The major advantage of using externally mounted temperature collars is that you can measure $T_b$skin of both animals in the field and in the laboratory without the need for surgery and recovery. A further advantage to using radio telemetry is that apart from body temperature, you can record additional measurements such as home range or distance traveled simultaneously.

During all body temperature measurements (implanted or external mounted sensors) it is obligatory to record the corresponding ambient temperature (T_a) continuously by using data-loggers or calibrated thermometers.

12.4 Ethical implications and legal aspects

Before embarking on a project it is important to establish the legal position with respect to animal experiments. Legislative requirements are different in different countries. Therefore, you should contact the ethics committee in your host country for further details concerning animal experiments and to ask if any special permitting is required at the proposed field site. In general, both you and your assistants will require permits or licenses to carry out specific procedures that are part of your project, such as injecting an animal, taking blood-samples, or implanting data-loggers or transmitters. You also need to check if you require a license to capture and mark the animals in question or when you want to bring them to the laboratory for the duration of your experiment. In the latter case you may need a whole set of housing requirements to show that the animals can be kept in a fit state for subsequent re-release. An additional factor to consider is transportation of blood samples and perhaps also isotopes between countries. You will need a permit to both export and import blood samples for the doubly-labeled water analyses. Finally, when working abroad you may need a CITES (Convention on International Trade in Endangered Species) permit that allows you to work with the animal you intend to study in case it is listed in Appendices 1–3 of the CITES legislation. The appendices are continually updated and you should contact the nearest CITES office to be sure you do not contravene the regulations (see Chapters 3 and 4).

In general, the use of any physiological methodology will normally involve several different legislative implications and if you do not start to obtain permissions early enough this may delay starting your project and may even significantly increase costs.

13

Primate behavioral endocrinology

Nga Nguyen

It is difficult to turn on the television, open a newspaper, or visit the doctor's office without hearing about one putative hormone–behavior relationship or another. That hormones affect behavior and that behavior, in turn, can affect hormones has been implicitly understood for centuries but the scientific study of hormone–behavior interactions (i.e., behavioral endocrinology) is only ~ 160 years old. Until recently, however, much of our knowledge of hormone–behavior interactions emerged from research and experimentation on a handful of laboratory species. Recent advances in non-invasive, hands-off techniques for measuring hormones have now made comparative research on wildlife, including primates, possible. This research has provided (and continues to provide) novel insights into human behavioral biology and ecology, and has contributed significantly to our understanding of human adaptation and evolutionary history. In this chapter, I provide a brief overview of the key historical and theoretical developments in wild primate behavioral endocrinology, summarize how (and in what manner) primatologists have used non-invasive techniques to monitor hormones in the wild to study the endocrinology of primate reproduction, social relationships, and stress, and review important methodological considerations for collecting, processing, and analyzing hormones in biological materials, as well as discuss the future directions of the field.

13.1 Major historical and theoretical developments in wild primate behavioral endocrinology

Primate behavioral endocrinology does not exist in a vacuum, but is part of the larger discipline of behavioral endocrinology, which has deep and far-reaching roots in modern biology. In this section, I provide brief overviews of the endocrine system, the history of behavioral endocrinology, and the strengths and weaknesses of field-based endocrinology research.

Primate Ecology and Conservation: A Handbook of Techniques. First Edition. Edited by Eleanor J. Sterling, Nora Bynum, and Mary E. Blair. © Oxford University Press 2013. Published 2013 by Oxford University Press.

13.1.1 The endocrine system

The term *hormone* was first coined in 1905 (Henderson 2005). Today, hormones are recognized as the blood-borne chemical messengers of the endocrine system, which, along with the central nervous and immune systems, is one of three closely integrated chemical communication systems in the vertebrate body (Nelson 2011). Some of the earliest experiments involving hormones suggested that they might be free from central nervous system regulation (Nelson 2011). However, we now know that the central nervous and endocrine systems are fully integrated and work together to monitor environmental perturbations and to coordinate animals' bodily functions and behaviors in response to these perturbations.

Hormones can travel from their point-of-origin endocrine gland to nearly every cell in the body, carrying with them information about the organism's state. In doing so, hormones coordinate behavior with bodily functions to minimize waste of energetic resources. Gonadal steroids, for example, ensure that animals will engage in costly mate-seeking behaviors only once they have produced adequate stores of gametes ready for fertilization (Adkins-Regan 2005).

A common misconception about hormones is that they direct animals to perform specific behaviors. However, hormones do not cause behavioral changes per se (Adkins-Regan 2005; Ellison and Gray 2009). Rather, they alter the probability that certain behaviors will occur in certain contexts by influencing organisms' sensory capabilities, central nervous system, or effector organs (i.e., muscles and glands; Nelson 2011).

Hormones circulate through the bloodstream to specific target organs/tissues that have specific receptors for them. The binding of a hormone to its receptor initiates a cascade of biochemical events within the target cell—events that can culminate in changes in cellular metabolism or gene expression and protein synthesis (Becker *et al.* 2002; Nelson 2011). That some hormones can modify gene action and function suggests that these hormones may be partly responsible for observed differences in behavioral and physiological development and life history among some closely related individuals, groups, and species (Ellison 2009).

13.1.1.1 Hormones and the epigenome

The idea that the phenotype is a product of both the genotype and the environment has been implicitly understood for generations, though the first written record of this "DNA is not destiny" concept dates to 1942 (Crews and McLachlan 2006). In that year, a report appeared describing how a single genetic strain of fruit flies, when exposed to different rearing environments, could give rise to different phenotypes. Waddington (1942) proposed that these phenotypic differences must arise from something other than mutations in the genetic code. Waddington argued that environmentally induced differences in gene action could occur among genetically identical individuals and that this process could give rise to phenotypic variability in nature. This concept would later form the foundation of the modern field of epigenetics (Crews 2011). Today, the term *epigenetic imprinting* (EI) is used to refer to environmentally induced changes in gene function that occur during development to produce changes in phenotype (without modification to the DNA) that can persist into adulthood (Crews and McLachlan 2006).

Hormones play a pivotal role in EI. A classic example of EI involves the role of sex hormones in sexual differentiation of brain and behavior. In male mammals, exposure of the developing brain to androgens during fetal development causes masculinization and defeminization of the parts of the central nervous system that govern the expression of sexually dimorphic behaviors (Phoenix *et al.* 1959). Once permanently structured (or "organized"), these sex differences can be stimulated (or "activated") later in life by the same or different hormones (Phoenix *et al.* 1959). Another classic example of EI involves the role of glucocorticoid hormone receptors in restructuring the offspring brain in response to variation in mothering behavior in rats. As adults, offspring of mothers that display high rates of infant licking and grooming display (1) enhanced cognitive abilities, (2) greater resistance to stressors, and (3) the type of maternal behavior they experienced as infants (Weaver *et al.* 2004).

Many other examples of endocrine-mediated EI exist (Gore 2008). EI is now widely recognized as a process that "not only shape[s] the subsequent behavioral phenotype of the individual but also modif[ies] the way the individual responds to adult experiences" (Crews and McLachlan 2006, p. S8).

Given the importance of epigenetic mechanisms to normal growth and development (Gicquel and Le Bouc 2006), concern has been mounting in recent decades about the potential for epigenetic processes to be disrupted by endocrine-disrupting chemicals (EDCs) (Crews and McLachlan 2006), which are now widely found in the environment (Kuster *et al.* 2005). EDCs are chemicals (often man-made) that mimic or block endogenous hormones. Many EDCs exhibit estrogenic activity and are often mistaken for estrogen by vertebrates. Endocrine disruption by EDCs is now recognized to have adverse effects on wildlife and people, and has been found to result in a range of developmental, reproductive, neurological, and immune disorders in affected populations (reviewed in Crews and McLachlan 2006).

Growing evidence suggests that some epigenetically mediated changes are heritable and may persist across many generations (Crews 2011). This has raised concerns about the transgenerational consequences of EDC exposure on humans and wildlife (Crews and McLachlan 2006; Gore 2008). Given our incomplete knowledge of the physiological processes that underlie normal growth and development in humans and wildlife and the growing prevalence of EDCs in the environment, these concerns highlight the importance of studying physiological processes in normal, healthy populations, so that we can identify and, if possible, treat diseased or dysfunctional populations in time (Crews and McLachlan 2006). Fortunately, advances in the past few decades in non-invasive techniques to monitor and assess hormone–behavior interactions in the wild have helped (and continue to help) augment our understanding of animal behavioral development and evolution.

13.1.2 A brief history of behavioral endocrinology

For centuries, researchers have implicitly recognized that hormones affect (and are, in turn, affected by) behaviors (Nelson 2011). However, scientific investigations into the nature of the hormone–behavior relationship date to only the last 160 years. In 1849, Arnold Berthold became the first practitioner of behavioral endocrinology when he demonstrated that a product of the testis was critical for immature male chickens to develop into normal adult roosters. Nearly a century later, Frank Beach (1948) would usher in the modern era of behavioral endocrinology with the publication of his highly influential book *Hormones and Behavior*, which synthesized the little that was known at the time about hormone–behavior interactions in animals.

In the 1970s, Robert Sapolsky began his now famous studies of the intersections between hormones, health, and social relationships among wild baboons in Kenya (reviewed in Sapolsky 1993). Sapolsky's pioneering studies effectively established an entirely new field of inquiry in primatology and would go on to inspire an entire generation of researchers interested in hormone–behavior interactions in wild primates. However, like landmark studies of other wildlife at the time (e.g., Wingfield *et al.* 2001), Sapolsky's research required capture and immobilization of animals for blood sampling, a procedure not always feasible or desirable under field conditions.

Fortunately, in recent decades, advances in more hands-off techniques for non-invasively sampling a diversity of behavioral, ecological, genetic, and physiological variables (including hormones) in wildlife have given field researchers previously unimaginable opportunities to "eavesdrop" on the social lives of their wild subjects without disruptions to the animals' lives (Altmann and Altmann 2003). Today, many field and laboratory researchers employ these techniques to study a range of hormone–behavior relationships in a wide variety of animals, including many non-human primates.

While many studies in behavioral endocrinology focus on examining proximate explanations for behavior, the discipline as a whole has much to contribute to our understanding of all four of Niko Tinbergen's explanations for behavior (i.e., phylogenetic, ontogenetic, functional, and proximate) (Marler 2005). For an excellent review of how behavioral endocrinology can help inform our understanding of Tinbergen's four explanations of behavior, the reader is advised to consult Gray and Ellison (2009).

13.1.3 Merits and shortcomings of field-based endocrinology

Field-based studies of hormone–behavior relationships have their merits and limitations. Unlike in the laboratory, many variables in the field are beyond researchers' control or measurement. In addition, in the laboratory cause–effect relationships can be more easily teased apart through careful experimentation while field-based endocrinology studies are usually non-experimental studies of covariation (Wallen and Hassett 2009). These studies typically evaluate the degree to which naturally occurring variation in a behavior of interest covaries with naturally occurring variation in a hormone of interest across individuals, groups, or populations (Wallen and Hassett 2009). Despite their shortcomings, field-based studies of hormone–behavior relationships offer a promising avenue for future research.

Studies of hormone–behavior interactions in wild populations are vitally important for verifying or validating hormone–behavior relationships established in captive populations for three reasons (Costa and Sinervo 2004; Fusani *et al.* 2005). First, captivity and provisioning can affect the expression of social behaviors, modify neural processes and brain structures, and influence the pattern of secretion of hormones in captive animals (reviewed in Costa and Sinervo 2004; Fusani *et al.* 2005). Second, laboratory studies have, of necessity, been confined to a handful of the tens of thousands of vertebrates found in nature. Field-based studies provide opportunities for remedying this deficiency. When conducted on populations in which experimentation is not feasible or desirable, field-based endocrinology studies can "provide meaningful leads to more mechanistic studies" that might later be more fully examined in captive populations (Wallen and Hassett 2009, p. 34). Finally, because of their greater ecological relevance, field studies offer important opportunities to study the physiological determinants or consequences of behavior within the selective environments in which these hormone–behavior relationships likely evolved (Whitten *et al.* 1998).

13.2 Applications of non-invasive techniques for monitoring hormones in wild primates

Thanks to recent methodological advances in non-invasive, hands-off techniques for collecting, preserving, and analyzing hormones, primatologists can now carry out studies of hormone–behavior relationships in the wild in ways that were previously unimaginable outside the laboratory (Whitten *et al.* 1998; Strier and Ziegler 2005). Below, I highlight the varied ways in which primatologists have employed these techniques in the past several decades to augment our understanding of primate reproductive ecology, social relationships, and stress physiology, with important implications for our understanding of human behavioral and social evolution.

13.2.1 Endocrinology of reproduction

Unlike most sexually reproducing animals, primates frequently mate outside the period of ovulation, often in the absence of visual signals of ovulation (Sillentullberg and Moller 1993). Field researchers, therefore, traditionally relied on behavioral or visual signals to identify the timing of ovulation in primates. However, these signals may be subject to selection and may therefore not always represent honest indicators of female fertility (Zinner *et al.* 2004). Fortunately, using non-invasive methods for measuring reproductive hormones in animal waste products (reviewed in Whitten *et al.* 1998; Strier and Ziegler 2005), primatologists can now more accurately monitor changes in female reproductive condition and more precisely predict the timing of ovulation and birth relative to behavioral and visual signals of female fertility through regular (near-daily or weekly) hormonal sampling of known individuals (e.g., Gesquiere *et al.* 2007).

Some of the earliest applications of these techniques focused on evaluating the degree to which behavioral and visual signals of estrus covaried with hormonal indicators of fertility, a topic which continues to be an important objective of field research on primate reproductive endocrinology today (Strier and Ziegler 2005).

In recent years, researchers have also begun to explore potential physiological causes and consequences of a range of reproductive phenomena. In topics ranging from the endocrinology of fetal loss (Beehner et al. 2006) and sexual signaling (Gesquiere *et al.* 2007) and conflict (Roberts *et al.* 2012), to reproductive suppression and skew (Bales *et al.* 2005) and variability in maternal behavior towards offspring (Nguyen *et al.* 2012), researchers are now combining observational sampling of behavior with techniques for measuring hormones in the field to gain deeper insights into the biological basis of reproductive behavior in primates.

An intriguing issue that has emerged in recent years concerns the influence of EDCs on primate reproductive ecology (e.g., Higham *et al.* 2007). Although some EDCs are found naturally in primate foods, their physiological and behavioral effects are poorly understood. Recently, researchers reported that during months of the year when baboon females consumed foods containing high amounts of EDCs, changes in the animals' reproductive physiology and behavior also occurred (Higham *et al.* 2007). Clearly, primates are not immune to the influence of chemicals in the environment that mimic female sex steroid hormones. Given the steady rise in man-made estrogenic and progestagenic compounds in the environment (and their long-life there) in recent decades (Kuster *et al.* 2005), these results raise concerns about the potential long-term effects of exposure to synthetic EDCs on humans and wildlife. By employing a mix of field and laboratory approaches, researchers can now begin to address these and other topics that have important implications for primate health and reproductive ecology.

13.2.2 Endocrinology of social relationships

Social relationships (both affiliative and agonistic) are the "glue" that hold animal societies together. Social relationships among primates were long believed to be free of neuroendocrine regulation. However, increasing evidence suggests that hormones play important roles in the organization or activation of the neuroendocrine mechanisms that underlie a range of social behaviors in wild primates (Ellison and Gray 2009).

13.2.2.1 Affiliation

The peptide hormones oxytocin and vasopressin have long been known to be important for the formation and maintenance of affiliative bonds in prairie voles, small, socially monogamous rodents (Young and Wang 2004). Research into the genetic and hormonal bases of affiliation in non-human primates lags far behind that of social rodents, in part because oxytocin and vasopressin (as peptides) are not excreted in feces, which limits their measurement in the field.

Socially bonded animals tend to exhibit strong attachment to one another (Bowlby 1969) and will typically exhibit behavioral or physiological distress when separated (Mendoza and Mason 1997). By studying attached animals that have become separated from one another in the wild, researchers may gain important insights into the biological bases of attachment and the adaptive value of these bonds in the selective environment in which these behaviors likely evolved.

For example, among wild baboons, females who lost a close relative exhibited short-term elevations in fecal glucocorticoids (steroid hormones typically associated with the stress response in vertebrates; Engh *et al.* 2006). The immediate distress occasioned by the death of a close associate was, however, in time alleviated among survivors by modifications to their social (grooming) network. While observations of behavioral responses to death in wild primates have increased recently (reviewed

in Fashing and Nguyen 2011), we still know little about the physiological determinants and consequences of primate responses to dying and death.

13.2.2.2 Agonism and social status

Conflict (intra-sexual and inter-sexual) is a pervasive feature of life in primate
and other animal societies (Aureli and de Waal 2000). In most species, males
exhibit higher rates of agonism (behaviors of aggression and submission) than
females and this sex difference is attributed to sexually differentiating hormonal
events during fetal development, sexual maturation, and even adulthood
(Glucksmann 1974).

There is now sufficient evidence to link the steroid hormone testosterone to
male aggression (Archer 2006). Indeed, testosterone is now widely believed to not
only organize (during development) but also activate (during adulthood) substrates
for a range of male reproductive behaviors, including aggression. However, testosterone only increases the probability of male aggression in the appropriate social
context. This view, known as the Challenge Hypothesis, argues that testosterone is
elevated and associated with aggressive behavior in males only during periods when
male–male competition over females is high (Wingfield *et al.* 1990). There is now
widespread support for the context-dependent role of testosterone in promoting
male aggression from research in many vertebrates, including primates (e.g., Muller
and Wrangham 2004; Archer 2006).

One way to minimize the high costs of aggressive conflict is through the
establishment and maintenance of stable dominance hierarchies (Aureli and de
Waal 2000). Perhaps no other topics in field primate behavioral endocrinology
have inspired more research than those concerning the physiological causes and
consequences of dominance (reviewed in Abbott *et al.* 2003). Although we have
made great strides in our understanding of the endocrinology of dominance in
recent decades, clear-cut answers to some of the field's most simple questions (e.g.,
"Which social ranks are most physiologically stressful to animals?"; "Are there
adverse physiological consequences of occupying stressful social ranks?") continue
to evade investigators (Sapolsky 2005). It is now clear that rank can mean different
things to different individuals (and at different points in their life) (Sapolsky 2005).

Interest in the physiological determinants and consequences of stressful social
status has not waned in recent years, but has instead grown steadily with the rise in
income inequality among humans and growing awareness that social status can
have significant impacts on individuals' health and quality of life (Sapolsky 2005).
Being socially complex models of human behavior, non-human primates remain

the ideal model system for studying these and other topics impossible (or unethical) to conduct in humans.

13.2.3 Endocrinology of stress

"Stress" can mean two things: (1) the internal/external and short-/long-term stimuli that disrupt homeostatic balance (i.e., "stressors") and (2) the behavioral and physiological coping mechanisms mounted in response to stressors (i.e., the "stress response"; Romero 2004). The stress response is often associated with the release of glucocorticoids (GCs) including cortisol, cortisone, and corticosterone. GCs are steroid hormones secreted from the adrenal cortex as part of the vertebrate stress response, a complex suite of behavioral and physiological events that help vertebrates mobilize energetic resources to cope with and respond to challenging or stressful stimuli (Sapolsky *et al.* 2000).

Elevations in GCs may be a consequence of heightened activity, arousal, or a sign of stress reactivity (Sapolsky *et al.* 2000). Moreover, recent evidence suggests that some GCs may be secreted not only in *response* to stressful events, but also in *anticipation* of them (Nguyen *et al.* 2008). Despite the recognized complexity of GC action in the body (Sapolsky *et al.* 2000), many researchers still narrowly view acute changes in GC levels solely as signs of physiological stress or as indicators of poor animal welfare. However, it is important to remember that short-term activation of the stress response is adaptive and helps vertebrates cope with fluctuating environmental conditions (Sapolsky *et al.* 2000). It is only when the stress response becomes a chronic condition that elevated GC levels can be viewed as pathogenic (Sapolsky *et al.* 2000), as prolonged activation of this response can deplete an organism's energetic reserves, with potentially adverse consequences for its health, immunity, neuroendocrine function, fertility, and survival (e.g., Sapolsky 2005; Blas *et al.* 2007).

Longitudinal monitoring of both GCs and behavior are needed to distinguish between acute (or adaptive) or chronic (or maladaptive) responses to environmental perturbations (Romero 2004). While measurement of GC levels provides a useful index of GC availability and activity in the body, GC-mediated behavioral and physiological responses also depend on the type of GC receptors present and the density and distribution of these receptors in the target tissues (Romero 2004), parameters that are virtually impossible to measure outside the laboratory.

It is now evident that investigators need to study more than one parameter when trying to draw conclusions about responses to (and consequences of) stressful events (reviewed in Moberg and Mench 2000). The increased use of corroborating evidence (e.g., adverse changes in behavior, metabolism, vigor, or immunity along with elevations in GCs; Moberg and Mench 2000) has the potential not only to correct past errors of interpretation of hormone–behavior interactions, but to also transform the field by providing important new insights into the full range of behavioral and physiological responses to stressful events.

13.3 Methodological considerations for field primate behavioral endocrinology

Hormones can be measured in a wide variety of biological materials (Table 13.1). Feces, by far the easiest of the materials to collect in the wild, remains the most commonly sampled material (Ziegler and Wittwer 2005), though urine may be the logical choice in some circumstances. In addition to choosing the sample material, researchers must also decide how often (and from whom) to sample, as well as choose from among an array of methods for sample collection, preservation, extraction, and analysis. In practice, the choices made will depend on the research question(s), logistics of the site, and species under study (Strier and Ziegler 2005). For example, studies of female reproductive physiology may require more intensive sampling (i.e., shorter inter-sample intervals) from a smaller number of individuals to fully capture the changes within and across different cycles. In contrast, studies comparing average hormone profiles across different environmental or social conditions may be amenable to sampling at longer intervals from a larger number of individuals.

Despite the ease with which hormones can now be monitored in the field, a number of methodological considerations must be taken into account to avoid common pitfalls and ensure meaningful, biologically relevant results that will lead to scientifically valid conclusions (Palme 2005). An excellent review of important points to consider when conducting studies of hormone–behavior interactions in the wild can be found in Palme (2005).

13.3.1 What endocrine values in different biological materials tell us

The lag time between hormonal stimulus and response (the interval between hormone release and action) can vary from seconds to days, with those responses that require gene transcription and protein synthesis (like most effects on behavior) needing the most time to take effect (Nelson 2011). Thus, some endocrine-mediated responses can occur hours or even days after the original hormone signal has disappeared from the blood. Hormones enter the urinary and gastrointestinal tracts from the bloodstream. Once metabolized, they will be excreted from the body in the urine and feces. Excreted hormones provide an integrative measure of

Table 13.1 Comparison of the advantages and limitations of the different biological materials from which hormones can be extracted.

Biological material	Hormones contained in material	Hormone concentrations in material reflect	Ease of collection	Impact of collection on animals	Ease of validation	Reference
Feces	Some **steroids** (or their metabolites)	Cumulative secretion and elimination of hormones over hours or days	Easy to collect repeatedly from terrestrial or arboreal animals	Little to none	Moderate to difficult	Ziegler and Wittwer 2005
Urine	Some **steroids** and **peptides** (or their metabolites)	Cumulative secretion and elimination of hormones over hours*	Moderately easy to difficult; can be collected most easily and frequently from arboreal animals	Little to none	Moderately easy	Knott 2005
Hair	Some **steroids**	Time averaged signal of hormonal activity; chronic hormone levels	Easy in humans; Difficult to obtain from wild animals, requiring capture and immobilization	Moderately to highly invasive	Moderately easy	Fourie and Bernstein 2011
Saliva	**Steroids** and some **peptides**	Moment-by-moment patterns of hormone secretion in blood, though at much lower levels	Easy in humans; Moderately easy in captive animals trained to give saliva, but difficult to obtain from wild animals	Moderately to highly invasive	Moderate to difficult	Heintz et al. 2011
Blood	**Steroids & peptides**	Moment-by-moment patterns of hormone production and secretion	Moderately easy in humans habituated to blood sampling, but can be difficult to obtain sufficient numbers of samples even from habituated subjects; Difficult to obtain from wild animals, requiring capture and immobilization	Moderately to highly invasive	Easy	Sapolsky 1993

* = urine has a shorter lag time than feces because animals urinate more frequently than they defecate (Whitten et al. 1998).

hormonal activity accumulated over several hours (in urine) or days (in feces) prior to excretion (Table 13.1; Whitten *et al.* 1998; Palme 2005).

The measurement of excreted hormones can help circumvent problems of data analysis and interpretation arising from the moment-to-moment variability in patterns of endocrine production and secretion typically seen in blood or saliva (Table 13.1). At the same time, diet can affect patterns of hormone secretion into urine and feces. To control for dietary differences between individuals and seasons, researchers should index hormonal values in urine (Munro *et al.* 1991) and feces (Wasser *et al.* 1993) following established protocols.

13.3.2 Preservation methods without electricity

Undoubtedly the best method for measuring hormones in a biological material is to analyze the material immediately after collection (Palme 2005). However, this is not logistically feasible at most field sites, so samples are usually stored until they can be transported to a laboratory for chemical analysis. Since hormones can be degraded by bacteria in the sample (Palme 2005), proper preservation in the field is essential for minimizing error or bias in hormone measurement caused by variation in sample storage condition or duration. Typically, urinary samples are frozen shortly after collection in freezers, or using liquid nitrogen or dry ice (Knott 2005) while fecal samples are dried by heat or lyophilization, or immersed in disinfecting chemical preservatives (e.g., alcohols or acids) (Ziegler and Wittwer 2005) before transportation to the laboratory for further processing. A reliance on energy-consuming freezers or lyophilizers or on access to liquid nitrogen or dry ice to preserve samples in the field has likely prevented measurement of excreted hormones at many remote field settings.

A small but growing number of alternative methods for preserving or processing samples in the field without freezing have emerged in recent years (Beehner and Whitten 2004; Knott 2005). These methods, which include drying urine on filter paper (Knott 2005) and extracting steroids from feces using solid-phase extraction cartridges in the field (Beehner and Whitten 2004), have the potential to greatly enhance the number and variety of species and populations represented in future field studies of hormone–behavior interactions.

13.3.3 Sample analysis considerations

Hormones must be extracted from some biological materials (e.g., feces, blood, saliva) prior to analysis to ensure accurate measurement of hormone levels without interference from the sample's matrix. Extraction functions to remove particulate matter that may interfere with the assay and to concentrate/isolate the hormone(s) of interest in a chosen solvent. Fecal steroids, for example, must be extracted from their solid matrix into a liquid solvent prior to assay (Palme 2005). Materials that are preserved in alcohol will immediately begin hormone extraction, so care must be taken to standardize methods for sample storage duration (and condition) and to avoid any loss of the chemical preservative (this may result in loss of hormones from the sample; Palme 2005).

Fecal extractions are typically done in the laboratory, though researchers are beginning to experiment with methods for extracting samples in the field, which avoids the necessity of freezing or drying samples in the field (Beehner and Whitten 2004). As a final step before analysis, some extracts must be further concentrated and purified using solid-phase extraction procedures (e.g., Beehner and Whitten 2004).

Once samples are ready for analysis, immunoassays are undertaken to measure the amount of the hormone of interest in the sample. Immunoassays can measure minute quantities of chemicals in a sample (Wild 2005), and are the preferred method for measuring the small amounts of hormones present in urine and feces. Competitive immunoassays measure the presence or concentration of an unlabeled substance (e.g., the hormone of interest) by observing how it competes with a labeled substance (e.g., a compound similar in structure to the hormone of interest labeled with radioactive isotopes or enzyme markers) for binding to an antibody (Wild 2005).

Radioimmunoassays (RIAs) are the more sensitive and thus popular of the two commonly used assays, although they cost more per sample and require special licenses for the use and disposal of radioisotopic tracers (Wild 2005). Enzymeimmunoassays (EIAs), on the other hand, are comparatively less expensive than RIAs and do not require any special licenses or many expensive reagents or equipment (Wild 2005). Traditionally, EIAs detected fewer of the hormones of interest in a sample than RIAs, though their popularity has grown in recent years as methodological refinements have made some EIAs more sensitive than in the past (Palme 2005). In practice, the choice of immunoassay will depend on the facilities and equipment available in the laboratory where the assays will take place as well as on the infrastructure available at the field site for sample preservation and processing, the type of material collected, and the number of samples.

It is critical that immunoassays be validated for the biological material chosen for each species to ensure that the results obtained from the assay reflect meaningful, biologically relevant phenomena rather than the influence of the applied sampling procedure(s) (Palme 2005). Given the diversity of validation techniques currently available (both chemical and biological), researchers should have little difficulty carrying out the necessary validations for the species and sample matrix and providing this information (or citing a published reference that provides this information) in published manuscripts.

13.4 Future directions in the behavioral endocrinology of wild primates

To date, most studies of hormone–behavior relationships in wild primates have relied on cross-sectional comparisons across groups or populations (Strier and Ziegler 2005). Although such studies provide important benchmarks and comparative data for future research, it is becoming increasingly evident that the personal and family history of an individual can affect its bodily functions which may, in turn, affect its behavior (e.g., Virgin and Sapolsky 1997; Nguyen *et al.* 2008; Onyango *et al.* 2008), possibly through modification of its epigenome (e.g., Weaver et al. 2004). As our understanding of the influence of age, sex, reproductive condition, environment, life experiences, family history, and so on, on hormone–behavior relationships grows, we must endeavor to more rigorously evaluate how (and in what manner) these variables can influence bodily function and behavior. The lessons that we learn from these future studies will not only shape our understanding of the biological basis of behavior itself, but will also add substantially to our knowledge of ourselves.

In this review, I have endeavored to highlight the major milestones and future directions in the field of primate behavioral endocrinology to inspire investigators interested in hormone–behavior interactions in wild primates to carry on the tradition of innovative and rigorous scientific research initiated by Robert Sapolsky and his predecessors. Field researchers face numerous logistical obstacles that are virtually absent from the laboratory environment. Nevertheless, the advantages of studying hormone–behavior interactions in the selective environments in which these interactions likely evolved clearly far outweigh their challenges.

14

Population genetics, molecular phylogenetics, and phylogeography

Mary E. Blair and Alba L. Morales-Jimenez

14.1 Introduction

Molecular genetic studies of primate ecology and conservation are growing in number due to advances in technology, economy, and feasibility including the improved ability to use low quality samples (Di Fiore 2003; Vigilant and Guschanski 2009). Molecular approaches allow our research questions to not only address patterns, but also the ecological and evolutionary processes behind those patterns, improving our empirical understanding of the biology of populations and species. Elucidating the processes that derive variation will also be essential to enable the conservation of natural populations that are capable of coping with continued environmental change (Frankham *et al.* 2002).

Major categories of genetic techniques in primate ecology and conservation include molecular phylogenetics—the study of evolutionary relationships among groups of organisms; phylogeography—the study of the historical processes that may be responsible for the contemporary geographic structuring of genetic signals within species or among closely related species; and population genetics—or the study of allele frequency distribution and change in populations. These techniques can be employed to answer or aid in answering a wide range of research questions related to primate ecology and conservation.

Our goal in this chapter is to provide a comprehensive synthesis and evaluation of molecular techniques in the study of primate ecology, evolution, and conservation, addressing how molecular tools can be used to answer the questions we list above in addition to many others. We discuss (1) how to obtain, preserve, and transport samples for genetic analysis, (2) laboratory techniques for DNA extraction, genotyping, and sequencing, and (3) data analyses relevant to research questions at the species level and the population level. Throughout we discuss

Primate Ecology and Conservation: A Handbook of Techniques. First Edition. Edited by Eleanor J. Sterling, Nora Bynum, and Mary E. Blair. © Oxford University Press 2013. Published 2013 by Oxford University Press.

the relevance of these methods to various research questions related to primate conservation as well as ecology and evolution.

14.2 Obtaining samples for genetic analysis

For sample collection and storage in the field, readers can refer to Chapters 3 and 4 of this volume for detailed guidelines. However, to collect samples for genetic analysis in particular, special consideration should be paid to the method of preservation in the field, especially for low-quality non-invasive samples such as feces. *RNA Later* solution (Ambion) is a particularly successful storage method for fecal samples. Feces can be stored in *RNA Later* at room temperature for several weeks, $-4°C$ for months or years, and at $-20°C$ indefinitely. Similar success has been reported using silica gel beads to dry fecal samples, or, for ape feces, a short storage period in 99% ethanol followed by drying in silica gel (Nsubuga *et al.* 2004). Although non-invasive sample collection makes it potentially possible to collect samples from individuals without ever seeing them, the fresher the sample the greater the success of extraction. Hair samples, with follicles intact, are often collected from elusive animals that make nests (i.e., gorillas, owl monkeys). These samples must be stored in a dry environment, such as small glassine envelopes or re-sealable bags, and can be stored at room temperature for months or at $-4°C$ for years (Bradley *et al.* 2008).

Several research, collection, and import/export permits as well as approved protocols (such as animal care, i.e., IACUC) are necessary to collect or transport biological samples, as explained in Chapters 3 and 4. However, many permits are not necessary for the transport of synthetic DNA, meaning any post-PCR (polymerase chain reaction) products.

Sampling strategy is very important to consider for genetic analyses, including the spatial scale of sampling in the field as well as sample size. For studies of population genetics and dispersal, there is general agreement that genetic samples should be taken at a spatial scale at least as large as the dispersal distance of the study species. However, researchers should also consider extended spatial scales given that they may not fully understand the scale of gene flow in their study species (Segelbacher *et al.* 2010). Also, a simulation study showed that it is more important to maximize the number of individuals sampled within a population rather than the number of populations sampled (Goudet *et al.* 2002). For phylogeographic and phylogenetic analyses, the number of samples and populations necessary to resolve relationships among groups depends on the amount of polymorphism and the extent of divergence. If most variation occurs among groups, then small sample sizes per group are appropriate. It is important to include multiple populations of closely related species, particularly if a non-recombining marker is being used (Baverstock and Craig 1996). For phylogenetic analysis the number of species that need to be included to obtain an accurate phylogeny is very important, as phylogenetic inference is susceptible to the number and phylogenetic distribution of the included species (Albert *et al.* 2009). It has been suggested that researchers should include replicates of sampled taxa one level below the level of inference (Baverstock and Craig 1996).

Researchers may also wish to consider using ancient or historical DNA samples, such as museum specimens, for genetic analysis. These types of samples are increasingly useful even for questions of ecology and conservation with advances in technology (e.g., Tracy and Jamieson 2011), or to boost sample size. To collect these samples from museum specimens one needs to apply for destructive sampling permission and also import/export permits if from an international museum. Samples can also be obtained from zoos, primate research facilities, or tissue repositories such as the Integrated Primate Biomaterials and Information Resource (IPBIR), the Gene Bank of Primates, and the European Primate Network (EU-PRIM-NET). Primate geneticists also often data-mine in NCBI's GenBank and other publically available resources. This can be especially useful for outgroup sequences in phylogenetic and phylogeographic analyses. However, it is important to check GenBank sequences for errors and potential nuclear pseudogenes listed as mitochondrial markers ("numts").

14.3 In the laboratory

The following section assumes intermediate knowledge of genetic analysis. For beginners, we refer readers to a basic and more comprehensive introduction such as Avise (2004).

14.3.1 DNA extraction

Manufactured DNA Extraction Kits (e.g., Qiagen, Aquagenomic) are now widely available as well as economically feasible for a variety of sample types. Phenol chloroform extraction is also widely used in places where commercial kits are unavailable or too expensive to import. The manufacturer's protocols are typically very successful for high-quality samples such as blood or tissue. However, certain sample types including non-invasively collected fecal samples and historical tissues often require modifications to these protocols to ensure success, as we outline below. Also, multiple negative extractions (extractions done parallel to the samples of interest but with no source of DNA) should be employed for both fecal samples and historical tissues to screen for contamination during the extraction process.

DNA extracted from feces is typically of low quality and thus some trouble-shooting is necessary. The success of DNA extraction from fecal samples may depend partly on the diet and the physiology of the focal species; researchers often have greater success extracting DNA from the feces of frugivorous or insectivorous primates as opposed to folivorous primates. When using DNA Extraction Kits for fecal samples, it often helps to modify the manufacturer's protocols to increase DNA yield, for example by extending the time period of cell lysis to as long as 24 hours, increasing the amount of proteinase K and extending incubation during protein digestion for up to 30 min, and extending the incubation step immediately before DNA elution to up to 30 min. The volume of the final elution can be smaller than suggested by the manufacturer to have a more concentrated DNA, from 200 ml to 75 ml.

For DNA extraction from historical tissues, a separate laboratory workspace is necessary that includes independent sets of reagents in order to reduce the risk of contamination from exogenous DNA and PCR amplified products. Standard historical DNA protocols (e.g., Wandeler *et al.* 2007) typically require soaking historical tissues in a bleach bath (10% bleach for 5 min) followed by several subsequent water washes prior to DNA extraction.

After extraction from low-quality samples, it is often useful to utilize a real-time quantitative PCR (qPCR) method to quantify the amount of primate DNA present in each sample (e.g., Morin *et al.* 2001). Typically, a concentration of at least ~0.5 ng/μl is necessary to amplify microsatellite markers successfully. For amplification of DNA for sequencing, a concentration of 10–100 ng/μl has been suggested as ideal. However, smaller concentrations could result in amplification if using Hot Start and/or increasing the number of cycles (see Section 14.3.3). If samples do not have a high enough concentration to proceed with PCR amplifica-tion of the marker of interest, the options are to re-extract the sample with double concentration at the elution step, or desiccate the DNA sample and re-suspend in a smaller volume of buffer.

14.3.2 Choosing appropriate markers

Before planning a study, it is important to consider the type of genetic markers one should use to inform a particular research question. In general, markers with faster rates of substitution (e.g., microsatellites or non-coding mitochondrial markers) are more appropriate for analyses within populations, while markers with slower rates of substitution (coding mitochondrial genes like *COX1* or *cyt-b*, nuclear introns, and Single Nucleotide Polymorphims [SNPs]) allow for the appropriate resolution for between population or species comparisons. Other very slowly evolving markers are appropriate for deep-time comparisons (exons).

Many markers have already been identified as informative for the study of genetic variation in a wide variety of primate taxa and populations. For example many SNPs have been identified and amplified across a wide range of old world monkeys (Malhi *et al.* 2011). A program is available to search all GenBank entries for Alu elements, a family of short interspersed elements (SINEs) caused by retrotransposons, that may be informative for particular primate taxa (Bergey 2011). Microsatellite markers have also been developed and published for a wide range of non-human primate taxa (Di Fiore 2003).

In many cases, and especially for rare or poorly studied species, it will be necessary to design and test primer sequences to facilitate the PCR amplification of variable markers of interest. For individual identification using microsatellite markers, it is important to test that a marker is polymorphic at the individual level in the species or population of interest. Primers can be designed using software like Primer3 or Primaclade for multiple taxa (Gadberry *et al.* 2005). In order to design primers it is important to use the genome of a close relative(s). To test the efficiency and dimerization probability of different primer pairs, the software Amplify3 can be used.

14.3.3 PCR amplification

PCR (polymerase chain reaction) amplifies specific regions of DNA to allow researchers to generate enough copies of a particular segment of DNA for subsequent manipulation and characterization. A typical PCR reaction includes deoxynucleotide triphostphates (dNTPs), $MgCl_2$, a buffer, a forward and reverse primer, and DNA polymerase enzyme. Sometimes trying different enzymes is necessary to get better results (e.g., AmpliTaq Gold [Invitrogen], HotStar [Qiagen], KOD [Toyobo]). If secondary compounds or inhibitors may be an issue, Bovine Serum Albumin (BSA) can be added to the reaction. Commercially manufactured PCR kits, mixes, or beads are often very successful, especially for multiplex microsatellite amplification (see Section 14.3.4), but they can be expensive. It is often cheapest to buy PCR reagents in bulk separately, which will also allow for easier manipulation of reagent concentration to optimize the performance of primers.

Researchers typically test a range of annealing temperatures to achieve optimum amplification of primers. Optimization is often necessary not only for newly designed primers but also for universal primers or primers isolated from species other than the focal taxon. To improve PCR amplification from low-quality samples, it helps to add more cycles to the PCR program or prolong the last step (extension), or to increase the amount of template DNA. Touchdown PCR, hot start PCR, or two-step PCR (Arandjelovic *et al.* 2009) methods are also quite successful for low-quality samples. After PCR amplification, gel electrophoresis is used to analyze the quality of the reaction. A good quality sample (positive control) should be amplified and run in the same gel as the samples of interest in order to compare their quality.

14.3.4 Microsatellite genotyping

Studies have suggested that between 10–20 (and probably closer to 20) micro-satellite markers should be used to distinguish between individuals within populations, especially in outbred populations (Csillery *et al.* 2006). Thus, it is often most economically viable to multiplex between two and five markers in a single PCR, if products will be visualized on an automated fluorescent sequencer (e.g., ABI 3730 DNA Analysis System). Markers can be multiplexed either using different fluorescent dye tags or, if they have non-overlapping size ranges, the same dye tags. Because alleles vary in the number of nucleotide repeats, they are typically scored by their size in base pairs or the number of repeats. Thus, PCR products can be assayed directly by either gel electrophoresis, where heterozygous individuals will show two PCR product bands of different sizes, or, if on a capillary sequencer, heterozygous individuals will show two peaks of fluorescence at different sizes. A size standard should be chosen that is appropriate for the size range of the markers or set of markers in a given PCR. On an automated sequencer, genotypes can be called automatically using software such as GeneMapper (ABI). It may be important to include a variety of markers with di-, tri-, and tetra-nucleotide repeat motifs, since allele-calling errors and stutter are common with di-nucleotide repeats.

A related and very useful technique for the study of primate ecology and conservation is a PCR-based sexing assay, which uses a single, multiplex PCR to amplify fragments of the amelogenin X gene and the Y-linked sex-determining region (SRY); heterozygous genotypes for this assay will thus be males, while homozygous genotypes will indicate females (Di Fiore 2005).

When using fecal or other low-quality DNA samples for microsatellite genotyping or sex determination, allelic dropout may occur, where a truly heterozygous individual is assayed as homozygous because one allele preferentially amplified over the other due to the low quality of the template DNA. Pre-amplification methods can reduce the likelihood of allelic dropout (Piggott *et al.* 2004), but as a precaution researchers should confirm homozygous genotypes several times (perhaps 7–10) when using low-quality template DNA. Also, before moving forward with data analyses of microsatellite genotypes even from high-quality samples, it is important to check whether any of the markers are in linkage disequilibrium, whether they deviate from Hardy–Weinberg equilibrium (HWE), or whether there might be any null alleles.

14.3.5 Cycle sequencing and sequence analysis

Before sequencing, the amplified product should be cleaned to eliminate primers, dNTPs, and to inactivate enzymes, which could potentially interfere with the sequencing reaction. There are several kits available for PCR cleanup (e.g., Qiagen PCR purification kit, ExoSap-IT protocol (Affymetrix)).

After cleanup, samples can be cycle-sequenced. A cycle sequencing reaction is a modified PCR reaction that uses only one primer (forward or reverse) in each reaction and includes fluorescently labeled ddNTPs that interrupt the extension of DNA when incorporated.

Ethanol precipitation is a technique to concentrate and purify DNA after the cycle sequencing reaction. Salt and ethanol are added to the cycle sequencing product to precipitate the DNA and separate it from other components by centrifugation. There are different protocols and reagents that can be used for this procedure; the most commonly used salt is sodium acetate. After the purification and concentration of the DNA, samples are re-suspended in formamide to be read by a sequencing analyzer.

14.3.6 Mitochondrial DNA: controlling for "numts"

One problem using mitochondrial DNA (mtDNA) is the possible amplification of nuclear insertions also known as "numts." The presence of numts has been identified in a variety of non-human primate species and can affect the phylogenetic analysis of mtDNA (Bensasson *et al.* 2001). Symptoms of numt contamination are ghost bands in the electrophoresis of PCR products, extra bands in restriction profiles, sequence ambiguities, frameshift mutations, stop codons, and unexpected phylogenetic placement. Because numts are usually short (100–300 bp), one way to avoid amplifying numts is to produce long-range amplicons before amplifying the marker of interest. Long-range PCR (LRPCR) amplification typically results in products 5–20 kb in length. There are several commercial kits that can be used for LRPCR, but high quality samples are often necessary for successful LRPCR amplification. Some numts can be thousands of bases in length, and so even when using LRPCR it is important to translate all mtDNA protein-coding genes to look for premature stop codons and frameshifts indicative of a numt sequence. The best way to ensure against the amplification of numts is to produce two overlapping LRPCR amplicons that together cover the entire mtDNA genome, ensuring circularity (Thalmann *et al.* 2004).

14.3.7 Sequence quality and base calling

Sequences can be retrieved from the sequencing analyzer, assembled, and studied for quality. Programs such as Sequencher (Gene Codes), Geneious (Bioware), or 4peaks (<http://www.mekentosj.com>) can be used to check the quality of the sequences; these software packages provide the researcher with the tools to evaluate the chromatograms and make edits when needed. Researchers typically also use these packages to trim away the sections of the sequence near the primer site and towards the end of the sequence, which are usually of very poor quality. Forward and reverse sequences can also be aligned to make sure the sequences are the same and check for any inconsistencies.

14.3.8 Sequence alignment

Once sequences are cleaned and edited, they can be aligned for future analysis using programs such as Clustal (Sievers *et al.* 2011) or MUSCLE (Edgar 2004). Clustal uses three steps: (1) it performs a pairwise alignment of all the sequences, (2) it uses the alignment scores to produce a neighbor-joining tree, and (3) it aligns the sequences sequentially, guided by the phylogenetic tree generated from step 2. MUSCLE uses a faster method to compute pairwise distances, which is convenient for larger sample sizes, followed by progressive alignment. Some researchers check the alignment given by the software and make corrections "by eye." This practice can cause problems when replicating a study, unless an explicit criterion is used (Morrison 2009) and therefore we recommend including the criterion used and/or the final alignment in the write-up of any study that includes "by eye" adjustments.

14.3.9 Next-generation sequencing

Next-generation sequencing technologies promise to revolutionize genomic research, allowing a typical researcher to produce exponentially greater amounts of sequence information than was previously possible. The price of using this technology is decreasing, and target enrichment techniques have been developed that may expand the utility of next-generation sequencing methods even for low-quality samples such as fecal DNA or historical DNA from museum specimens (e.g., Maricec *et al.* 2010; Mamanova *et al.* 2010). Next-generation technology in particular may allow researchers to incorporate studies of adaptive genetic diversity to explore emerging research questions (see Section 14.5). As multilocus analyses have become the norm, next-generation sequencing is very likely to become the standard genetic analysis technique in the near future for applications of molecular methods to ecology and conservation. However, before widespread adoption of these techniques, we must negotiate current limitations such as the cost and use of different genome library preparation methods and also data storage and downstream bioinformatics.

We refer readers to several recent review articles and online guides for the details of different next-generation sequencing pipelines used for population genetics and phylogenetic analyses (Shendure and Ji 2008; Mamanova *et al.* 2010; Glenn 2011; SEQanswers wiki <http://seqanswers.com/>). However, some important issues to consider include: (1) which platform (e.g., Illumina versus 454) would be the most appropriate for your research question and loci of interest (Glenn 2011); (2) what library preparation and target enrichment strategy would be most appropriate for your research question (Mamanova *et al.* 2010); and (3) how will you manage the downstream bioinformatics (e.g., Hird *et al.* 2011)? Another crucial issue is how to manage costs while negotiating the trade-off between missing data (a lack of sufficient read coverage per individual and/or locus) and the number of loci or individuals you can analyze in a given sequencing run.

14.4 Data analysis at the species-level

14.4.1 Phylogenetics and systematics

Systematics is the biological field that detects, describes, and explains the diversity of the biological world. Systematics involves two different tasks: taxonomy and phylogeny. Taxonomy involves the description of organisms as well as their assignation to a category under the Linnaean system. Phylogeny studies the relationships among organisms using mutual similarities (phenetics) or shared evolutionary ancestry (cladistics). Cladistics is widely accepted because it infers relationships on the basis of homology rather than homoplasy.

In theory, taxonomy should reflect the phylogenetic relationships of a group. However, at least 22 species concepts have been proposed (Mayden 1997). Under the biological species concept (BSC; Mayr 1963), species are interbreeding populations that are reproductively isolated from other groups. However, in mammals and other animals, reproductive isolation is often difficult to determine or observe, and in some groups of primates hybridization is common between species of the same genus and sometimes even between genera. The phylogenetic species concept defines species as the smallest diagnosable cluster of individual organisms with a parental pattern of ancestry and descent, separated from other clusters by a unique combination of character states (Cracraft 1983). Using genetic data, phylogenetic species can be defined through a combination of phylogenetic tree inference (described below) and population aggregation analysis (Davis and Nixon 1992), which groups taxa together based on the presence of shared, fixed characters.

Species-targeted conservation is the focus of influential conservation legislation and has facilitated fundraising and public awareness. Species may be targeted for several reasons (e.g., as flagships—see Chapter 16) and many conservation actions are based on the species approach including the design of protected areas, action plans, and listing (see Chapter 16). Listing of endangered species and action plans are conservation strategies that intend to prioritize those taxa that are most threatened. These strategies rely on the available taxonomic information and, if there are conflicting classifications, it is possible that some species could be ignored when setting conservation priorities and accurate analyses of their risk of extinction will be impossible.

Phylogenetic inference can give us hypotheses about the different lineages in a group of organisms and, if following the PSC, reciprocally monophyletic lineages may be considered species. However, phylogenetic inferences or trees need to be used with caution as they may be reflecting the evolutionary history of a particular gene and not necessarily of the species. Some gene trees do not show the evolutionary history of the species because of the presence of paralogy, incomplete lineage sorting, or horizontal transfer. Therefore a gene tree is telling the story of the gene history rather than the species history. Using multiple genes may improve the likelihood of inferring species history from multiple gene trees. However, the use of multiple genes can result in many different topologies and there is no straightforward way to choose among them and infer the species tree. It is important to examine different topologies in a geographic context and consider the possible evolutionary processes influencing each one.

Specific strategies for phylogenetic tree inference are widely debated, and we refer readers to other books for more detail (Felsenstein 2004; Hall 2011). After achieving a molecular sequence alignment, one must choose an appropriate model of molecular evolution for the data set (e.g., using software such as jModelTest; Posada 2008). Under the chosen model of sequence evolution, one can begin phylogenetic inference using maximum likelihood or Bayesian inference (see Table 14.1 for software to implement each of these and also maximum parsimony). It is very common to run several phylogenetic inference methods on the same data set to check for congruence across methods. Some researchers also use distance-based approaches such as UPGMA or Neighbor-Joining, which are simpler algorithms that do not require testing of different models of sequence evolution, since they incorporate models into the formulas for distances. Distance-based methods have some strict assumptions that could result in topological errors. For maximum parsimony and likelihood analyses, measures of nodal support are achieved through bootstrapping, typically initiated from a random starting tree. For Bayesian inference, clade credibility values are generated using a Markov chain Monte Carlo (MCMC) process; thus, it is extremely important to ensure that a Bayesian analysis runs long enough to reach convergence. Software such as Tracer (Drummond *et al.* 2006) can be used to infer convergence, but it is also helpful to initiate independent runs with simultaneous heated and cold chains and chain swapping, especially for difficult data sets. Taxon instability analysis can help to determine if a particular taxon might be driving low nodal support (Mesquite; Maddison and Maddison 2011).

Bayesian relaxed clock approaches are increasingly used to estimate divergence dates among taxa (e.g., BEAST; Drummond *et al.* 2006). These approaches take into account rate variation across lineages, improving divergence date estimates compared to other methods. However, the major limitation to the accuracy of these estimates is still the fossil-based calibration of the "molecular clock," which is intrinsically limited by the fossil record. Researchers can improve estimations of uncertainty by using date ranges for calibration, and perhaps also by enabling tip dates, which is possible in BEAST and allows for the incorporation of historical samples within a statistical framework.

In addition to defining species for accurate assessments of extinction risk and conservation priority setting, a key usage of molecular-based species identification is in "conservation forensics." DNA barcoding, or using a short genetic marker to identify a species, has been proposed as a method to help agencies charged with the enforcement of wildlife trade legislation, such as CITES, by identifying species from even small or degraded bits of tissue confiscated as bushmeat or traditional medicines (see Chapter 18). A recent study used universal primers for *COX1*, a mtDNA marker, as a DNA barcode to successfully distinguish between commonly hunted and traded African and South American mammals and reptiles, including several primates (Eaton *et al.* 2010).

14.5 Data analysis at the population-level

14.5.1 Phylogeography

Phylogeography is the study of the historical processes that may be responsible for the geographical distribution of genealogical lineages within genera, species complexes, or especially within species. Phylogeography bridges molecular phylogenetics and population genetics, focusing on the historical or phylogenetic components of population structure, or phylogeographic structure, across species' ranges. We discuss some methods for phylogeographic analysis that are particularly useful for ecology and conservation below, but for more detail and background on phylogeographic methods we refer readers to Avise (2000) and Hickerson *et al.* (2010).

A particularly important application of phylogeography is defining biologically meaningful units of diversity within species. Different levels of genetic evidence will suggest different temporal scales of phylogeographic structure; for example, high bootstrap support (or high clade credibility in Bayesian analyses) for an intraspecific phylogroup at one genetic marker is evidence for structure. However, similar phylogeographic structure across *several* unlinked genetic markers suggests a genomically extensive phylogenetic split, probably representing deeper, meaning older, structure.

Intraspecific phylogroups are sometimes called Evolutionarily Significant Units (ESUs), defined as monophyletic groups of genetically differentiated populations within a larger monophyletic species (Ryder 1986; Vogler and Desalle 1994). ESUs are often characterized by reciprocal monophyly at mtDNA loci with significant divergence of allele frequencies at nuclear loci (Moritz 1994). It has been suggested that ESUs should be defined more broadly using the concepts of genetic and ecological exchangeability, which are rejected when there is evidence for ecological or genetic differentiation between populations (Crandall *et al.* 2000). Thus it is important to take a phylogeographic approach and consider ESUs in a geographic context to determine if geographic barriers might exist between phylogroups that could support a lack of ecological exchangeability. Indeed, ESUs are typically spatially oriented towards known or suspected Pleistocene refugia (Avise 2004).

Confirming the distinctiveness of ESUs within a species is essential to inform conservation management as these units are historically isolated lineages that cannot be recovered if lost. There are many examples where molecular analyses identified deep phylogeographic separation between some populations that were previously considered panmictic, and in some cases these analyses came too late to prevent the extinction of unique populations (Avise 2004; DeSalle and Amato 2004).

Comparative phylogeography examines phylogeographic structure across a range of multiple co-distributed species. If similar patterns of structure are found across many species, and especially if these patterns fit with non-molecular data including morphological, behavioral, or geological differences, then we can infer very deep phylogeographic structure that may drive regional patterns of endemism. Such information can be used to target conservation efforts towards each unique pocket of phylogeographic diversity, sometimes called an area of genetic endemism (e.g., for macaques and toads on Sulawesi; Evans *et al.* 2003). It is also very

Table 14.1 *Software packages and their use for molecular phylogenetics, phylogeography, and population genetics.*

Program	Utility and description
Sequencher	Sequence assembly, analysis and editing (Gene Codes)
Geneious	Sequence assembly, analysis and editing (Bioware)
4peaks	Sequence assembly, analysis and editing (<http://www.mekentosj.com>)
Clustal	Multiple sequence alignment (Sievers *et al.* 2011)
MUSCLE	Multiple sequence alignment (Edgar 2004).
PRGmatic	Next-generation sequencing; downstream bioinformatics (Hird *et al.* 2011)
jModelTest	Phylogenetics; testing models of DNA evolution (Posada 2008)
PAUP*	Phylogenetics; distance-based, maximum parsimony, and maximum likelihood (Swofford 2002)
MrBayes	Phylogenetics: Bayesian inference (Ronquist and Huelsenbeck 2003)
RAxML	Phylogenetics; maximum likelihood (Stamatakis *et al.* 2005)
PHYML	Phylogenetics; maximum likelihood (Guindon and Gascuel 2003)
Tracer	Bayesian approaches; testing for convergence (Drummond *et al.* 2006)
Mesquite	Phylogenetics; taxon instability analysis (Maddison and Maddison 2011)
BEAST	Phylogenetics; Bayesian inference and Bayesian relaxed clock divergence dating (Drummond *et al.* 2006; Drummond and Rambaut 2007) and Phylogeography; phylogeographic diffusion model (Lerney *et al.* 2009)
S-DIVA	Phylogeography; dispersal–vicariance analysis (Yu *et al.* 2010)
Phylomapper	Phylogeography (Lemmon and Lemmon 2008)
ARLEQUIN	Population genetics; general (allele and genotype frequencies, population subdivision, gene flow and genetic distances, assignment tests; Excoffier *et al.* 2005)
KINSHIP	Population genetics; pedigree relationships and parentage assignment (likelihood-based; Goodnight and Queller 1999)
CERVUS	Population genetics; parentage assignment (likelihood-based; Marshall *et al.* 1998)
STRUCTURE	Population genetics; population subdivision and recent migration (Bayesian; Pritchard *et al.* 2000; Falush *et al.* 2003)
TESS	Population genetics; population subdivision and spatial analysis (Bayesian; Chen *et al.* 2007)
BAYESASS	Population genetics; recent gene flow inference (Bayesian; Wilson and Rannala 2003)
IMa2	Population genetics; multi-population estimation of historical effective population sizes and gene flow among populations (Hey 2010)
BOTTLENECK	Population genetics; tests for genetic signatures of a recent population bottleneck (Piry *et al.* 1999)
GIMLET	Population genetics; population size estimation for capture-mark-recapture studies (Valiere 2002)
Capwire	Population genetics; population size estimation for capture-mark-recapture studies (Miller *et al.* 2005)
CIRCUITSCAPE	Population genetics; isolation by resistance (McRae 2006; McRae and Shah 2009)
CDPOP	Landscape genetics; spatially explicit individual-based population modeling in complex landscapes (Landguth and Cushman 2010)
R	Numerous packages are available for population genetics, phylogenetics, and phylogeography (e.g., Geneland for landscape genetics Guillot *et al.* 2005; R Core Development Team 2011)

important for conservation plans to incorporate species with unique phylogeographic histories that do not follow typical patterns.

There are many software packages available to test different models of biogeographic history (e.g., for the ancestral origin of a group in a particular region, by dispersal, or by vicariance) and visualize phylogeographic data in a spatial context (Table 14.1). S-DIVA (Yu *et al.* 2010) is a software package that has made improvements over earlier programs for dispersal–vicariance analysis by accounting for uncertainty in parameter optimization and phylogeny. Phylogeographic diffusion models can also be tested under Bayesian inference (BEAST; Lerney *et al.* 2009) or likelihood-based methods (PhyloMapper; Lemmon and Lemmon 2008). These programs sample time-scaled phylogenies while accommodating phylogenetic uncertainty and alternative hypotheses about ancestral origins, spatial dynamics, and dispersal patterns.

Statistical phylogeographic approaches are strengthened when considering a wide range of plausible models. This can become cumbersome under a likelihood approach, but a family of methods called approximate Bayesian computation (ABC) are "likelihood-free," bypassing the computational difficulties of calculating likelihood functions. ABC methods simulate data from a coalescent model using parameter values randomly drawn from a prior distribution under an array of biogeographic hypotheses. Using sets of summary statistics, one can choose the simulated data set and therefore the biogeographic hypothesis that most closely matches the observed data set (Hickerson *et al.* 2010; Hickerson and Meyer 2008).

14.5.2 Population genetics and molecular ecology

In population genetics, we examine genetic diversity among individuals and groups in a population. Molecular ecology refers to the application of population genetics to answer research questions related to ecology, including: How are individuals related within social groups? How is reproduction partitioned among members of social groups? How many individuals are there? Has population size changed in the recent or historical past? How are populations connected to each other via dispersal, and how are dispersal patterns influenced by the local ecology of landscapes? For population genetics and studies of relatedness within and among populations and groups, microsatellite markers and SNPs are most often utilized because they allow for identification at the individual level. Hypervariable markers of the mtDNA genome are also useful to study patterns along matrilines.

Several estimators of relatedness can be used to compare patterns across groups and populations (e.g., Lynch 1988; Queller and Goodnight 1989). Numerous unlinked loci with high levels of heterozygosity yield the best estimates of pairwise relatedness. Examining these patterns can generate and test predictions based on socioecological hypotheses about primate social relationships and affiliative behavior towards kin (e.g., see Chapter 9). Of particular interest to researchers studying primate behavioral ecology at long-term field sites are paternity determinations, which can be made using genetic data through parentage exclusion analyses or likelihood-based parentage inference (see Table 14.1; Chakraborty *et al.* 1988; Marshall *et al.* 1998). Likelihood-based parentage inference is typically preferred over exclusion analyses because it allows for some genotyping errors and the possibility that some potential parents may not have been sampled.

A key analysis for both ecological and conservation research questions is genetic capture-mark-recapture (Vigilant and Guschanski 2009; and see Chapter 2, Section 2.4). Several software programs have been developed to incorporate genetic data into population size estimates (Table 14.1) and some allow for studies where sampling has taken place in one session. Thus, when extensive sample collection is possible, genetic censusing can be much more efficient than traditional census techniques (see Chapter 2). A more recent study has captured spatial information in a genetic capture-recapture analysis of wildcats to estimate population density across a landscape (Kery *et al.* 2011).

An important question, both for conservation management and ecology, is how the size of a population may have changed over time. Signatures of a recent population bottleneck can be inferred by comparing observed and expected heterozygosity excess (implemented in the program BOTTLENECK, Table 14.1). If heterozygozity is higher than expected across a majority of markers in a population, it may have recently experienced a genetic bottleneck (Piry *et al.* 1999). Coalescent simulation-based Bayesian methods can also be used to infer historical population sizes and population splitting times (e.g., Approximate Bayesian Computation or IMa2, see Table 14.1). Researchers working with historical DNA can also make direct comparisons between current and historical population genetic structure and effective population size (e.g., Tracy and Jamieson 2011).

Measuring dispersal, population structure, and connectivity among populations is another extremely important use of population genetics for the study of primate ecology as well as conservation management. Traditionally, researchers have estimated population genetic structure using summary statistics such as Wright's F-statistics. F_{ST}, for example, measures inbreeding due to the differentiation among sub-populations relative to the total population, specifically by measuring deviations of observed from expected values of heterozygosity. The value of F_{ST} ranges from 0 to 1, with a higher value corresponding to greater population structure. For F-statistics and other approaches to examine population genetic structure such as analysis of molecular variance (AMOVA), researchers must *a priori* define hierarchical levels within their data set (e.g., population, subpopulation, group).

Individual-based Bayesian models are particularly useful for studies of potentially structured populations because they allow the inference of structure with greater precision and without *a priori* information about population membership. Bayesian clustering approaches group individuals into populations of random mating individuals that minimize Hardy–Weinberg and gametic disequilibrium across the data set (Beaumont and Rannala 2004). As in Bayesian phylogenetic inference, Bayesian clustering methods search likelihood space using MCMC, so burn-in periods and other parameters must be carefully set to ensure convergence and avoid reaching local maxima instead of the global optimum. Individuals of unknown natal localities can also be assigned to their most likely population of origin, allowing the estimation of migration rates and even dispersal distances, assuming the population of origin has been sampled. Bayesian analyses have several advantages over F_{ST}-based analyses in studies of population structure because they identify individuals, can incorporate genotyping error, and many Bayesian clustering software programs now also explicitly include geographic information (see Table 14.1). However, different programs require accepting different assumptions, and the use of more than one program may increase the accuracy of this approach (Excoffier and Heckel 2006; Latch *et al.* 2006). Different assumptions also apply to programs that use Bayesian clustering methods to identify recent migrants or historical rates of gene flow (e.g., BAYESASS will only correctly estimate migration rates if $F_{ST} > 0.10$; Faubet *et al.* 2007). Another simulation study showed that several Bayesian clustering programs perform poorly when there is strong isolation-by-distance (IBD) in the data set (Safner *et al.* 2011). IBD refers to the process whereby populations differentiate because of impeded gene flow due only to an increase in geographic distance.

A general drawback to Bayesian approaches is the necessity of specifying prior parameter distributions, a problem similar to that of specifying populations *a priori* in frequency-based methods. Researchers attempt to bypass this issue by specifying non-informative or uniform priors, often the default for many clustering software programs, which hold little or no prior information about the parameters. However, priors should be specified when there is concrete prior knowledge about parameters, and can be particularly useful when testing whether any individuals are migrants to their supposed populations. When using priors, researchers should systematically examine the effects of different priors on the parameters estimated by the model.

As primates are found in increasingly heterogeneous and human-dominated landscapes, incorporating spatial information with genetic data is essential to better understand population processes and inform conservation management in these

Box 14.1 Landscape genetics and least-cost modeling

Landscape genetic approaches use spatially explicit models to examine how landscape features affect the spatial distribution of genetic variation. Least-cost modeling is a common method employed in the landscape genetic approach, which builds on the IBD framework. IBD analyses examine the correlation between measures of genetic distance among populations and the geographic distance separating those populations, typically using Mantel tests for matrix correspondence (Smouse *et al.* 1986).

In least-cost modeling, landscape heterogeneity is incorporated into calculations of "least-cost" dispersal routes using prior knowledge of species dispersal ability and habitat preferences combined with detailed information of landscape features using Geographic Information Systems (GIS) software (e.g., Fig. 14.1). Least-cost behavioral distances could also be calculated by taking into account social structuring and other social barriers to movement. Least-cost geographic distances are then compared to Euclidean linear geographic distances regarding how well they correlate with genetic distances, typically using Mantel tests. If genetic distances correlate more strongly with least-cost distances than with Euclidean measures, one can infer that landscape heterogeneity has some effect on the distribution of genetic variation. It is important to use partial Mantel tests for these comparisons, because least-cost and Euclidean distances are not independent of one another (Cushman and Landguth 2010). A disadvantage to using Mantel tests, however, is the inherent difficulty in choosing among closely related models or models with only slightly different Mantel's *r*-values, which may be within a reasonable margin of error of one another. Software is available to utilize these methods and other methods in landscape genetics with increased sensitivity (Table 14.1; see Balkenhol *et al.* 2009 for a review of statistical approaches). Of particular interest is an approach called isolation-by-resistance, which incorporates aspects of graph theory and electrical circuit theory (i.e., electronic resistance) to predict spatial genetic structure in complex landscapes (McRae 2006). For any study of landscape genetics, detailed GIS or remote sensing data will be necessary.

Some recent studies using the least-cost distance approach do show that landscape features may influence spatial genetic structure in some primates. In *Rhinopithecus bieti*, non-Euclidean distances incorporating the presence or absence of habitat gaps explained genetic variation better than Euclidean distances, although the landscape composition of those gaps was not considered (Liu *et al.* 2009). In sifakas (*Propithecus tattersalli*), large rivers but not roads are important barriers to gene flow (Quemere *et al.* 2010). In Central American Squirrel Monkeys (*Saimiri oerstedii citrinellus*), commercial oil palm plantations in the Central Pacific of Costa Rica have a moderate effect on genetic distances between individuals, while other matrix habitats such as cattle pastures or residential areas do not (Blair 2011; Blair and Melnick 2012).

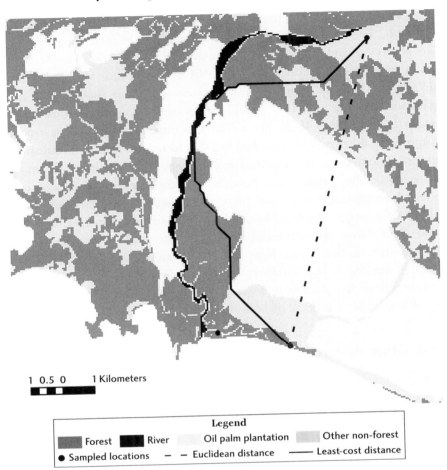

Fig. 14.1 Example of least-cost paths between sampled groups, where oil palm plantations were given a cost of 1000. For a python code to create cost distances in ArcGIS software [ESRI], see Etherington (2010).

complex arenas. Landscape genetic approaches are increasingly used to understand the influence of landscape patterns on dispersal patterns and population genetic structure. This methodological approach uses spatially explicit models to examine how landscape features affect the spatial distribution of genetic variation (see Box 14.1; Manel *et al.* 2003).

Finally, population genetics can be used to investigate patterns of sex-biases in dispersal, which represents an important factor in primate socioecological models and conservation management in terms of ensuring successful transfers, translocations, and reintroductions. Sex-biased dispersal affects the population genetic structure of different genetic markers differently based on their patterns of inheritance. The direction of sex-biased dispersal can be inferred using a combination of mtDNA and autosomal or Y chromosomal markers. For example, if only females disperse or disperse farther, mtDNA and autosomal loci should show similarly low levels of structure, Y chromosomal markers should show more significant structure, and IBD should be stronger among males than among females (Di Fiore 2009). If only males disperse or disperse farther, genetic structure is expected in the maternally inherited mtDNA loci, while little to no structure is expected in autosomal or Y chromosomal markers, and IBD should be stronger among females than among males. Also, the more dispersing sex should show lower corrected assignment index scores (AI_c), which represent the probabilities that individuals' genotypes occurred by chance in a population. However, simulation studies have shown that tests based on AI_c can only reliably detect sex biases in dispersal when the bias is quite large, and with exhaustive sampling (Goudet *et al.* 2002). A pattern of high structure across all markers and strong associations between genetic and geographic distances among both males and females would suggest low levels of dispersal by both sexes.

14.6 Other research questions

There are numerous other potential applications of genetic data to studies of primate ecology and conservation, more than we have space to discuss here. These applications could include genetic analysis of diet and seed dispersal through the collection of fecal samples, disease and host–parasite interactions (see Chapter 4), or determining appropriate pairs for captive breeding programs (see Chapter 17). Also, with the advent of next-generation sequencing and the possibility of rapid genome sequencing for large samples of individuals, studies of adaptive genetic diversity will become more feasible, such as the hereditary aspects of complex individual behaviors like temperament, personality, or cognitive ability (behavioral genetics), or the phenotypic plasticity of life history traits, longevity, and aging (ecological genomics).

Acknowledgments

We thank the editors and anonymous reviewers for their comments on this manuscript. We also thank our colleagues at the Molecular Primatology Laboratory at New York University. MB thanks D. Melnick, J. Munshi-South, and A. Goncalves da Silva for helpful input on population and landscape genetic approaches.

15

Demography, life histories, and population dynamics

Olga L. Montenegro

15.1 Introduction

Information on primate demography and life histories is fundamental for understanding primate evolution, ecology, and conservation issues. Primate life history data are used for testing evolutionary and socioecological hypotheses (Jones 2011; Wich *et al.* 2007), and even theories for the evolution of human life history (Ingicco *et al.* 2011; Wich *et al.* 2004). Primate population dynamics provide insights into the influence of social factors on classical population growth models, which is of academic interest. Also, primate conservation strategies seriously need demographic information to understand the synergistic effects of population decline and genetic deterioration, and to better predict outputs of diverse management alternatives. This chapter briefly reviews the main techniques used to address primate life history, demography, population dynamics, and extinction risk. The information below concerns both free-ranging primate populations and captive individuals.

15.2 Determination of demographic parameters

15.2.1 Sex and age determination

In species with sexual dimorphism (i.e., apes, baboons, macaques), the sex of adults is straightforwardly determined by distinct observable characteristics, but in other species, sex determination might require close examination of restricted individuals. Sex of infants and juveniles is difficult to assess in most species (Robbins *et al.* 2009). Precise age is also difficult to assess and age classes are used instead. Table 15.1 illustrates some commonly used methods to assign sex and age class in selected primate species.

Primate Ecology and Conservation: A Handbook of Techniques. First Edition. Edited by Eleanor J. Sterling, Nora Bynum, and Mary E. Blair. © Oxford University Press 2013. Published 2013 by Oxford University Press.

Table 15.1 *Commonly used methods to assign sex and age class in selected primate species.*

Species	Age determination criteria	Capture required	Source
Lemur catta (Ring-tailed lemur)	Weight, tooth eruption, and attrition and sexual characteristics.	Yes	Sauther *et al* (2002), Cuozzo and Sauther (2006)
Propithecus verreauxi verreauxi (Sifaka)	Tooth wear analysis from dental casts of captured individuals.	Yes	Richard *et al.* (2002)
Theropithecus gelada (Gelada baboon)	Size, coat color, muzzle shape, teeth replacement, canine eruption, cape and whiskers development, paracallosal skin (in females).	No	Dunbar (1980)
Pan troglodytes (Chimpanzee)	Individual recognition from birth. For adult females of unknown age, estimated age of her oldest offspring plus 13–15 years.	No	Nishida *et al.* (2003)
Gorilla gorilla (Western gorilla)	Behavioral and morphological markers (i.e., suckling and riding ventrally or dorsally in infant, head shape and size relative to body size, fur color, sexual swellings in females, other secondary sexual characters in both females and males). Age classes assigned to individuals of unknown age by comparison with photos and videos of gorillas of known age.	No	Breuer *et al.* (2010)
Pongo abelii (Sumatran orangutan)	Focal animal samples. Comparison with photographs of orangutans of known age. Developmental trajectory of all known-age animals, using photos taken at different ages, and recording ages at which changes occurred.	No	Wich *et al.* (2004)
Presbytis thomasi (Thomas' langur)	Weaning, age at first reproduction in females, and descent of testicles and loud calling in males, defined age classes (infants, subadults and adults).	No	Wich *et al.* (2007)

15.2.1.1 *Age determination from direct observation*

Age determination from direct observation requires an understanding of the changes in size, physical appearance, and behavior that an animal undergoes while growing. For example, characteristics such as size, coat coloration, muzzle shape, teeth replacement, canine eruption, pelage and whisker development, physical growth, and color of paracallosal skin in females are commonly used to assign approximate age to individual gelada baboons (Dunbar 1980). Long-term studies usually require several different observers throughout time, making standardization of age criteria necessary. Currently, studies on gorillas (Breuer *et al.* 2010) and orangutans (Wich *et al.* 2004) use photographs and records of changes occurring through development to standardize age class determination criteria. In combination with physical characteristics, behavior may be helpful to distinguish between age classes. For example, in Thomas' langurs (*Presbytis thomasi*), Wich *et al.* (2007) recognized adult males based on descent of testicles and emission of loud calls.

15.2.1.2 Age determination from captured individuals

15.2.1.2.1 Tooth emergence and wear

Dentition is one of the most useful characteristics for age determination in mammals. This is because patterns of tooth emergence and replacement show a consistent pattern throughout an animal's growth. The sequence of tooth emergence and attrition is very useful not only for primate life history but also for zooarchaeological, and paleoprimatological research (Ingicco *et al.* 2011; Kelley and Schwartz 2010). Definition of age classes based on dentition requires an understanding of the dental formula for both deciduous and permanent teeth as well as the timing of tooth replacement sequence and wear (Box 15.1). The dental formula is a standardized way to describe the number of each type of tooth (incisor, canine, premolar, and molar) characteristic of mammalian dentition, in both the maxilla (upper tooth row) and mandible (lower tooth row). Dental formulas vary between Strepsirrhine and Haplorrhine (Platyrrhini and Catarrhini) primates (Fig. 15.1).

The sequence of permanent teeth eruption varies among primate species, making it necessary to use or develop appropriate guides. For example, Ingicco *et al.* (2011) proposed a method for age determination for Colobine primates, based on the sequence of eruption of both deciduous and permanent teeth. For juveniles, stages were defined based on the eruption of each new tooth, while for adults they used the gradient of wear on each cusp. The emergence of M3 (third upper molar) was defined as the boundary between juvenile and adult animals. Their absolute age designation to the life stages was based on growth data (Bolter 2011). With this method it is important to consider the sex of the individual. In the above example, even though in both males and females M3 erupts at around 3 years of age, female age at first reproduction is 4 years, while males are sexually mature at 5 years (Bolter 2011). These differences are important in demography and life history studies because chronological ages do not usually coincide with maturation timing in either sex.

> **Box 15.1** Using dental emergence and wear as criteria for age determination
>
> Primate skulls deposited in natural history museums or found in the field from natural mortality or from hunted individuals provide important information on age or life stage. Tooth emergence and replacement are useful in estimating the age of immature animals. In order to assign a primate skull to a relative age or a life stage (i.e., infant, juvenile, adult) the following information needs to be considered:
>
> - Knowledge about tooth morphology.
> - Understanding of dental formula of deciduous and permanent teeth for the given species. An excellent source for both types of primate dental formulas is Swindler (2002).
> - Understanding the eruption sequence for the given species: The first upper molar (M1) is the first permanent tooth to erupt in primates, but the sequence of the other permanent teeth varies among species (Swindler 2002).
> - Knowledge of correct timing of dental eruption and replacement.
> - Recording of the origin of the skull under observation (free-ranging or captive individuals): Captive animals exhibit accelerated development, including tooth emergence (Smith and Boesch 2011; Zihlman *et al.* 2004). Such differences affect estimation of chronological age.
> - Understanding of the possible effects of diet on tooth wear patterns.

In wild ring-tailed lemurs, age classes were recognized based on several combined criteria, including eruption or upper canines, overall dental wear on occlusal surfaces, and dentine exposure, following dental attrition scores defined for this species (Sauther *et al.* 2002; Cuozzo and Sauther 2006).

Dental casts are very useful, not only for age determination, but also for studies on dental health and its association with habitat quality (Sauther *et al.* 2002). Tooth impressions are easily taken from anesthetized individuals, both in the field and in captive conditions (see Chapter 3 for capture protocols), or from museum specimens. Developing a reference collection of dental casts is a useful tool in primate population research. Studies using mark-recapture procedures with multiple recapture occasions allow estimating of individual rates of tooth wear. For example, Richard *et al.* (2002) used dental casts from recaptured sifakas (*Propithecus verreauxi verreauxi*) in a population of more than 200 marked animals.

Another way of obtaining tooth information is by X-rays. A modern advance in this method is digital morphology and high-resolution X-ray computed tomography. A Digital Morphology library is currently available from The University of Texas Digital Morphology Group, which provides 2D and 3D imagery of

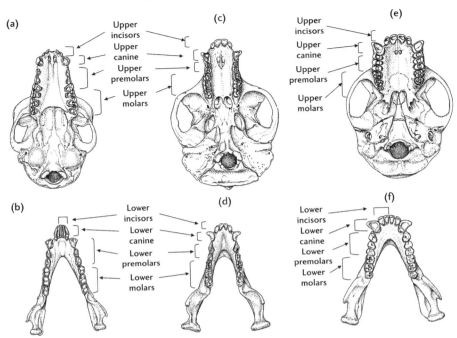

Fig. 15.1 Dental formulas for three primate species (drawings are not to scale): ring-tailed lemur (*Lemur catta*) I 2/2, C 1/1, P 3/3, M 3/3 (a–b); De Brazza's monkey (*Cercopithecus neglectus*) I 2/2, C 1/1, P 2/2, M 3/3 (c–d); and brown capuchin (*Cebus apella*) I 2/2, C 1/1, P 3/3, M 3/3 (e–f). Note that dental formula for the ring-tailed lemur and brown capuchin are the same, but tooth morphology is different, especially for the lower incisors and canine.

biological specimens. Their website contains images of several primate species (and many other groups) of different sex-age classes (growth series). Many of the scanned specimens at that digital library are of New World monkeys, for which tables of tooth emergence and wear are currently lacking. A primate skull of unknown age could, at minimum, be assigned to one out of two age-classes (juvenile or adult) if images for that particular species are available. Another method that uses teeth for age determination, most commonly utilized in archaeological research, is the histological analysis of enamel and dentine growth lines (Smith 2006).

15.2.1.2.2 Body weight and sexual characteristics
Increments in body size with growth, as well as sexual maturation signs, in combination with tooth eruption and wear are commonly used for age and sex

determination in captured animals. For example, the dental attrition scores used for ring-tailed lemurs were combined with weight, testicle descent (in males), and nipple length (in females), to assign every captured animal to one of four age-classes (Sauther et al. 2002; Cuozzo and Sauther 2006). Usually adults, and even juvenile animals older than one year, are difficult to differentiate between more than one age category based on body weight because there is overlap between body weights of animals from different cohorts (Richard et al. 2002). Consecutive weight measurements of growing individuals could be used to construct regression lines, from which the age of other immature individuals could be estimated. This method was used by Neyman (1979) to estimate age and birth dates of cotton-top tamarins. The use of this procedure is limited to growing individuals, and is affected by small sample sizes.

A new field method has been adapted for taking remote measurements on body features in primates, based on parallel laser metrics with digital cameras (Rothman et al. 2008). This method could be useful for estimating and monitoring changes in body size in growing animals, and has the potential to be used for relative age estimations in the field in a non-invasive manner.

15.2.2 Sex ratio

15.2.2.1 Population sex ratio

In general terms, sex ratio is the proportion of males to females in a population. However, several definitions of sex ratio need to be considered when estimating this parameter. The male to female ratio at conception is the *primary sex ratio*, the sex ratio at birth is the *secondary sex* ratio, and the adult sex ratio is the *tertiary sex ratio*. In primates, these sex ratios usually differ considerably. While sex ratio at birth may be close to 1:1, several factors can skew adult sex ratio in favor of one sex. Those factors include differential mortality between males and females, one-sex-oriented dispersal, and maturation differences between sexes. In many primate species, females reach their sexual maturity earlier than males of the same age. Also, primate population dynamics are more influenced by the sex ratio of actual reproducing adults than by the general population sex ratio, because of their social systems.

15.2.2.2 Operational sex ratio

Operational sex ratio (OSR) is the ratio of sexually active males to receptive females at a given time (Emlen and Oring 1977). OSR usually differs from adult sex ratio because not all adults are ready to mate at a given time. Adult females, for example, are not ready to mate when nursing a newborn or during the anestrous portion of their reproductive cycle. In primates, the estimation of OSR depends on whether breeding is seasonal or not. Mitani *et al.* (1996) presented a way to estimate OSR for 18 primate species, including both seasonal and non-seasonal breeders. Their estimation required knowledge of the following variables:

- adult sex ratio (males: females),
- breeding season length (in days),
- inter-birth interval (in years),
- length of estrous period (in days), and
- number of cycles females typically have before conception.

Other studies have used a simpler method for estimating OSR. For example, Takahashi (2001) estimated OSR for one unprovisioned, habituated troop of Japanese macaques (*Macaca fuscata*) as the number of estrous females per troop male. Estrous females were identified by observation of mating, presence of vaginal plugs, and coloration of faces and genitals. In other Cercopithecines, timing of ovulation can be estimated by the changes occurring in the perineal turgescence during the ovarian cycle. Also, ovulation can be determined by female hormonal profiles from fecal samples (Ziegler *et al.* 2000). However, when this method has not been successful, knowledge of gestation length has proved useful for determining time of ovulation, by counting backwards from the day of parturition (Ostner and Kappeler 2004).

15.2.3 Estimating population vital rates

15.2.3.1 Fertility rates

Fertility rate is the number of successful offspring produced in a period of time. Associated terms are fecundity (the potential reproductive capacity) and birth rates (the number of individuals born, either alive or not, in a period of time). Since there may be confusion between these terms, it is necessary to clarify which is the intended meaning when using them (Alberts and Altmann 2002). Often in primate populations, this parameter is estimated as the number of offspring divided by the total number of female-years observed during the study (Ostner and Kappeler 2004; Wich *et al.* 2007; Robbins *et al.* 2009). Fertility rates are estimated for the whole population, as well as separately for each social group within a population (Robbins *et al.* 2009). Although fertility rates are usually estimated from females, male fertility rates also can be estimated by paternity determination, using DNA from male fecal samples (Altmann *et al.* 2010).

In estimating fertility rates, it is necessary to first identify whether the population has a birth flow or a birth pulse pattern. For structured populations (such those of primates) with a birth flow pattern, the fertility rate is calculated as follows (Caswell 2001):

$$F_j = \left(\frac{1+p_1}{2}\right)\left(\frac{F_j + p_j F_{j+1}}{2}\right)$$

where F_j is fertility (number of offspring per individual) of individuals in the j-category at time t, and p_j is the survival of individuals in the j-category from time t to time $t+1$.

For structured populations with a birth pulse pattern, the way the fertility rate is estimated depends on whether the observation was made before or after the breeding season. The following are the equations to calculate fertility rates in those two cases (Caswell 2001):

Fertility rate for pre-reproductive census:

$$\hat{F}_j = F_j.$$

Fertility rate for post-reproductive census:

$$F_j = p_j F_j.$$

In captive primates, fertility rates are estimated directly from the institution's breeding records, and at the inter-institutional level, from species studbooks. In those cases, contraception and infant removal practices need to be taken into account because they influence birth rates.

15.2.3.2 Survival and mortality rates

Survival rate is measured as the proportion of animals that remain alive between two censuses. Likewise, mortality rate is the number of animals that die during two observational periods. Distinguishing between dispersal and mortality when an individual disappears from a group is a common problem in primate demography studies (Robbins et al. 2009). In marked populations, deaths can be determined from evidence of identifiable remains of marking equipment in the field or in predator feces (Ostner and Kappeler 2004).

Analytically, researchers have estimated yearly survival rate as $S = 1 -$ yearly mortality (Wich et al. 2004, 2007). When using mark-recapture methods, current techniques incorporate distinction between survival and detection probabilities. Such methods are based on individual mark-recapture histories, and are facilitated by computer software, such as MARK (White and Burnham 1999), one of the most currently popular in the subject (also see Chapter 14 for genetic mark-recapture methods).

15.2.3.3 Dispersal rates

Dispersal refers to the movement of an individual from its natal site or from its original home range to another location with no return, at least in the short term. Permanent change of individuals from one social group to another is also regarded as dispersal in primates, and it must be differentiated from temporary inter-group transfer, where individuals may return to their original groups. Dispersal should also be distinguished from migration, which implies long distance travels, usually in groups. When an animal disappears from a monitored group, dispersal to another group could be determined by searches in neighboring areas for other groups or solitary animals. The dispersal rate is usually calculated from those animals that disappear from a group but are seen alive at least once afterwards (Ostner and Kappeler 2004). Dispersal can be also estimated by capture-recapture methods (Richard *et al.* 2002), radio-tracking (Crofoot *et al.* 2010), and even genetic data (Tung *et al.* 2010, and see also Chapter 14 for genetic methods to measure dispersal).

15.3 Life history characteristics

Age at first reproduction, gestation length, inter-birth interval, length of lactational period, and lifespan are life history traits of high importance in primate demographics and sociobiology. Together, these life history characteristics limit the number of offspring a female can produce over her lifespan, and directly influence population dynamics. Primate life history patterns vary among species and the correlates of such variation are of great interest among primatologists. An updated database on primate life histories is being constructed and will be an important data resource (Strier *et al.* 2010).

15.3.1 Age at first reproduction

Determination of age at maturity requires individual-based data. Females born during the study years provide direct information of age at first reproduction. In captivity, age at first reproduction has been obtained from studbooks. Kaplan–Meier analyses have been used on large data sets, such those for captive orangutans (Anderson *et al.* 2008). Kaplan–Meier analysis is a standard survival method used in medical research.

15.3.2 Gestation length

Gestation length refers to the time lag between fertilization and parturition. Time of fertilization may be indirectly inferred by sexual behavior. Pregnancy may be

Fig. 15.2 Two red howler monkeys (*Alouatta seniculus*) in a tropical lowland forest of Colombia. Note the pregnant female at the right. Photo by Jorge Contreras.

evident only in an advanced stage (Fig. 15.2), making it difficult to assess a precise gestation length. However, in free-ranging primates, excreted steroids in feces have been used for pregnancy assessment (Wasser *et al.* 1988). Most of the information on gestation length in primates comes from captive animals. Diagnosis of pregnancy in captivity is usually done by non-human primate pregnancy test kits in addition to observing behavior (Hodgen *et al.* 1976).

15.3.3 Inter-birth interval

Inter-birth interval (IBI) is the elapsed time between successive births. When estimating IBI, infant mortality before weaning needs to be considered, because of its effects on shortening the interval. In the field, estimation of IBI also depends on whether the population has a birth pulse or a birth flow breeding pattern. In captivity, IBI may be influenced by management protocols, such as contraception programs and infant removal. Kaplan–Meier analyses and interval data analysis are commonly used to estimate IBI, both in free-ranging populations (Wich *et al.* 2004) and in captivity (Anderson *et al.* 2008).

15.3.4 Lactational period

This parameter can be inferred from infants' age at weaning, and it is better seen as a range rather than an absolute time. In observational studies, age at weaning is inferred by infant and female behavior. Also, in restricted females, enlarged nipples and presence of milk could be detected as signs of lactation. Cessation of nipple contact is an indication of weaning (Wich *et al.* 2007). Age at weaning is estimated with interval-censored data analysis. This technique is used when the event or process of interest (in this case, weaning) occurred at a time that is unknown, but between two known observation times (in this case, two consecutive censuses).

Two main indirect methods for estimating weaning age come from predictive relationships resulting from inter-specific comparisons (Humphrey 2010). The first approach is based on the threshold of body weight achieved before weaning, which in many species has been estimated to be one third of adult body size. The second approach is dental development. Emergence of the first permanent molar has been found to be highly correlated with weaning time in many, but not all, primate species.

15.3.5 Lifespan

Information on maximum lifespan in primate species comes mainly from captive animals. In the wild, maximum age at death is more difficult to estimate. In general terms, addressing maximum longevity would require following a cohort until its last member dies. Longitudinal long-term studies have provided estimations of longevity for several species, mainly apes and baboons, but even the longest studies may still have not reached the maximum age at death (Wich *et al.* 2004). Also, to estimate a species' variance of maximum lifespan would require many replicate populations (Bronikowski and Flatt 2010), which is very difficult for primates, because of their typically long lifespans.

15.4 Population dynamics

Population dynamics refers to changes in both size and structure of a population over time. Understanding such dynamics is very important for primate conservation because it reveals if a population is increasing, decreasing, or remaining stable through time. Studies on primate population dynamics require either long-term monitoring or periodic censuses. Changes in population size are estimated by either comparing the number of individuals in a population over time or by modeling population growth with life table analysis or matrix analysis.

15.4.1 Population growth rate

The main demographic factors affecting population size are the number of new individuals added to the population by births and immigration, and the number of individuals lost by deaths or emigration. The simplest classical model of population growth assumes no migration, leading to the estimation of population size at a time $t + 1$ as follows:

$$Nt + 1 = Nt + B - D$$

where, $Nt + 1$ is population size at a time $t + 1$, Nt is the population size at a previous time t, B is the number of births, and D is the number of deaths. Per capita rates of births (b) and deaths (d) are estimated as:

$$b = \frac{Number\ of\ births}{Number\ of\ individuals\ per\ time}$$

$$d = \frac{Number\ of\ deaths}{Number\ of\ individuals\ per\ time}$$

Per capita rate of population growth is estimated using estimation of b and d as:

$$r = b - d$$

And the change in population size (dN) over time (dt), can be calculated as:

$$\frac{dN}{dt} = rN$$

The value of r indicates whether the population size remains stable ($r = 0$), is increasing ($r > 0$) or is decreasing ($r < 0$). This rate is also called the intrinsic rate of increase and is given as the number of individuals produced per individual per time in a population. Another measure of population growth rate is the finite rate of increase (λ). This measure is obtained by comparing population size in two times t and $t + 1$, as follows:

$$\lambda = \frac{Nt + 1}{Nt}$$

Where $Nt + 1$ is the population size at time $t + 1$ and Nt is the population size at a previous time t. When population sizes are equal at both times t and $t + 1$, population remains stable ($\lambda = 1$), but when $Nt + 1 > Nt$, population is increasing ($\lambda > 1$), or when $Nt + 1 < Nt$, population is decreasing ($\lambda < 1$). The rate of population growth measured in this manner represents the proportional change of population size between two time periods. Lambda is also written as a population change factor:

$$Nt + 1 = \lambda Nt.$$

As seen here, intrinsic rate of increase (r) and finite rate of increase (λ) are two forms of expressing a population growth rate, the former as a per capita rate and the latter as a proportion or percentage of change in population size. These two terms are associated in the following manner:

$$r = ln\lambda$$

$$\lambda = e^r.$$

In primate populations, the above equations have been used, with a few modifications, to take into account dispersal migration. For example, Robbins *et al.* (2009) estimated monthly growth rates of a gorilla population by multiplying the initial population size by an increase factor ($1 + r_m$) and by adding an adjustment factor that accounted for monthly immigrants or additional habituated animals. Estimation of λ as the proportion N_{t+1}/N_t assumes that population size is known at times t and $t + 1$. When population size is unknown, and it is estimated by count data, detection probability (β) needs to be taken into account (Williams *et al.* 2001). If β_t is the detection probability at time t and β_{t+1} is detection probability at time $t + 1$, and those two probabilities are not equal ($\beta_1 \neq \beta_{t+1}$), estimation of λ is as follows:

$$\lambda = \frac{Nt + {}^*\beta t + 1}{Nt \, {}^*\beta t}.$$

15.4.2 Population growth models

The simplest technique to model population growth is the exponential growth model, whose main assumptions are that: (1) the population is closed, (2) the population grows with no limits, and (3) birth and death rates are constant and independent of population size. This model predicts population size as $N_{t+1} = e^{rt} N_t$. Since animal populations do not follow the above assumptions, the use of this model is limited to short periods of time. The logistic population growth model (density-dependent growth model) incorporates limits to population growth (carrying capacity, K). In this model, changes in population size over time depend on population size: $dN/dt = rN\,(1 - N/K)$. The above two models assume unstructured populations. However, for primates, and most animal populations, age-structured population models (such as life tables and matrix models) are more appropriate for describing population growth, because they take into account age-specific vital rates.

15.4.2.1 Life tables

Life tables summarize age-specific reproduction and survival data in a population. They are based only on females and their daughters because population growth is limited by female fertility. The basic information needed to build a life table includes the number of individuals in each age class, and their fertility—(b) and mortality—(d) specific rates. Such data are used for estimating other parameters such as age-specific survival, generation time, and net reproductive rate for a population (Box 15.2). Also age at first reproduction (a) and age at last reproduction (ω), can be inferred from a life table.

A life table is also used for the estimation of net reproductive rate (R_0) and generation time (T), as follows (see definition of terms in Box 15.2):

$$R_0 = \sum lx^* mx.$$

Net reproductive rate (R_0) represents the average number of females produced by females over their entire lifespan. R_0 is also called replacement rate, and it is another way to examine whether a population is remaining stable in size ($R_0 = 1$), increasing ($R_0 > 1$) or decreasing ($R_0 < 1$). R_0 is also used to estimate generation time (T):

Box 15.2 Life table parameters

Life tables summarize age-specific data on female fertility and mortality and allow estimation of the parameters that explain population growth. The following presents the common notation and main parameters in a life table, and the way they are calculated (Lemos-Espinal *et al.* 2005):

x = age (in years)

X = age class (interval from age x to $x + 1$)

Age 0 = newborns

$N_{t,X}$ = Number of individuals in age-class X at time t

S_X = Number of individuals that survive from age x to age $x + 1$

Dx = Number of individuals that die from age x to age $x + 1$. It is calculated as $N_x - N_{x+1}$

l_X = Survivorship probability from birth to the start of age class X. It is calculated as N_X/N_o

d_X = Death probability from age x to age $x + 1$. It is calculated as $l_x - l_{x+1}$

q_X = Age class X specific mortality rate. It is calculated as d_X/l_X, or D_x/N_x

p_X = Survivorship probability from age x to age $x + 1$. It is calculated as $1 - q_X$

b_X = Number of female offspring expected from a female in age class X. It is also written as m_X

$$T = \frac{\sum Xlx * mx}{R_0}.$$

Generation time (T) is the lag time between a female birth and her daughter's birth, considering the mean of her daughters. In other words, T is the mean time in which a generation is replaced by the next one. In populations with overlapping generations, generation time is estimated as the mean age of reproduction.

Net reproductive rate (R_0) and generation time (T) allow the estimation of population growth rates, such as the intrinsic rate of population growth (r) and finite rate of population growth (λ). For example, knowing R_0 and T, r can be estimated as:

$$r = \frac{\ln R_0}{T}.$$

Information from life tables is used to estimate how much population size changes over time. For example, the number of newborns produced from time t to time $t + 1$ (N_{t+1}) is given by the number of females that survive from time t to time $t + 1$ ($Ntx * px$) multiplied by their fecundity (mx), as follows:

$$Nt + 1 = \sum Ntx * px * mx.$$

If age-specific vital rates remain constant over time, the above equation may be used to project population sizes in future years. This method shows that after a few years λ remains constant, and that the proportion of each age stabilizes over time, leading to a stable age structure.

Many life tables of primates come from captive or colony breeding populations (Gage 1998). For wild primate populations, information is available for a few species under long-term observational studies (Richard *et al.* 2002; Nishida *et al.* 2003; Wich *et al.* 2004; Wich *et al.* 2007). Such long-term studies, however, are not available for most primate populations.

Although life tables are very useful for population dynamic studies, there are several important issues that need to be considered. The main assumption in a life table is that age is the only driver affecting vital rates. However, vital rates may also change due to environmental factors. Also, in primate populations, other parameters such as size and social status are better predictors of vital rates than age (Alberts and Altmann 2002).

Alternative methods for studying populations are matrix models, which provide a link between individual life cycle and population dynamics (Caswell 2001). There are two basic matrix models: age-structured matrix (Leslie matrix), and stage-structured matrix (Lefkovitch matrix).

15.4.2.2 Age-structured matrix model

A matrix is an arrangement of rows and columns containing values of demographic processes in a population. In a Leslie matrix, a population is structured by age, and the only two demographic parameters that model population growth are age-specific fertility (values on top row of the matrix) and age-specific survival rates (values in the subdiagonal of the matrix). The remaining cells of the matrix are zeros (Fig. 15.3a). In the matrix, age-specific survival rate is also understood as growth (G), because it refers to the number of individuals from age x that survive and grow to age $x + 1$.

Fig. 15.3a shows a hypothetical population structured in ten age classes. Every value a_{ij} of the matrix represents the contribution of age-class j to age-class i. For example, the value for F_{13} is the contribution of age-class 3 to age-class 1 through fecundity, and the value for G_{21} is the contribution of age-class 1 to age-class 2 through growth (individuals of age-class 1 that survived and grew to age-class 2). From Fig. 15.3a, it can be inferred that ages at first and last reproduction are age-classes 3 and 8 respectively. Projection of population size and structure at time t + 1 ($N_{t + 1}$) is estimated by multiplying the L matrix by the age-abundance vector Nt:

$$Nt + 1 = L * Nt.$$

If matrix L is multiplied repeatedly by consecutive age distribution vectors, a stable age distribution is drawn, after initial fluctuations. Once a stable age distribution is reached, multiplying the matrix L by the age distribution vector (w) is the same as multiplying the matrix by a scalar number. This scalar number is called the dominant eigenvalue of the matrix and it is an estimate of the population growth rate λ. It is also possible to estimate the reproductive value, v, which represents the relative contribution of each age class to future generations. The reproductive value vector (v) is obtained by repeatedly multiplying the transposed matrix by the age abundance vector, until age distribution stabilizes. Values in vector v are standardized, letting the first value equal to one. Vectors w and v are the dominant right and left eigenvectors associated with the matrix (Caswell 2001). In relatively stable environments, annual matrix models could be used to obtain an average matrix, which accounts for annual differences in estimations of λ. However, in environments with cycling changes, it is more appropriated to use periodic matrices (see Caswell 2001).

15.4.2.3 Stage-structured matrix model

There are cases in which developmental stage is a better predictor of vital rates than chronological age (Caswell 2001). In those cases, a stage-structured matrix model is

(a)

$$
\begin{array}{c}
 \\ 1\\2\\3\\4\\5\\6\\7\\8\\9\\10
\end{array}
\begin{array}{c}
\begin{array}{cccccccccc}
 & & & & & \text{L} & & & & \\
1 & 2 & 3 & 4 & 5 & 6 & 7 & 8 & 9 & 10
\end{array}\\
\begin{pmatrix}
0 & 0 & F_{13} & F_{14} & F_{25} & F_{61} & F_{71} & F_{81} & 0 & 0\\
G_{21} & 0 & 0 & 0 & 0 & 0 & 0 & 0 & 0 & 0\\
0 & G_{32} & 0 & 0 & 0 & 0 & 0 & 0 & 0 & 0\\
0 & 0 & G_{43} & 0 & 0 & 0 & 0 & 0 & 0 & 0\\
0 & 0 & 0 & G_{54} & 0 & 0 & 0 & 0 & 0 & 0\\
0 & 0 & 0 & 0 & G_{65} & 0 & 0 & 0 & 0 & 0\\
0 & 0 & 0 & 0 & 0 & G_{54} & 0 & 0 & 0 & 0\\
0 & 0 & 0 & 0 & 0 & 0 & G_{87} & 0 & 0 & 0\\
0 & 0 & 0 & 0 & 0 & 0 & 0 & G_{98} & 0 & 0\\
0 & 0 & 0 & 0 & 0 & 0 & 0 & 0 & G_{109} & 0
\end{pmatrix}
\end{array}
*
\begin{array}{c}
n_t\\
\begin{pmatrix} N_1\\N_2\\N_3\\N_4\\N_5\\N_6\\N_7\\N_8\\N_9\\N_{10}\end{pmatrix}
\end{array}
=
\begin{array}{c}
n_{t+1}\\
\begin{pmatrix} N_{1(t+1)}\\N_{2(t+1)}\\N_{3(t+1)}\\N_{4(t+1)}\\N_{5(t+1)}\\N_{6(t+1)}\\N_{7(t+1)}\\N_{8(t+1)}\\N_{9(t+1)}\\N_{10(t+1)}\end{pmatrix}
\end{array}
$$

(b)

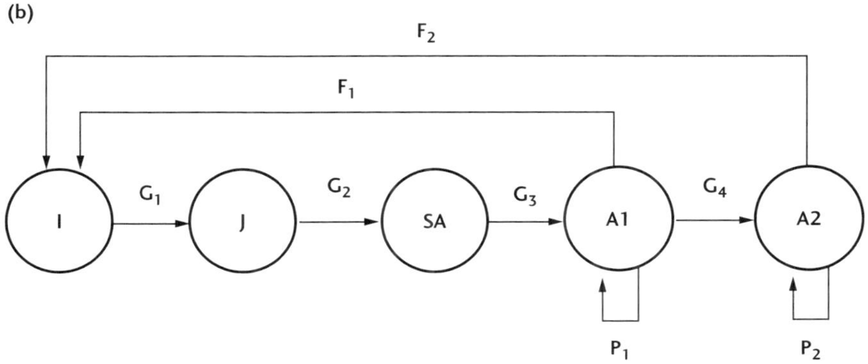

(c)

$$
\begin{array}{c}
 \\ \text{I}\\ \text{J}\\ \text{SA}\\ \text{A1}\\ \text{A2}
\end{array}
\begin{array}{c}
\begin{array}{ccccc}
\text{I} & \text{J} & \text{SA} & \text{A1} & \text{A2}
\end{array}\\
\begin{pmatrix}
0 & 0 & 0 & F_1 & F_2\\
G_1 & 0 & 0 & 0 & 0\\
0 & G_2 & 0 & 0 & 0\\
0 & 0 & G_3 & P_1 & 0\\
0 & 0 & 0 & G_4 & P_2
\end{pmatrix}
\end{array}
*
\begin{array}{c}
n_t\\
\begin{pmatrix} N_I\\N_J\\N_{SA}\\N_{A1}\\N_{A2}\end{pmatrix}
\end{array}
=
\begin{array}{c}
n_{t+1}\\
\begin{pmatrix} N_{I(t+1)}\\N_{J(t+1)}\\N_{SA(t+1)}\\N_{A1(t+1)}\\N_{A2(t+1)}\end{pmatrix}
\end{array}
$$

Fig. 15.3 Matrix population models. (a) Leslie matrix for a hypothetical population structured in ten age classes. Vector n_t indicates the number of individuals in each age class, and vector n_{t+1} represents the projected number of individuals in each age class at time t + 1. (b) A life cycle graph for a hypothetical population, structured in five developmental stages (infants, juveniles, subadults, young adults or adults 1, and adults 2) G_1 to G_4 represent the contribution of a particular stage to the next one through survival and growth to the next stage, F_1 and F_2 are the contributions of both adults 1 and adults 2 to infants through fecundity and P_1 and P_2 are their permanence in the same stage. (c) Stage-structured matrix model for the hypothetical population illustrated in (b). Vector n_t represents the number of individuals in each age class and vector n_{t+1} represents the projected number of individuals in each stage at time t + 1.

very useful because the population may be structured by size classes or developmental stages. A stage-classified population can be illustrated with a life cycle diagram, which is a schematic representation of the different age classes or stages connected by different demographic processes within a population (Fig. 15.3b). This graphic method helps to visualize the relative contribution of each size class or stage to other classes by growth (G), survival (P), and fertility (F). For the hypothetical example of Fig. 15.3b, let us suppose that numbers 1 to 5 represent five stage classes in a primate population as follows: infants (1), juveniles (2), subadults (3), young adults (4), and adults (5). Infants contribute to juveniles by growth (G_1). The amount of that contribution can be calculated by estimating the number of infants that reach the juvenile stage. Likewise, juveniles contribute to the subadult class by growth (G_2). Young adults contribute to adult class by growth (G_3), but they also contribute to infant age class by fertility (F_1), and to the young adult class by survival and remaining in the same stage (P_1). Adults contribute to the infant age class by fertility (F_2) and to adults by survival and remaining in the same stage (P_2).

15.4.2.4 Sensitivity and elasticity analyses

An advantage of using matrix population models is the possibility of performing prospective perturbation analyses that provide a measurement of the impact of small changes in vital rates on the population growth rate λ (Caswell 2001). Sensitivity (S_{ij}) of each a_{ij} value of a population projection matrix measures the absolute effect on λ of an absolute change in each element of the matrix. The following equation is used to calculate the sensitivity (s_{ij}) of each vital rate (a_{ij}) in a population matrix (Caswell 2001):

$$S_{ij} = \frac{\partial \lambda}{\partial a_{ij}} = \frac{v_i w_j}{<w, v>}.$$

In this equation, w and v are the dominant right and left eigenvectors respectively of the population matrix. A sensitivity matrix is built with the S_{ij} values and it is used to examine the response of λ to perturbations in each a_{ij}. The vital rate (a_{ij}) with the largest sensitivity (s_{ij}) would be the one with the largest absolute effect on population growth.

Although sensitivity analysis helps to identify the vital rate of highest effect on λ, it leaves two main problems unsolved. The first is that vital rates (a_{ij}) are measured at very different scales. The second is that sensitivity does not allow examination of the relative contribution of each vital rate to λ. An analytical approach to address these problems is elasticity analysis (de Kroon et al. 1986).

Elasticity analysis is the standardization of the sensitivity (s_{ij}) values. Calculation of the elasticity (e_{ij}) for each element of the population matrix is done as follows (de Kroon *et al.* 1986):

$$e_{ij} = \left(\frac{a_{ij}}{\lambda}\right)(s_{ij}).$$

An elasticity matrix is built with the e_{ij} values. An important feature of such a matrix is that the sum of all e_{ij} is one, making it possible to compare the relative importance of each vital rate to population growth (see Lawler *et al.* 2009 for an example with Verreaux's sifaka).

15.5 Modeling extinction risk

15.5.1 Population viability analyses

Population viability analysis (PVA) is the use of quantitative techniques to predict possible trajectories of one or several populations and their probability of persisting over time. This tool is used for species of conservation concern, and the outputs are used to rank competing conservation or management alternatives. In fact, the IUCN Red Lists often use PVAs for their assessments and conservation plans (see Chapter 16). Morris and Doak (2002) classify PVA into four main types: count-based PVA, demographic PVA, multi-site PVA, and spatially-explicit individual-based models. The choice of any of these PVA types would depend upon the quantity and quality of available information of one or several populations.

Count-based PVA is based on estimations of population size, typically with annual censuses. The simplest of this type of PVA is the count-based density-independent model, in which several estimations of λ are obtained from the basic equation $\lambda = N_{t+1}/N_t$. Due to environmental variation over time, vital rates are expected to vary as well, making λ variable over a continuous range of values. One way to measure the average stochastic population growth rate is to convert the annual estimations of λ to a log scale and estimate their mean (μ) and variance (σ^2). These two last parameters would describe the normal distribution of future log population sizes. Morris and Doak (2002) present several methods to estimate μ and σ^2 and confidence intervals. These estimates are used to build an extinction time cumulative distribution function (CDF) as a metric of population viability. For this analysis, population viability is defined as the probability of the population ever reaching a quasi-extinction threshold (the minimum number of individuals below which population has very low chances of recovery). The equation proposed by Morris and Doak (2002) allows calculating the probability of the population reaching the quasi-extinction threshold, at different times, given

Table 15.2 *Some software used for population dynamics and PVA analyses.*

Software	Use	Free?	Source
ALEX	PVA, spatially-structured populations	Yes	Possingham and Davies (1995)
MARK	Mark-recapture analysis	Yes	White and Burnham (1999)
NetLogo	Individual-based spatially explicit modeling	Yes	Wilensky (1999)
demogR	Leslie matrix model, life tables	Yes	Jones (2007)
RAMAS	Metapopulations, Leslie matrix-based model	No	Akçakaya (1997)
VORTEX	PVA—individual-based model	Yes	Lacy et al. (1995)

an initial population size. This method could be used for endangered primate species whose populations have count data for several years but with very little or non-existent demographic data.

Other types of PVA are based on demographic data. Those models involve stochastic projection matrix models, or individual-based models. The basics of projection matrix models were explained in Section 15.4. Since those matrices allow projected estimations of λ, the tendency for population increase or decrease can be examined over the long term under a stochastically varying environment. Computer simulations enable the obtainment of an extinction time cumulative distribution function (CDF), and examination of whether population size has fallen below the quasi-extinction threshold each year (Morris and Doak 2002).

Multi-site PVA is used for assessment of extinction risk for multiple populations. This type of PVA requires knowledge of site-specific demographic information, correlations in population growth rates or vital rates across sites, and movement rates between sites (Morris and Doak 2002). This approach is used for both spatially structured populations and metapopulations. All the PVAs mentioned here are facilitated by computer computations (Table 15.2). When comparing the performance of several software packages, Brook et al. (2000) found that individual-based and matrix-based packages differ in their risk assessment, because the latter ignore stochasticity in the sex ratio and recommend modeling only the female component of the population when using matrix-based packages. Most available PVAs for primate species have used individual-based models, using the software Vortex (i.e., Manansang et al. 2005; Holst et al. 2006).

15.5.2 Spatially explicit individual-based models

Spatially explicit models are used to incorporate a spatial context to the ecological phenomena under study. Population models involving spatial features include metapopulation and spatially structured population concepts. Extinction risk evaluation for this type of population was mentioned above as multi-sited

PVA. These models can also include landscape features at regional scale by using remotely sensed data and geographic information systems (GIS).

Individual-based modeling (IBM), also known as agent-based modeling, is a new technique for understanding biological (and other) systems, based on the characteristics and behaviors of the individuals, or agents, which are the building blocks of these systems. In the study of population dynamics, individual-based modeling is most useful for those cases when classical population models are not feasible, because of a lack of long term data sets, and/or when demographic rates are affected by individual behavior and not only by population size or environmental factors (Grimm and Railsback 2005). Individual-based modeling has been used increasingly and recent reference texts and software are available (Grimm and Railsback 2005; Railsback and Grimm 2011). Agent-based modeling has also been used for primate behavioral studies (Bryson *et al.* 2007; King *et al.* 2011b; see also Chapters 4 and 9 for applications to studying epidemiology and group ranging and inter-group encounters, respectively). Also, a spatially explicit IBM model was used to understand the impact of hunting and the importance of un-hunted areas for an exploited primate species (Wiederholt *et al.* 2010).

15.6 Conclusion

To advance in our understanding of primate ecology and evolution, as well as to make informed conservation decisions, more demographic studies are needed. Primate demographic studies require long-term data sets, because of the long lifespan of most primate species. Although analytical tools for population dynamics have been available for a long time, the complex individual behavior of primates violates many of the assumptions of such methods. Matrix models are a valuable approach, and could be used more extensively for modeling primate populations. Risk assessments are based on population dynamics and can be seen as a practical extension of demography. The better the demographic information that is made available for primate populations, the more effective conservation decisions can be.

Acknowledgments

I wish to thank the editors Dr. Mary Blair, Dr. Eleanor Sterling, and Dr. Nora Bynum, for inviting me to write this chapter and for all their support during this process. I also appreciate the careful reading from the editors and the external peer reviewers, whose comments helped me to improve early versions of this chapter. Illustrations in Fig. 15.1 were created by Marcela Morales, and the photograph Fig. 15.2 was kindly provided by Jorge Contreras.

16

Determining conservation status and contributing to *in situ* conservation action

Mary E. Blair, Nora Bynum, and Eleanor J. Sterling

16.1 Introduction

Conservation *in situ* refers to maintaining and enhancing wild populations and ecosystems. In this chapter, we provide an overview of how to determine the conservation status of your study organism(s) and how to contribute to conservation action *in situ*. First, we discuss international and national conventions and lists of threatened species where you can look to determine what is known so far about the conservation status of your study population. Then, we synthesize how to determine population status and identify threats to populations and their magnitude. Finally, we discuss tactics for *in situ* conservation action, including how to connect with local and national officials and agencies engaged in conservation activities, incorporating human dimensions including local knowledge in the development of conservation efforts, and contributing to species action plans and other active management activities.

16.2 Overview of national and global conventions and lists that include primates

Several national and global conventions on biodiversity conservation include primates and are useful in guiding determination of conservation status as well as the development of plans for conservation action to protect wild primate populations *in situ*. We provide an overview here of where you can look to determine what is known so far about the conservation status of your study organism(s), while also discussing the limitations of using global systems to assess individual species' conservation status, as data quality and availability vary greatly across species and regions (see Box 16.1).

Primate Ecology and Conservation: A Handbook of Techniques. First Edition. Edited by Eleanor J. Sterling, Nora Bynum, and Mary E. Blair. © Oxford University Press 2013. Published 2013 by Oxford University Press.

The International Union for the Conservation of Nature (IUCN)'s Species Survival Commission (SSC) works to coordinate international conservation of endangered taxa. The IUCN is an international Union of Members including over 80 states, 140 government agencies, and 800 non-governmental organizations. The SSC is an international network of more than 7500 volunteer experts organized into more than 120 Specialist Groups, Authorities, and Task Forces, including the Primate Specialist Group (PSG, <http://www.primate-sg.org>). The PSG heads the development of the global strategy for primate conservation as well as regional and taxon-specific action plans. The overall goal of the PSG is to maintain "the current diversity of the Order Primates, with a dual emphasis on: (1) ensuring the survival of endangered and vulnerable species wherever they occur and (2) providing effective protection for large numbers of primates in areas of high primate diversity and/or abundance" (Oates 1996, p.vii). The PSG works to evaluate the conservation status of all identified primate species and subspecies to contribute to the IUCN Red List of Threatened Species^{TM} (<http://www.iucnredlist.org/>, hereafter referred to as the Red List), the most comprehensive evaluation of the status of over 52,000 species across the globe, including information on taxonomy, geographic range, habitat and ecology, and threats to survival (IUCN 2012). The 2004 World Conservation Congress mandated development of uses for the Red List in national legislation and international conventions, as well as in conservation planning and scientific research (IUCN 2005).

The current definitions of the Red List categories are noted in Table 16.1. The threat classification system has been revised several times since the Red List was initiated over 40 years ago. The first Red List assessments relied on the opinions of experts, but recent assessments use data-driven criteria to help provide objective estimates of extinction risk. In particular, quantitative data are used to place a taxon in one of three "threatened" categories, Critically Endangered, Endangered, or Vulnerable (IUCN 2001; IUCN 2011; Table 16.1). Taxa are evaluated against all the criteria and listed under the highest threat category for which any one of the criteria is met. Experts compile and review the primary data needed to allocate a taxon into a category. For primates, the Red List includes assessments of species, subspecies, and sometimes subpopulations if certain criteria are met (IUCN 2011). For other taxonomic groups, such as plants, varieties may be listed as well.

The Red List is most useful to set taxon-based priorities for conservation action where extinction risk, as delineated by the Red List threat classification categories, can be used to set priorities among taxa and focus limited resources on actions that will do the most to advance towards the goals of international conservation strategies. However, a researcher needs to think critically when interpreting the Red List categories and threat assessments; the effectiveness of taxon-based, priority

Table 16.1 *A summary of the IUCN Red List of Threatened Species[TM] threat categories and criteria for classification in those categories, adapted from IUCN (2011).*

Red List category	Criteria for classification
Extinct (EX)	Exhaustive surveys of known or expected habitat throughout historic range have failed to record an individual, and none survive in cultivation, captivity, or as a naturalized population well outside the past range.
Extinct in the Wild (EW)	Exhaustive surveys of known or expected habitat throughout historic range have failed to record an individual, but the taxon is known to survive in captivity, or as a naturalized population well outside the past range.
Critically Endangered (CR)	Evidence for: population size < 50 mature individuals, < 250 mature individuals with evidence of continuing decline, or an observed, projected, or suspected reduction in population size of $\geq$ 80% over a 10 year or 3 generation period ($\geq$ 90% if causes of reduction are clearly reversible, understood, and have ceased), OR an extent of occurrence less than 100 km^2 or area of occupancy less than 10 km^2 with evidence of severe fragmentation, continued decline, or extreme fluctuations, OR quantitative analysis showing the probability of extinction in the wild is at least 50% within 10 years or 3 generations.
Endangered (EN)	Evidence for: population size < 250 mature individuals, < 2500 mature individuals with evidence of continuing decline, or an observed, projected, or suspected reduction in population size of $\geq$ 50% over a 10 year or 3 generation period ($\geq$ 70% if causes of reduction are clearly reversible, understood, and have ceased), OR an extent of occurrence less than 5000 km^2 or area of occupancy less than 500 km^2 with evidence of severe fragmentation, continued decline, or extreme fluctuations, OR quantitative analysis showing the probability of extinction in the wild is at least 20% within 20 years or 5 generations.
Vulnerable (VU)	Evidence for: population size < 1000 mature individuals, < 10,000 mature individuals with evidence of continuing decline, or an observed, projected, or suspected reduction in population size of $\geq$ 30% over a 10 year or 3 generation period ($\geq$ 50% if causes of reduction are clearly reversible, understood, and have ceased), OR an extent of occurrence less than 20,000km^2 or area of occupancy less than 2000km^2 with evidence of severe fragmentation, continued decline, or extreme fluctuations, OR quantitative analysis showing the probability of extinction in the wild is at least 10% within 100 years.
Near Threatened (NT)	Evidence fails to qualify taxon for CR, EN, or VU but the taxon is likely to qualify in the near future.
Least Concern (LC)	For widespread and abundant taxa that fail to qualify for CR, EN, VU, or NT.
Data Deficient (DD)	Inadequate data are available to make an assessment of the risk of extinction.
Not Evaluated (NE)	The taxon has not yet been evaluated against the criteria.

setting exercises is limited by both the quality of the demographic data used to assess the status of a taxon (see Box 16.1) and taxonomic uncertainty. Without a sound definition of taxonomic units (e.g., species and subspecies), accurate analyses of threats and conservation status are impossible (Mace 2004). For example, Population Viability Assessments (PVAs) require a reliable number of individuals remaining to accurately model potential future viability (see Chapter 15, Section 15.3). If several populations are thought to represent one species, when in reality they are two or more species, their population size will be overestimated and their risk of extinction underestimated; such a misrepresentation led to the extinction of several unique lineages of the New Zealand tuatara (Daugherty *et al.* 1990; DeSalle and Amato 2004).

The issue of taxonomic uncertainty has been historically challenging in primates as a group (e.g., see Blair *et al.* 2011). At the root of this problem is the lack of consensus regarding how to define species, with researchers typically debating the biological species concept (Mayr 1963), where species are interbreeding populations reproductively isolated from other groups, against the phylogenetic species concept (Cracraft 1983), where species are defined as the smallest diagnosable clusters of individuals with a parental pattern of ancestry and descent (see Chapter 14, Section 14.3.1 for more discussion of species concepts and molecular genetic methods to determine taxonomic units). In many animals and in primates in particular, reproductive isolation is difficult to determine in part because of the common occurrence of natural hybridization among close relatives with overlapping geographic distributions (Bynum *et al.* 1997; Bynum 2002; Zinner *et al.* 2011). By contrast, hybrids are not an issue in the phylogenetic species concept, which designates species as the smallest definable units with evolutionary significance. An increased application of the phylogenetic species concept has resulted in what some authors consider "taxonomic inflation" of many groups, and has affected the determination of conservation status (Isaac *et al.* 2004). Preuss's red colobus (*Procolobus preussi*) makes for an interesting example here. The IUCN Red List now recognizes it as a species distinct from *Procolobus pennantii* and *badius*. It used to be considered a subspecies of *pennantii* (and still is by some) or *badius*, and was listed as Endangered. But, now that it is considered a full species (and is still declining), it is listed as Critically Endangered, and therefore merits more conservation attention (IUCN 2012). It is important for researchers to understand the taxonomic history of their study population(s) and how that history may have affected any changes in conservation status; this information is typically included in the "taxonomic notes" of the IUCN Red List assessment.

Although the Red List is intended to be a global assessment, there are guidelines for its application at national and regional levels (IUCN 2003), and there are now many national and regional red lists (for example, the National List of Endangered Wildlife in Costa Rica, MINAE 1997). These lists are particularly important where endemic species have only been assessed at regional or national levels and have not yet undergone a global assessment (Rodrigues *et al.* 2006). It is also important to recognize that for a given taxon, their global Red List category may change when evaluated at the national or regional scale. For example, even a widespread species may have a small local population in a country if that country is at the edge of the species' range.

A particularly important national list is that of the US Endangered Species Act (ESA), passed in 1973 to provide protection for species at risk of extinction. Of course there are no native non-human primates in the USA, but 590 foreign species are listed, and many of them are primates. These foreign species are listed in order to regulate their trade in and out of the USA, but their listing also brings important publicity to the species, which can help to garner public engagement for international conservation efforts (see Section 16.4).

Another important list used to set taxon-based priorities for primates at global, regional, and national levels is the Top 25 Most Endangered Primates list, generated by the PSG, Conservation International, and the International Primatological Society (e.g., Mittermeier *et al.* 2009). Started in 2000, the list is updated through the expert opinions of primatologists attending the biennial International Primatological Society Congress. The list represents a consensus of 25 primate species "considered to be amongst the most endangered worldwide and the most in need of urgent conservation measures" (Mittermeier *et al.* 2009, p.1). The list takes into account not only Red List categories but also the representativeness across the list for important regional areas to global primate conservation, and the representation of particularly important threats. For example, the Javan slow loris (*Nycticebus javanicus*) replaced *Loris tardigradus nycticeboides* on the 2008–2010 list in order to highlight "a crisis threatening all the Asian lorises," meaning the pet and traditional medicine trade; "the Javan slow loris, representing the plight of all, is evidently the hardest hit of any of the lorisiformes in this respect" (Mittermeier *et al.* 2009, p. 5).

In light of the tremendous threat of trade to wild primate populations (see Chapter 18), another important international list for conservation prioritization is that of the Convention on International Trade in Endangered Species of Wild Fauna and Flora (CITES), which started in 1975 after a resolution adopted in 1963 at an IUCN meeting. CITES is an international agreement between 175 governments to ensure that international trade does not threaten the survival of wild animals and plants. Taxa are listed in one of the three CITES appendices (<http://www.cites.org/>): Appendix I includes taxa for which international trade is prohibited by CITES due to their highly endangered status, Appendix II

Box 16.1 The aye-aye: a case study in assessing the conservation status of a rare and elusive primate (by Erin McCreless and Eleanor J. Sterling)

Ideally, baseline knowledge of a species' population size and trends is readily available, and can be used to determine conservation status and develop management plans. However, for some species—particularly for rare or elusive animals that are difficult to survey in the wild—even basic population data may not exist. This is the situation for the aye-aye (*Daubentonia madagascariensis*), a nocturnal and little-studied lemur whose history on the IUCN Red List illustrates the challenge of implementing the Red List categories and criteria when data are scarce.

The only surviving member of its family, the Daubentoniidae, the aye-aye's nocturnal and largely solitary habits make it poorly suited to traditional primate census techniques (see Chapter 2). Most of our knowledge about aye-aye distribution is based on incidental sightings, and its population size, abundance, and trends are virtually unknown. The species was presumed extinct in 1935 but was redis-covered in the 1950s, though still thought to be restricted to a few small popula-tions on the east coast of Madagascar (Petter and Peyrieras 1970). Consequently many conservation strategies and several protected areas emerged from concern over aye-aye conservation. However, a review in the mid-1990s revealed that the aye-aye is actually quite widely distributed and adaptable (Sterling 1994a). Aye-ayes are now known to occur along Madagascar's eastern and northern coasts and in the central and northern areas of the west coast, and inhabit a wide variety of habitats, from primary rainforest and deciduous forest to secondary growth, dry scrub forest, and cultivated lands (Sterling and McCreless 2006). Nevertheless, a scarcity of data regarding population abundance and trends across the species' range leaves its conservation status in question.

The aye-aye was classified as Endangered in its first assessment in 1986 based on the opinion of experts. Despite clear understanding of its widespread distribution coming to light ten years later (Sterling 1994a), for a variety of reasons, including the uniqueness of the taxon and ongoing conservation strategies that featured the aye-aye as a "flagship" species (see Section 16.4.2) as well as little knowledge on its population sizes or trends, it continued to be listed as Endangered for another decade. In 2008, its status was changed to Near Threatened. The most recent IUCN Red List assessment states that populations are thought to have declined by 20–25% over the last 24 years due to habitat loss and exploitation—resulting from beliefs held towards aye-ayes by Malagasy people (IUCN 2012).

A shift in IUCN Red List status to a lower threat category is often considered a conservation success, but for the aye-aye the status change simply reflects increased recognition of knowledge about the species' distribution. However, little to nothing has changed in our overall understanding of aye-aye abundance, population trends,

Box 16.1 *Continued*

or threats from hunting since the 1990s. Population trend estimates are speculative at best—there is no mainland population that has been well-sampled enough to understand basic population sizes—and should be improved to determine whether aye-ayes are truly threatened and what conservation actions, if any, are needed. Traditional line-transect surveys are not practical for this animal, but indirect survey methods often used to study apes and other primates could supply valuable information about the species' distribution (see Chapter 2). Aye-ayes leave distinctive toothmarks on *Canarium* seeds, wood, and bamboo, and leave these items scattered around the forest floor after feeding (Duckworth 1993; Sterling 1994b; Goodman and Sterling 1996). The nests they build in trees, for resting during the day, can also serve as indicators of their presence (Sterling 1993a; Ancrenaz *et al.* 1994). Local human communities often take note of aye-ayes, both because the animals eat cultivated plants and because of the beliefs surrounding them (Andriamasimanana 1994; Sterling and Feistner 2000; Sterling 2003; see Section 16.4.2), though this information may be difficult to translate into population estimation data. Finally, new tools for analysis of low-quality DNA samples (see Chapter 14) offer the possibility of using genetics to identify the individual animals residing in an area.

While indirect survey methods can improve our knowledge of elusive species like the aye-aye, neither traditional nor indirect surveys may ever be able to provide the reliable population estimates required to confidently assign these animals to IUCN Red List categories. Under a more strict interpretation of the categories, aye-ayes perhaps should be listed as Data Deficient rather than Near Threatened. The most recent IUCN Red List guidelines (IUCN 2011) note that the Data Deficient category is meant to highlight taxa for which sufficient data are lacking to make a sound status assessment. However, there is strong pressure in the guidelines to categorize taxa as Data Deficient only when there is no alternative, because taxa placed in this category are less obviously targets for conservation action (IUCN 2011). Thus, although precise information on rare and elusive taxa such as the aye-aye is usually lacking, the IUCN advocates the use of projections, assumptions, and inferences in order to place these taxa in the appropriate category of threat, while clearly documenting any assumptions or other uncertainties associated with the information utilized to assign the category. In other words, the IUCN guidelines urge scientists to err on the side of caution, suggesting that they list taxa as threatened when data are ambiguous (while acknowledging any ambiguity). The reasoning is that if a species is threatened, inaction due to inadequate information may result in population extirpation or the species going extinct. By contrast, other scientists worry that listing a species as threatened without good justification "waters down" any listing process, potentially distracting attention from other species in need of immediate attention. The more information we have about species, the more efficient we can be in setting our priorities. The case of the aye-aye illustrates a limitation inherent in using a single, standardized, global system to assess species' conservation status, when data quality and availability varies greatly across species and regions. This case underscores the need to both intensify research efforts for little-known species, and to think critically when interpreting the IUCN Red List categories and threat assessments.

includes taxa for which trade is regulated by CITES but not prohibited, and Appendix III includes taxa that are regulated by a member government of CITES. At the global level, CITES listing has emerged as an important action to call attention to primate populations that are heavily traded. For example, in June, 2007 all *Nycticebus* (slow loris) species were transferred from Appendix II to Appendix I of CITES based on the increasing and unsustainable demand for slow lorises in international trade (Nekaris and Nijman 2007).

Regardless of your focal population or the topic of your field study, it is important to check the Red List, CITES appendices, and national and regional lists for information about your study population. These lists and other comprehensive references such as *All the World's Primates* (Rowe and Myers 2011; <http://alltheworldsprimates.org/>) are valuable resources for available population status and threats information, and they can inform you about any gaps in knowledge that your study could fill.

16.3 Determining conservation status

Successful primate conservation strategies and optimal allocation of funds for conservation efforts require accurate diagnosis of populations threatened with extinction. Resources are often limited, and if we conserve a population that is not threatened, or ignore a population that needs to be protected, we may waste valuable resources by allocating them incorrectly. An accurate diagnosis of threatened populations should follow a series of questions about the population (Cowlishaw and Dunbar 2000):

1. Is the population declining and what is the evidence for that decline?
2. Are the other known populations of this species also in decline?
3. Could the decline lead to extinction?
4. What appears to be contributing to this decline?

At the base of all these questions are issues of definitions of rarity. We often focus on rarity as a target for conservation action. We think of rare species as those with low population sizes only found in one small geographically restricted area. However, there are multiple ways of defining rarity (Hartley and Kunin 2003; Kunin and Gaston 1993) that we need to keep in mind when managing populations. For instance, widespread species such as the pig-tailed macaque *Macaca nemestrina* (IUCN 2012) may have naturally low individual population sizes and if we ignore local extinction of several populations, a widespread species may ultimately not be able to maintain enough connectivity to support reproduction and may become threatened with extinction.

Understanding the overall distribution and population abundance of your focal species is a good first step in assessing conservation status. It is useful to look both at extent of occurrence—the area contained within a minimum convex polygon around all known, inferred, or projected sites currently occupied by the taxon—as well as area of occupancy—the area within the extent of occurrence that is actually occupied. Taxa generally do not occur evenly throughout an area, and the latter fine-scale analysis helps to exclude areas of unsuitable or unoccupied habitats when thinking about conservation action (see Chapter 6). To determine the area of occupancy for a taxon, researchers can refer to gazetteers or databases including the IUCN Red List. These analyses are quite sensitive to scale, as species mapped at high spatial resolutions will appear rarer than those mapped at coarser spatial resolutions. This translates into lower extinction risk values for species with the least available information (Hartley and Kunin 2003).

The next step is to look at any previous research on the population size or distribution in your area of interest, determining methods used to estimate population size and, if they are comparable between years, trends in population size. It is useful to know if the data from these studies were observed, estimated, projected, inferred, or suspected, as this will help you to assess comparability. If this information is not available, or if previous estimates are not comparable, it may be worthwhile to begin surveys of the species under study following techniques as noted in Chapter 2 to at least estimate population size in the present. Genetics and demographic modeling (especially population viability analysis) can also help to infer population size and trends (see Chapters 14 and 15 on genetic methods to estimate population size and on demographic modeling including PVA, respectively), and are particularly useful when information from the field on population size is incomplete. Demographics may be especially important to consider; Strier and Ives (2012) found an unexpected increase in fertility rates and in adult male mortality due to an expansion of habitat and a behavioral shift with increased use of the ground in a recovering population of muriquis, emphasizing the importance of understanding demographic rates in endangered species recovery.

Based on this assessment, if the population appears to be in decline, it is useful for you to think through if the decline is exponential, linear, or accelerated, as this can give you a sense of the urgency of the situation. A population viability analysis or a population–habitat viability analysis can help you to quantitatively assess the potential for extinction (see Chapter 15). In particular it can be useful to assess the extent of available habitat (Chapter 6) and estimate the proportion of habitat that has some level of protection using a resource such as the World Database on Protected Areas (<http://www.wdpa.org/>) and in-country sources. In-depth ecological work on life history characteristics of a population is also invaluable for assessing population persistence and it is also necessary to undertake a population viability analysis (Chapter 15).

You can undertake a threats assessment to identify possible explanations for a population decline (e.g., The Nature Conservancy's Threat Ranking System; Salzer 2007). Threats can be divided into direct threats, such as habitat loss or fragmentation, unsustainable hunting, predation or competition from invasive species, or pollution, as well as indirect, or underlying, threats such as rising human population and consumption rates, weak governance, lack of enforcement of existing regulations, and so on. Methods described in Chapters 6, 8, and 18, including bushmeat and market surveys, remote sensing to assess habitat change, and interviews with local communities will be useful to assess a threats' status and trends over time. Threat analyses methods can then facilitate the determination of which of these threats seem to have the largest effect on the population. Once you have mapped out your threats you can identify which individuals, institutions, and groups contribute to the threats as well as their potential resolution and work to develop a plan to invest conservation action based on this strategic analysis. Ideally you will set measurable targets and metrics for determining success for an initiative. Sometimes with cryptic populations, for instance with Bonobos where it is next to impossible to track animal populations (Grossmann *et al.* 2008), your target may be reducing threats or measuring a proxy target such as the expanse of high-quality habitat. However, if you use habitat as a surrogate for species decline, you need to have good documentation of the relationship between habitat quality and species persistence.

Recent assessments of muriquis (*Brachyteles*) represent informative examples of efforts by researchers to assess the conservation status of endangered primates, focus research projects on continued monitoring and assessment, and inform conservation priorities and actions. A population viability analysis, and a population and habitat viability analysis for both species of *Brachyteles* (*B. arachnoides* and *B. hypoxanthus*) were conducted in the 1990s (Strier 1993; Rylands *et al.* 1998). These analyses clarified where there were gaps in the information available about muriqui conservation status (as reviewed in Strier 2000) and subsequently led to a series of surveys to improve the quality of data available to assess muriqui status. These analyses also sparked the establishment of new long-term research sites and the Committee for the Conservation of the Muriqui (Oliveira *et al.* 2005), garnering national and international attention on the threat of extinction to these unique species. With updated data from ongoing research, the Red List status of *B. hypoxanthus* changed from Endangered in 1996 to Critically Endangered in 2000 (*B. arachnoides* has remained Endangered). Population viability models are periodically repeated with updated data (e.g., Mendes *et al.* 2005) to inform priority actions for conservation and coordinated research programs to continue population monitoring, and to study the behavioral ecology, population genetics, and disease ecology of these endangered primates.

16.4 Conservation action

16.4.1 Action plans

The first step in planning conservation action for a primate population is to develop a conservation action plan, sometimes called a conservation management plan. As mentioned above, the IUCN SSC PSG produces action plans for species or groups of species. These plans summarize available knowledge on a particular species or species group and lay out a program of actions for their conservation. Getting in touch with the regional coordinator at PSG for your geographic area of interest will be important to determine whether a species action plan has already been developed for your species of interest, to what extent it has been implemented, and how you can help. In most cases, an action plan has not yet been published, but may be in development—and in almost all cases, there will be gaps in information that your study could help fill.

Other primatologists working on your taxon or region of interest are also excellent contacts to coordinate the development of an action plan. An excellent place to search for primatologists is the University of Wisconsin-Madison's Primate Info Net (<http://pin.primate.wisc.edu/>). You can also examine the list of authors for the most recent IUCN Red List assessment for your taxon of interest, and of course local and regional universities in your geographic area of interest. Local and national officials and agencies where you acquire your research permit (see

Chapters 3, 4, and 8) are a good starting point to find out whether there are local or national conservation plans existing or in development for your species or region of interest and, if not, whether they may be interested in working with you to develop one. This is especially important for species that may be listed as threatened at local or national levels but not at an international level (i.e., not on the Red List). Other key contacts to help you connect with ongoing conservation activities in an area are conservation-related non-governmental organizations working in your region of interest.

As mentioned above, action plans typically summarize available knowledge on a particular species or species group to present a detailed assessment of status and threats, and lay out a suite of recommended actions to reduce threats and improve the conservation status of regional primate populations or particular species. These actions can include both *in situ* and *ex situ* conservation activities (see Chapter 17 for more detail relating to *ex situ* conservation activities, e.g., Species Survival Plans). Often species action plans include a diversity of recommendations; for example the Asian Primate Action Plan (Eudey 1987) included recommended actions relating to trade controls, education and training, social and human dimensions of conservation, captive breeding and reintroduction, protected area establishment and management, and research and surveys. Different actions are recommended depending on the conservation context of a situation. For example, protected area management and establishment may be major action items in one region, while education, research, and surveys may be the most important actions in other regions. For a more detailed discussion and a historical perspective on a variety of conservation actions and tactics, we refer readers to Cowlishaw and Dunbar (2000).

16.4.2 Incorporating local knowledge into conservation efforts

The success of conservation efforts will often depend on consideration of local knowledge, customs, and values throughout the design, implementation, and evaluation of conservation projects. Conversely, ignoring human dimensions of ecosystems and biodiversity can result in conservation efforts that fail in their biodiversity goals and/or result in conflict with local populations (as reviewed in Sterling *et al.* 2010). In order to incorporate human dimensions into conservation action, the ecological and conservation science research used to inform conservation action needs to be multidisciplinary in nature, drawing from the social sciences as well as the natural sciences. Several chapters in this volume present multidisciplinary techniques to incorporate local knowledge, customs, and values in ecological and conservation research and practice (e.g., Chapter 8 and 17). We argue here that multidisciplinary approaches are critical not only to improve the effectiveness and sustainability of conservation actions, but also to improve our understanding of the ecology of non-human primate communities and the systems in which they live.

An example of a multidisciplinary research effort that greatly informed conservation comes from Madagascar, where primatologists gathered knowledge from local community associations, enabling them to survey several previously unknown sites with populations of greater bamboo lemurs (*Prolemur simus*; Ravaloharimanitra *et al.* 2011). Their efforts vastly increased our understanding of the distribution of this Critically Endangered taxon.

Similarly, community-based conservation projects aim to directly integrate local communities into conservation projects and to balance the needs of both wildlife and humans. However, these projects may involve communities in a myriad of different ways, depending on the conservation context and desired outcomes. Typically, programs may involve: (1) direct benefit-sharing schemes, where local people gain directly from conservation activities in the form of cash income or employment (e.g., as researchers, game scouts, or tour guides), (2) indirect benefit-sharing schemes, where local people gain indirectly through a share of profits from conservation income being used on local development projects (e.g., building hospitals or schools), or (3) local governance schemes, where local people directly manage and monitor resources (Cowlishaw and Dunbar 2000). Increasingly, conservation project personnel are working to involve local communities throughout the full life cycle of the endeavor, from planning to implementation to evaluation and adaptive management.

General recommendations for designing community-based conservation projects are difficult to generate because of the specifics of a given conservation context and desired outcomes. However, we can say that overall the question should not be whether or not one should attempt to integrate local communities and knowledge into a conservation project, but rather **how** to attempt to do so. A review of such projects suggested that they should operate at the scale of landscape mosaics, should make trade-offs between conservation and development explicit, and should be firmly rooted in local social processes, considering local knowledge, customs, and values throughout the project (Sayer 2009).

Ecotourism is a conservation strategy that often includes integration of local communities and has risen in prominence over the past several decades. The International Ecotourism Society (TIES) has defined ecotourism as "responsible travel to natural areas that conserves the environment and improves the well being of local people" (<http://www.ecotourism.org>). For primates, the potential for successful ecotourism as well as its pitfalls have been exemplified by the Mountain Gorilla Project in the Parc National des Volcans (PNV) in Rwanda. From 1978, when tourism was introduced at the site, through to the end of the 1980s, revenue from tourism resulted in successful improvements to park protection, an increase in the recruitment rates of the mountain gorilla population, and an expansion of the program to include environmental education initiatives in local communities as well as a veterinary program (Butynski and Kalina 1998). However, civil war in the 1990s caused demand for tourism to shrink and resulted in closure of the PNV from 1997–1999, revealing tourism as a precarious income source. More recently, revenue from tourism has increased again to surpass previous highs, but local communities around the PNV do not seem to be reaping the benefits of the program, with only 6% of generated revenue reaching them, although policies are in place to attempt to address this imbalance (as reviewed in Rogers 2011). Recent research has also documented disease transmission between the mountain gorillas and humans, including tourists and researchers, further complicating cost–benefit analyses of the program (as discussed in Cranfield 2008).

Another key example of a tourism project is the Napo Wildlife Center (NWC, <http://www.napowildlifecenter.com/>). The NWC is a wildlife refuge and an award-winning ecotourism resort in Yasuní National Park in Ecuador. The leaders of the Quechua Añangu Community, an indigenous community living within the park, started the tourism project and sought initial funding through foreign investment for the construction of a lodge and infrastructure. After the initial investment, the Añangu have owned and operated the NWC since 2007, and the success of the project has enabled the financing of local schools for the community as well as a full-time doctor. Yasuní supports primate populations from at least ten primate species (Bass et al. 2010), and its overall biodiversity has inspired the Yasuní-ITT Initiative by the Ecuadorian government, which seeks to leave the oil fields in the park untapped in exchange for international compensation (Finer et al. 2010).

Future ecotourism projects should consider local knowledge, customs, and values throughout the design, implementation, and evaluation of the project. They should also adhere to best practice guidelines that emphasize a cautious approach that balances economic and conservation objectives and is backed up by impact studies and risk assessments (Macfie and Williamson 2010). In particular, see Chapter 4 for a discussion of key considerations about the risk of human to non-human primate disease transmission (and vice versa) at ecotourism destinations.

A related issue to be aware of is that approaches focused solely on the conservation of primate populations may ignore or even deter the conservation of other flora and fauna that form the ecological community in which primates live. Because non-human primates often inspire fascination in humans, primates are often effective "flagship species," attracting the attention of humans (and funding) towards a threatened community or ecosystem. However, it is important to consider how the fascination of humans with primates or a particular primate taxon might have unintended consequences on the conservation of other species, or on area-based or habitat-based conservation actions (Rose 2011).

16.4.3 How researchers can help, just by doing research

In many cases, primatologists can contribute to the conservation of primate populations just by continuing to do their research. We discussed in the introduction how to choose a research site and whether an established or new site might be best for a particular project. An important function of long-term research sites is that they serve as refugia for threatened and particularly unsustainably harvested species. For example, in Tai National Park, Cote d'Ivoire, a survey showed significantly higher primate and duiker encounter rates in long-term research areas as compared to adjacent areas within the national park. Conversely, evidence of poaching was significantly lower within the research area (Campbell *et al.* 2011b).

In addition, primatologists are increasingly interested in understanding the ecology and behavior of primate populations living in human-dominated environments as humans continue to develop and modify primate habitats. In this volume, there are many case studies relating to studying primate populations in human dominated environments (e.g., Box 14.1, Chapters 3, 8, and 18). This research can directly feed into plans for how to conserve primate populations in human-dominated areas (e.g., in the matrix and not only in protected areas). This is especially important for non-forest-dwelling species with habitat requirements that closely overlap those of humans (e.g., Bonnet macaques, Singh *et al.* 2011a; chacma baboons, Hoffman and O'Riain 2012).

Finally, researchers are well positioned to contribute directly to the capacity development of the next generation of primatologists and of local communities of engaged citizens. Perhaps the simplest way to do so is through partnering with local scientists and by hiring field assistants, which could include local community members as well as undergraduate and graduate students from local and regional universities near your field site and from your home institution. Long-term field sites also have the potential to offer longer-term employment and increased financial stability to local communities. In addition, researchers can take advantage of freely available and adaptable teaching and capacity development resources on topics related to ecology and conservation (e.g., materials for all grade levels from the Ecological Society of America such as the EcoEdNet repository <http://www.ecoed.net/>; resources for teachers such as the Network of Conservation

Educators and Practitioners <http://ncep.amnh.org>; and community and resource professional resources such as those available through the Conservation Gateway <http://www.conservationgateway.org/>).

16.5 Conclusion

In this chapter, we have provided an overview of how to determine the conservation status of your study organism(s) and how to contribute to conservation action *in situ*. In particular, we have emphasized the importance of developing conservation action plans in collaboration with local researchers, managers, and communities in recognition that the success of conservation strategies will often depend on consideration of local knowledge, customs, and values. In order to incorporate human dimensions into conservation action, the ecological and conservation science research used to inform conservation action needs to be multidisciplinary in nature, drawing from the social sciences as well as the natural sciences. We argue here that multidisciplinary approaches are critical not only to improve the effectiveness and sustainability of conservation actions, but also to improve our understanding of the ecology of non-human primate communities and the systems in which they live.

Acknowledgments

We thank two anonymous reviewers for their thoughtful comments on an earlier draft of this manuscript. We also thank Connie Rogers for her guidance and inspiring discussion about community-based ecotourism focused on primate populations.

17

Captive breeding and *ex situ* conservation

Dean Gibson and Colleen McCann

17.1 Introduction

The IUCN red data book currently lists 415 primate species of which 37 are critically endangered, 86 are endangered, and 78 are vulnerable to decline (IUCN 2012). These represent nearly half of the primate order (Vie *et al.* 2009; Rowe and Myers 2011). The continued acceleration of threats to species diversity has resulted in many populations being actively managed by humans. The intensive management of populations by zoos can play a major role in a species' recovery and conservation (Baker *et al.* 2011). But in order to do this, populations must be genetically robust, demographically viable, and sustainable in perpetuity (Lees and Wilcken 2009). In an evaluation of the world's vertebrates, Hoffmann *et al.* (2010) found that captive breeding played a role in the recovery of 17 of 68 threatened species, one of which was a primate species, the golden lion tamarin (*Leontopithecus rosalia*). While *ex situ* conservation such as captive breeding can be combined with *in situ* conservation actions, captive breeding is presently absent from, or at best plays an insignificant role in most conservation efforts and government policies (Conde *et al.* 2011). Therefore, the question of whether or not *ex situ* efforts via captive breeding can play a role in assisting *in situ* primate conservation warrants a more thorough investigation.

This chapter presents a summary of current zoo management efforts toward *ex situ* sustainability (defined as a methodology for providing assurance colonies of primate species for *in situ* needs) and conservation support. We also provide an overview of available historic and current captive primate population census data and include several examples of success and failure at maintaining survivorship and successful reproduction in captive populations. Additionally, we present two unique and contrasting case studies that highlight these challenges.

Primate Ecology and Conservation: A Handbook of Techniques. First Edition. Edited by Eleanor J. Sterling, Nora Bynum, and Mary E. Blair. © Oxford University Press 2013. Published 2013 by Oxford University Press.

17.1.1 Brief synopsis of *ex situ* primate survivorship and evolution of zoos

In the late 19th century, the majority of wild primates captured for *ex situ* collections died while still in transit (Wallis 1997). For the 1 in 25 estimated to have survived capture and the long journey to captivity, *ex situ* life was cut short due to transmission of human disease, inappropriate (e.g., solitary) housing, and the lack of expertise in animal husbandry (Wallis 1997). Since that time, zoos progressed with improvements in primate survivorship and propagation, but were still considered to be wildlife consumers rather than producers well into the 1970s (Perry *et al.* 1972; Stevenson 1983). By the 1980s, captive breeding for many primates was routinely successful; however, maintaining population growth and genetic diversity became the new challenge (Flesness 1986). Through the 1990s zoos continued to struggle with successful population management, while also emerging as respected conservation organizations with many noteworthy contributions to *in situ* primate conservation (Wiese and Hutchins 1997). Today's zoos range from roadside menageries to highly developed conservation parks (West and Dickie 2007). The world's leading zoos focus their efforts on conserving wildlife (Lees and Wilcken 2009) via applied field research, habitat protection, and conservation planning assistance, while increasing the awareness of the millions of zoo visitors worldwide by providing conservation education messages about the precarious relationship between humans and the natural world (Wiese and Hutchins 1997; Conway 2011).

The mission of modern zoos is centered on three main objectives: education, research, and species conservation; and the living animal collection is the cornerstone to achieving those objectives (Fig. 17.1). As such, strategies must be in place for the maintenance of both animal collections within individual zoos and throughout the zoo community. Zoos develop collection plans as roadmaps for determining which species are maintained over time, and based on objectives that meet their education and conservation goals. The long-term sustainability of animal populations in zoos is dependent on regional collaborative efforts and sound population management that ensures survivability and propagation (Box 17.1).

The Animal Record Keeping System (ARKS), International Species Information System (ISIS), and studbooks for individual species populations available to zoo managers are the tools used in population analyses to determine current and future growth potential and sustainability. Population management methods are aimed at retaining as much of the genetic variability of the *ex situ* population as possible while meeting demographic needs. *Ex situ* populations that meet these

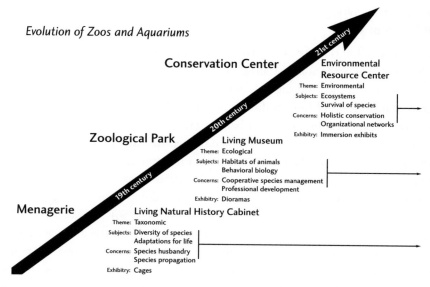

Evolution of Zoos and Aquariums

Conservation Center

21st century

Environmental
Resource Center
Theme: Environmental
Subjects: Ecosystems
Survival of species
Concerns: Holistic conservation
Organizational networks
Exhibitry: Immersion exhibits

Zoological Park

20th century

Living Museum
Theme: Ecological
Subjects: Habitats of animals
Behavioral biology
Concerns: Cooperative species management
Professional development
Exhibitry: Dioramas

19th century

Menagerie

Living Natural History Cabinet
Theme: Taxonomic
Subjects: Diversity of species
Adaptations for life
Concerns: Species husbandry
Species propagation
Exhibitry: Cages

Fig. 17.1 Evolution in the mission of zoos. The horizontal arrows indicate that professional capacities of concern and subjects communicated to the public in earlier phases of zoo development are now vital services to conservation. Reprinted with permission from the Chicago Zoological Society (IUDZG/CBSG 1993).

criteria can serve as valuable assurance populations for *in situ* species conservation efforts.

17.2 Primate population data sets and sustainability

The intensive management of populations by zoos can play a major role in species' recovery and conservation but, for this to occur, populations must be genetically robust, demographically viable, and sustainable. In this section, we provide an overview of available captive primate population census data and the sustainability of captive primate populations. The results of six available captive primate population analyses that span the years 1972 through 1986 are listed in Table 17.1. Making conclusive comparisons is difficult due to the varied parameters and data sets used by each author. However, as a component of their overall review of the data, two authors did report on population viability based on the Perry *et al.* (1972) criteria: defined as "a living population size of 100 of which 50% are captive born." The results of Perry *et al.* (1972), Stevenson (1983), and Flesness (1986) validate a positive trend toward the establishment of sustainable captive primate populations

Box 17.1 *Ex situ* cooperative breeding programs*

Cooperative breeding programs[1]

On their own, animal collections in individual zoos and aquariums are typically too small to be of much value to long-term conservation. Cooperative regional or international *ex situ* breeding programs are required to form larger, viable populations (WAZA 2011). Cooperative breeding programs provide the following:

- Demographically and genetically viable populations for *ex situ* species programs;
- Animals for public educational and/or exhibit initiatives;
- Fundraising opportunities for species conservation initiatives; and
- Research opportunities to gain species biology and husbandry knowledge.

In order for *ex situ* populations to serve all of these functions, populations must be viable and sustainable over the long term. This requires that they be demographically stable, genetically healthy, reproductively self-sustaining, and of sufficient size in order to maintain genetic diversity.

There are three such cooperative programs currently in place: the Species Survival Plan (SSP®), the European Endangered Species Program (EEP), and the Global Species Management Plan (GSMP)**.

Species Survival Plan[2]*

The Species Survival Plan (SSP®), an Association of Zoos and Aquariums (AZA) copyrighted breeding and conservation program, is designed to maintain a healthy, self-sustaining, genetically diverse and viable, and demographically stable population of a species in human care; and to guide zoo and aquarium-based efforts towards *in situ* species conservation. Each SSP has a qualified species coordinator who, along with a management committee and expert advisors, is charged with managing the species for AZA institutions. SSP Programs are managed by their corresponding Taxon Advisory Group (TAG), and charged with developing a comprehensive population studbook with a record of all historical and current individuals in the population; a breeding and transfer plan that identifies population management goals; and recommendations that consider each animal's social and biological needs as well as its transfer and breeding feasibility. All recommendations are designed to maintain or increase a genetically diverse and demographically stable population. Additional functions include:

- Collaborating with other institutions/agencies to ensure integrated conservation initiatives;
- Increasing public awareness of wildlife conservation issues;
- Developing and implementing *ex situ* and *in situ* education strategies;
- Developing *in situ* re-introduction programs, if possible;
- Providing a discussion forum for topics applicable to the species; and

Box 17.1 *Continued*

- Providing species-specific information for the development of Animal Care Manuals detailing the best husbandry practices for maintaining species in *ex situ*.

Sustainability of zoo-based populations in AZA

In 2011, the AZA implemented a change in the structure of its population management strategies in order to increase the long-term sustainability of zoo-based species programs. This re-evaluation of species programs included an analysis of sustainability for each taxonomic group (using ZooRisk, Vortex, and other available population analysis software), identification of achievable goals for attaining population sustainability, assessment of current impediments to achieving sustainability, and the development of recommendations for modifications to SSP programs to increase population sustainability. Population sustainability of each managed species program was assigned one of three levels: Green, Yellow, or Red based on a set of genetic and demographic criteria.

GREEN SSP Programs:

- These populations are currently sustainable for the long term.
- The population is demographically sustainable for greater than 100 years or greater than 10 generations.
- The population is able to maintain greater than 90% gene diversity over this time.

YELLOW SSP Programs:

- The population currently cannot retain > 90% gene diversity for greater than 100 years or more than 10 generations.
- Factors affecting sustainability may include:
  - ❖ too few individuals in the population;
  - ❖ insufficient space available for maintaining a sustainable population;
  - ❖ lack of husbandry and breeding expertise;
  - ❖ low gene diversity;
  - ❖ poor demographics.

RED Programs

- The population is not currently sustainable.
- The population has fewer than 50 individuals.
- The program is not designated as an SSP program.
- The program is managed as an official AZA Studbook if the Taxon Advisory Group (TAG) recommends the species in the Regional Collection Plan (RCP).

In 2010, at the time of the program review, seven primate species met the criteria for the Green SSP program level: Golden lion tamarin (*Leontopithecus rosalia*), cotton-top tamarin (*Saguinus oedipus*), western lowland gorilla (*Gorilla gorilla gorilla*), siamang (*Symphalangus syndactylus*), chimpanzee (*Pan troglodytes*),

Sumatran orangutan (*Pongo abelii*), Bornean orangutan (*Pongo pygmaeus*), and bonobo (*Pan paniscus*). Programs are evaluated on a quarterly basis and in the latest population review (2011) three additional primate populations have met the sustainability criteria: black-and-white ruffed lemur (*Varecia variegata*), squirrel monkey (*Saimiri sciureus*), and the guereza (*Colobus guereza*).

European Endangered Species Program[3]

The European Endangered Species Program (EEP) is the population management program for species maintained by the European Association of Zoos and Aquaria (EAZA). Each EEP has a coordinator who is assisted by a species committee. The coordinator collects information on the status of all the animals kept in EAZA zoos and aquariums of the species for which he or she is responsible; produces a studbook; carries out demographical and genetic analyses; produces a plan for the future management of the species; and provides breeding and transfer recommendations to participating institutions.

Global Species Management Plan[1]

In 2003, WAZA adopted a procedure for establishing inter-regional programs, which may relate to a number of species for which international studbooks have been established. These programs, called Global Species Management Plans (GSMP), are those officially recognized and endorsed by WAZA.

A GSMP can provide important links between *ex situ* management of a species and *in situ* conservation initiatives. This is particularly valuable when both the *ex situ* and *in situ* populations are small and fragile, requiring intensive management of the species in order to avoid extinction. In such cases, all available resources and concerted efforts are applied to a unified conservation goal. In many cases, well-managed *ex situ* populations may contribute to the conservation of a species *in situ*, as ambassadors to raise funds and awareness for habitat and population protection, as well as by providing assurance populations for re-introduction initiatives when applicable. The Javan (or silvery) gibbon (*Hylobates moloch*) program is a recognized GSMP by WAZA.

* Full description of cooperative breeding programs and various facets of population management for zoo-based species programs is available from the WAZA, AZA and EAZA websites:

[1] WAZA.ORG <http://www.waza.org/en/site/about-waza/who-we-are>.

[2] AZA. ORG <http://www.aza.org/about-aza/>.

[3] EAZA.NET <http://www.eaza.net>.

** Additional regional associations and comparable *ex situ* species programs exist in some form for other geographic areas.

*** For the purposes of this chapter, we provide a more in depth description of the AZA species program as an example of the population management structure and processes for maintaining *ex situ* breeding programs.

Table 17.1 Summary table of primate species census analyses and those utilizing Perry et al. (1972) criteria.*

Author/date	Data used	Perry et al. criteria	Results	Additional or other analysis	Results
Perry et al. (1972)	IZY Rare Census through 1970 (species listed by IUCN as Rare or Endangered)	Yes	Zero primate species met Perry et al. criteria; *Eulemur mongoz* might have if data were complete, *Leontopithecus rosalia* and *Pongo pygmaeus* populations were promising	No	
Pinder and Barkham (1978)	IZY Rare Census through 1975 (species listed by IUCN)	No		Yes	26 mammal species determined to be self-sustaining, 2 were primate species: *Eulemur mongoz* and *Macaca silenus*.
Stevenson (1983)	IZY Rare Census through 1979 (IUCN listed and all species)	Yes	8 IUCN listed primate populations met Perry et al. criteria— (*Eulemur macaco, Eulemur mongoz, Varecia variegata, Saguinus oedipus, Leontopithecus rosalia, Macaca silenus, Macaca sylvanus, Pongo pygmaeus*)	Yes	Additional estimates/analysis increased number to 17 species. *150 institutions reporting to ISIS (most from USA)
Schmidt (1986)	ISIS data through 1983, IUCN listed species	No		Yes	20 primate populations had over 30 individuals; only 15 of these were increasing.
Flesness (1986)	ISIS data through 1984; ISIS data fraction plus 1981 IZY extrapolation (IUCN and all species)	Yes	24 species (ISIS data only) met Perry et al. criteria (IUCN listed species: *Eulemur macaco, Callimico goeldii, Leontopithecus rosalia,*	Yes	An additional 3 species met Perry et al. criteria with ISIS + IZY (*Saguinus imperator, Macaca sylvanus, Microcebus murinus*;

| Gibbons (1995) | ISIS data through June, 1994 | No | Pongo pygmaeus; Non listed species: Lemur catta, Varecia variegata, Galago senegalensis, Aotus trivirgatus, Cebus apella, Saimiri sciureus, Ateles geoffroyi, Callithrix jacchus, Cercopithecus diana, C. neglectus, Erythrocebus patas, Macaca fuscata, M. nigra, Papio hamadryas, Mandrillus sphinx, Colobus guereza, Hylobates lar, Symphalangus syndactylus) | Reports on 1994 primate population figures, lumps 1989–1994 captive bred numbers together. | Perry et al. criteria with extrapolations (Callithrix pygmaea, Saguinus fuscicollis, Cercocebus torquatus, Chlorocebus aethiops, Cercopithecus mitis, Miopithecus talapoin, Macaca fascicularis, M. mulatta, Papio cynocephalus, Theropithecus gelada); 9 species were determined to be borderline (Eulemur mongoz, Cheirogaleus medius, Gorilla gorilla, Pithecia pithecia, Pan troglodytes, Callicebus moloch, Miopithecus talapoin, Macaca nemestrina, Semnopithecus entellus). 200 institutions from 14 countries reporting to ISIS at this time. | Most species are unlikely to be genetically sustainable, only 23 have population sizes greater than 100. 17 species meet the criteria for genetically self-sustaining populations. Only a few species account for the majority of individual primates and captive births within geographic groups. This unequal distribution is noted as a cause for sustainability concerns. |

* Perry et al. criteria: defined as "a living population size of 100 of which 50% are captive born."

by demonstrating that while no species in 1972 met the basic criteria for a sustainable captive population, 46 species might have met these criteria in 1986.

We investigated whether a positive trend toward basic sustainability potential has continued, using the most up-to-date information available from the 2009 International Species Information System (ISIS) data, which represents 780 institutions in approximately 80 countries (see <http://www.ISIS.org>; Bingaman-Lackey, pers comm.; Table 17.2). A caveat of our comparison is that this data set is considerably larger than previous data sets (e.g., Stevenson 1983; Flesness 1986). Also, to facilitate direct comparison across data sets we used the Perry *et al.* (1972) criteria, which do not incorporate demographic and genetic analyses, such as those used by today's AZA management programs. Such analyses are equally necessary for determining accurate population viability.

Of the populations represented in the 2009 ISIS data set, 55 species (28%) meet the basic "sustainability" criteria first established by Perry *et al.* (1972), representing an increase of 31 species compared to the last analysis by Flesness (1986). However, we also compared our results to Flesness's extrapolated data set (see Table 17.1) where 46 species (40%) were identified as potentially meeting the Perry *et al.* (1972) criteria. In this comparison, the 55 species (28%) in 2009 is an increase of only 9 additional potentially sustainable species over 25 years. Thus, on a percentage basis, we have fewer potentially sustainable primate populations in 2009 than we had in 1986. This is irrespective of having triple the number of reporting institutions and 72% additional species (115 species in 1986 compared to 198 species in 2009). While many aspects of captive management have improved and have been successful over the last 25 years, these numbers indicate some cause for concern about the current status of sustainable captive primate populations.

17.2.1 Current trends

How have *ex situ* primate populations fared over the years? In addition to the genetic and demographic parameters influencing population sustainability, other species-specific variables, both intrinsic and extrinsic, can bolster or plague the sustainability of a given population. Examples of North American primate populations that have experienced continued success in captivity throughout the years include: cotton-top tamarins (*Saguinus oedipus*), golden lion tamarins, guerezas (*Colobus guereza*), black-and-white ruffed lemurs (*Varecia variegata*), and western lowland gorillas (*Gorilla gorilla gorilla*). While species-specific characteristics dictate the husbandry and management practices in each of these captive breeding programs, the common denominator for their success has been the availability of founder individuals, sufficient knowledge of their reproduction requirements, and the maintenance of appropriate social groupings. Coupled with strong institutional

Table 17.2 *Primate population census data set (1971–2009).*

Species	Perry et al. (1972)			Stevenson (1983)			Flesness (1986)			This chapter		
	IZY[1] 1971	# Captive born[2] in 1971	% Captive born in 1971	IZY 1979	# Captive born in 1979	% Captive born in 1979	ISIS[3] 1984	# Captive born in 1984	% Captive born in 1984	ISIS 2009	# Captive born in 2009	% Captive born in 2009
PROSIMIANS												
Lemur catta	73	28					579	529	91	2484	2406	97
Eulemur macaco			38*	111		80	495	442	89	239	223	93
Eulemur fulvus				223		80				139	114	82
Lemur m. mayottenus				171		60						
Eulemur rufus	43	15	35*	70		80				79	79	100
Eulemur rubiventer										136	126	93
Eulemur albifrons				54		80				135	130	96
Eulemur mongoz	167	64	38*	134		64	46	21	47	111	105	95
Varecia variegata				147		80	388	363	94	800	779	97
Varecia rubra										577	559	97
Cheirogaleus medius				25		75	61	58	95	41	41	100
Microcebus murinus				85		81				145	144	99

(*continued*)

Table 17.2 *Continued*

Species	Perry et al. (1972)			Stevenson (1983)			Flesness (1986)			This chapter		
	IZY[1] 1971	# Captive born[2] in 1971	% Captive born in 1971	IZY 1979	# Captive born in 1979	% Captive born in 1979	ISIS[3] 1984	# Captive born in 1984	% Captive born in 1984	ISIS 2009	# Captive born in 2009	% Captive born in 2009
Galago senegalensis										85	72	85
Nycticebus pygmaeus										171	136	80
NEW WORLD MONKEYS												
Aotus trivirgatus							142	84	59	31	25	81
Cebus apella							121	69	57	763	602	79
Cebus capucinus										148	103	70
Cebus xanthosternos										133	113	85
Saimiri sciureus							409	209	51	1116	762	68
Saimiri boliviensis										712	671	94
Pithecia pithecia				52		46				354	338	95
Ateles geoffroyi							236	137	58	470	362	77
Ateles fusciceps										345	282	82
Cacajao calvus	38	4	11	27		38				8	2	25
Alouatta caraya										234	220	94
Callimico	16	6	38	84		68	131	121	92	410	405	99

Leontopithecus rosalia	76	39	51	134	82	394	384	98	457	442	98
Leontopithecus chrysomelas									352	341	97
Saguinus oedipus				427	70	242	193	80	1068	1033	97
Saguinus bicolor									152	146	96
Saguinus imperator						64	39	61	411	404	98
Saguinus midas									328	294	90
Saguinus labiatus						185	161	87	167	163	98
Callithrix jacchus									689	606	88
Callithrix geoffroyi				36	60				374	361	97
Callithrix pygmaea									674	627	93
OLD WORLD MONKEYS											
Cercopithecus diana						127	79	62	110	108	98
Cercopithecus neglectus						126	98	78	208	197	95
Erythrocebus patas						127	93	73	270	201	74

(continued)

Table 17.2 Continued

Species	Perry et al. (1972)			Stevenson (1983)			Flesness (1986)			This chapter		
	IZY[1] 1971	# Captive born[2] in 1971	% Captive born in 1971	IZY 1979	# Captive born in 1979	% Captive born in 1979	ISIS[3] 1984	# Captive born in 1984	% Captive born in 1984	ISIS 2009	# Captive born in 2009	% Captive born in 2009
Chlorocebus aethiops										177	129	73
Colobus guereza							256	184	72	587	572	97
Trachypithecus auratus										227	198	87
Macaca fascicularis										305	225	74
Macaca mulatta										391	244	62
Macaca nemestrina										312	181	58
Macaca tonkeana										127	117	92
Macaca silenus				288		80	205	162	79	461	433	94
Macaca sylvanus				607		80	62	52	84	374	266	71
Macaca fuscata							217	187	86	732	653	89
Macaca nigra							181	126	70	248	222	90
Papio hamadryas							105	70	67	1442	1060	74
Mandrillus sphinx							191	145	76	549	527	96
Theropithecus gelada										156	155	99

APES

Nomascus gabriellae						114	93	82
Nomascus leucogenys						160	121	76
Hylobates lar			228	127	56	450	360	80
Symphalangus syndactylus			122	76	62	324	274	85
Pan paniscus	21	4	19	32	45	171	140	82
Pan troglodytes	152	28	677		63	1196	837	70
Pongo pygmaeus	539		341	228	67	584	464	79
Gorilla gorilla gorilla			275	133	48	761	647	85

[1] *International Zoo Yearbook (IZY) Census of Rare Animals in Captivity began in 1962 (Stevenson 1983) and was last printed in 1998.*

[2] *"# Captive born" is the total number of living captive born animals in the population (born in prior years and year indicated).*

[3] *ISIS census data began in 1975 and continues to this day.*

* No. captive born not reported by Madagascar.

commitments to work collaboratively towards maintaining the species in each region, these factors have contributed to the viability and future sustainability of the populations.

Conversely, other captive populations have experienced a dramatic decline in numbers and institutional support. Again, the failures can be attributed to both intrinsic and extrinsic factors. One example that illustrates this is the endangered lion-tailed macaque (*Macaca silenus*), which experienced perhaps the greatest decline of any captive population in North America. The decline was initiated as a result of disease issues, but ultimately a lack of institutional commitment led to an unsustainable captive population (Carter 2005). In the 1980s, with the start of the formal captive breeding programs (SSP and EEP), the lion-tailed macaque became one of the pioneering cooperative breeding programs for an endangered primate. In the 1990s, concern over Herpes B virus emerged when a long-term population manager was no longer in place. Thus, the lack of a champion in place for the population, combined with a growing concern over disease issues, likely led to a steady decline in institutional support for the species, which resulted in a cessation of population growth. Despite increased knowledge of disease risk assessments and the implementation of personal protective measures for primate caretakers, the lion-tailed macaque population continues to decline in numbers. According to the 2010 studbook, the last birth in the population occurred over six years ago and only two females in the population are under the age of 10 years (Carter 2010). Based on these conditions, it is estimated that more than half of the current holders of the species in North America will no longer be able to maintain this species in their zoo. If this trend is not reversed, the population will phase out of AZA zoos through attrition.

Other population failures can be attributed to reproductive challenges, for instance with the Diana monkey (*Cercopithecus diana*) and Goeldi's monkey (*Callimico goeldii*). Diana monkeys historically have had poor reproductive success in North America largely due to a lack of successful breeding of mated pairs. This appeared to be a result of the poor breeding performance by some individuals negatively impacting the growth of this small, fragile population (Whittaker 2008). In contrast, the Goeldi's monkey population was one that was relatively large and stable early on in its establishment in zoos. This early success led to the employment of reversible contraception to manage the growth of the population. Several females in the population that were selected not to breed were given melengestrol acetate (MGA) contraceptive implants. In subsequent years, after population breeding goals changed and the MGA implants were removed, reproduction in the treated females was not fully restored. It was discovered that the MGA implants lead to reproductive pathologies that likely impacted the reversibility of the contraception (Murnane *et al.* 1996; Asa and Porton 2009). This was a devastating setback for this population and what emerged was a cautionary lesson on the experimental nature of reproductive manipulations with captive species.

Lastly, some populations have ultimately not succeeded as a result of a lack of available space needed to expand the population for growth and sustainability. For instance, some species, such as the mongoose lemur (*Eulemur mongoz*), had breeding moratoriums in place for several years due to the lack of additional institutions needed to hold the species. This is despite the fact that the species is of conservation concern, and husbandry and reproductive techniques have been successfully achieved.

Thus, the continued challenge of population sustainability lies with the species that have not reached a basic level of success—and the answer to why this occurs is often unique to each species and the historical factors impacting their population growth.

17.3 Recreation, education, and conservation

While maintaining *ex situ* populations for re-introduction may be the inspiring goal of many zoos, it is a fair conclusion that most *ex situ* primate populations will not directly assist with *in situ* conservation efforts. However, *ex situ* primate populations, regardless of size and sustainability, serve a conservation purpose through their enormous public appeal, which gives them an equally important ambassador role. Simply put, the educational value of primates in today's zoos cannot be overstated.

Many zoos have sophisticated education programs as well as recreation options such as daily animal shows, animal feedings, or presentations featuring animal ambassadors. Experiencing a live animal up close inspires awe and wonder while providing an opportunity to present important educational and conservation messages. These are valid and important roles for zoos (Lacy 2010), which along with aquariums are visited by approximately 700 million people around the world on a yearly basis (WAZA 2011). It is evident that program animals are engaging and generate excitement and interest from zoo visitors. Research has demonstrated that zoo visitors exposed to animal ambassadors were more likely to ask questions, experienced higher rates of learning and retention, and spent more time watching exhibit animals (Povey and Rios 2002). Primates serve as popular exhibit animals and they are excellent ambassadors for the ecosystems they inhabit, and as a result, play a key role in zoo conservation education efforts (Mittermeier 1997).

The bridge between conservation and recreation is education (Western 1986), and the expertise of zoo educators is a valued conservation resource (Hutchins and Conway 1995). Techniques and materials developed by zoo educators are often donated by zoos and used by field conservation projects for *in situ* outreach and educational programs. Conservation education, while important to the urban zoo visitor, is even more so to communities within countries where wild primate species still exist. As an example of the reach of education and conservation in zoos, over the last ten years AZA zoos formally trained more than 400,000 science teachers and supported more than 3700 conservation projects totaling $90,000,000 annually in more than 100 countries (AZA 2011).

Additionally, members of WAZA collectively contribute $350 million per year for conservation projects, which include education and research programs in the wild, making them the third major contributor to conservation worldwide after The Nature Conservancy and World Wildlife Fund global network (Conde *et al.* 2011). Not only do zoos design and deliver effective environmental education programs, support wildlife research, and provide funds, manpower, and expertise to support conservation efforts (Lees and Wilcken 2009), they also cooperate to educate politicians and influence political decisions concerning conservation and wildlife issues (Hutchins and Conway 1995).

Table 17.3 *A Sample of primate research conducted in zoological settings.*

Primate taxa	Research category	Reference
Primates	Welfare	Kagan and Vesey 2010
Primates	Genetics	O'Brien 2006
Lemuridae	Nutrition	Villers *et al.* 2008
Lemuridae	Re-introduction	Britt *et al.* 2004
Daubentoniidae	Morphology	Krakauer *et al.* 2002
Indridae	Reproduction	Brockman *et al.* 1995
Loridae	Husbandry	Fitch-Snyder *et al.* 2008
Callitrichidae	Population Management	Ballou *et al.* 2010
Callitrichidae	Reproduction	Baker and Woods 1992
Cebidae	Life history	Fernandez-Duque 2012
Cebidae	Husbandry	Jens *et al.* 2012
Cercopithecinae	Physiology	Beehner and McCann 2008
Cercopithecinae	Behavior	Melfi and Feistner 2002
Colobinae	Life history	Shelmidine *et al.* 2009
Hylobatidae	Systematics	Geissmann and Orgeldinger 2000
Pongidae	Health	Murphy *et al.* 2011
Pongidae	Behavioral Management	Stoinski *et al.* 2004

17.4 Research

Zoos provide a unique opportunity to conduct research that enhances our understanding of a species' biology and behavior. Often these types of data are difficult to obtain under field conditions where close monitoring of specific behaviors may not be as feasible, and thus may be more conducive in a captive setting. Research conducted in zoos can be categorized as "basic research" which yields key information on a species' biology or theoretical framework, or "applied research" that can produce data to help inform a practical situation. The challenges of captive research design aside (e.g., small sample sizes, confounding variables, and independence of data points), data obtained from *ex situ* studies can be applied to both *in situ* and *ex situ* research questions. Table 17.3 provides a sample of the types of primate research conducted in zoos to highlight the breadth of research topics covered. There are a multitude of primate studies being conducted in zoos and, as a result, important contributions are being made to our understanding of various aspects of a species' biology (Hosey *et al.* 2009). However, for the purposes of this chapter we provide a limited sample of studies with the aim of illustrating the significant value of primate zoo collections for scientific research. For a more thorough review of the types of research conducted on zoological collections refer to Hosey *et al.* (2009).

17.5 Re-introduction

The ideal *in situ/ex situ* conservation partnership has been aptly expressed by Bob Lacy, "the *in situ* population is the goal of the conservation program, while the *ex situ* population is a temporary way station to help assure that the *in situ* population will persist" (Lacy 2010, p. 28). Zoos have made tremendous efforts with many species to achieve this goal, some of which were through the technique known as re-introduction, defined by IUCN as "an attempt to establish a species in an area which was once part of its historical range, but from which it has been extirpated or become extinct" (IUCN 1998, p. 2). Therefore, the purpose of a re-introduction project should be to re-establish a self-sustaining wild population and to maintain the viability of that population (Baker 2002).

   Successful re-introduction involves establishing institutional and governmental cooperation, securing protected habitat, mitigating threats to survivorship, and finally, securing commitments to the demanding logistics of both pre- and post-release monitoring. The World Conservation Union Species Survival Commission (IUCN/SSC) Re-introduction Specialist Group (RSG) provides many resource guidelines for the re-introduction of species into nature (IUCN 1998), as well as

Table 17.4 *Primate re-introduction case studies.*

Species	Location	Ranking	Citation
Varecia variegata variegata	Eastern Madagascar	Successful	Soorae 2008
Hylobates spp.	Indonesia	Successful	Soorae 2008
Pan troglodytes	Robondo Island, Tanzania	Successful	Soorae 2008
Gorilla gorilla gorilla	Bateke Plateau, Congo and Gabon	Successful	Soorae 2008
Nycticebus coucang	Sumatra, Indonesia	Partially successful	Soorae 2008
Ceropithecus mona	Southern Nigeria	Partially successful	Soorae 2008
Leontopithecus rosalia	Atlantic Coastal Forest, Brazil	Highly successful	Soorae 2010
Pan troglodytes schweinfurtheii	Conkouti-Douli National Park in Congo	Successful	Soorae 2010
Pongo abelii	Sumatra, Indonesia	Successful	Soorae 2010
Pongo pygmaeus	East Kalimantan, Indonesia	Partially successful	Soorae 2010

documents detailing best practices for specific taxa, including primates (e.g., Beck *et al.* 2007). In 2008 and 2010 the IUCN/SSC RSG invited managers of re-introduction projects to rank their projects into one of four categories: highly successful, successful, partially successful, and failure (Table 17.4)—and highlighted the lessons learned in order to provide insights into the re-introduction process and increase the chances of success of future re-introduction initiatives (Soorae 2008, 2010).

The golden lion tamarin (GLT) re-introduction project is perhaps the most well-documented primate re-introduction project and serves to highlight potential difficulties with the re-introduction process. This project became feasible in the mid-1980s only after the emergence of a self-sustaining captive population, the establishment of a protected area in Brazil, and the commitment of a long-term field study (Kierulff *et al.* 2002). While this re-introduction is generally viewed as highly successful, research has shown (as with many species except ungulates) that captive-bred animals reared in caged environments lack proficiency in wild survival skills and behavior (Beck *et al.* 1994, 2002; Hoffmann *et al.* 2010). Although the re-introduction program evolved to attempt to improve survival skills, the GLT project data are consistent with other re-introduction data in concluding that specific survival skill training did not promote survivorship for captive-born animals (Beck *et al.* 2002).

Another option for *ex situ* populations to partner with *in situ* conservation efforts is through restocking, reinforcement, or supplementation, which is defined as "the addition of individuals to an existing population of conspecifics" (IUCN 1998, p. 2). This method of re-introduction was tested on a prosimian primate species which, at the time, was thought to be more "hard wired" or innate for survival skills. The restocking of the black-and-white ruffed lemur in Madagascar was similar to the golden lion tamarin case in that a protected area was available as well as a sustainable captive-bred population and commitment for long-term study. The main difference was a project goal of assessing re-introduction feasibility for prosimians along with increasing the genetic diversity of a small isolated *in situ* population through a restocking methodology. In spite of taxonomic differences and amount of free ranging experience, project results were similar, with low survivorship rates for *ex situ* captive-born individuals. This project, while successful on many levels (e.g., increasing the gene diversity of the wild population as the captive-bred animals reproduced successfully with their wild counterparts), also unfortunately demonstrated once again that animals raised in zoo environments should be avoided as release candidates as they lack many critical survival behaviors (Britt *et al.* 2002).

Behavior and survival skill differences between *ex situ* and *in situ* primate populations are clearly evident. This is not surprising given that most primate species have significant dependency periods on adults for learning the necessary social and survival behaviors to exist and reproduce in a wild environment. As generations of primates are born and maintained in an *ex situ* environment, learned *in situ* survival and behavior skills are inevitably lost. Although modern zoos often invest heavily to provide natural environments with naturalistic enclosures, it is important to note that these environments do not necessarily function naturally (Lacy 2010) for the species held within them with respect to facilitating the necessary skill set for release into the wild. In addition, while free-ranging habitats do provide opportunities to develop many important foraging and locomotor skills, successfully mimicking wild conditions in *ex situ* environments for exhibit purposes or for survival skill training of captive primates should be a priority for zoo management and exhibit design if re-introduction is the end goal.

17.5.1 Alternatives to re-introduction

As most of our *ex situ* captive populations are unlikely to be candidates for re-introduction, *ex situ* husbandry knowledge and management resources should be transferred and applied to captive primate populations in the country of the species' origin where re-introduction and restocking efforts are a more realistic option. Acclimatization is likely to be quicker and survival higher with animals raised in an environment similar to where they will be released (Conde *et al.* 2011). Also, within-region transfers are more cost effective, logistically simpler, and generally benefit from established administrative structures and lines of communication (Lees and Wilcken 2009). Current examples include Nigeria's Pandrillus project and the re-introduction/restocking of drills (*Mandrillus leucophaeus*) currently planned for Cameroon's Limbe Wildlife Sanctuary (Gadsby, pers. comm.), as well as Vietnam's Cuc Phuong Primate Sanctuary and the re-introduction/restocking of two langur species, the Hatinh langur (*Trachypithecus hatinhensis*) and the red-shanked douc (*Pygathrix nemaeus*) (Vogt and Forster 2010). Translocation does not involve the use of *ex situ* captive bred animals, although it requires a skill set (immobilization, transport, health survey) that zoo biologists can apply to *in situ* primate conservation efforts. Additionally, translocations of wild-born individuals are considered to be a more efficient use of conservation resources and ultimately have greater success than re-introductions of captive bred animals (Britt *et al.* 2002; Kierulff *et al.* 2002). Some examples include the translocation of baboons (*Papio papio*) in Kenya, rhesus macaques (*Macaca mulatta*) in India (Strum and Southwick 1986; Strum 2005), and black howler monkeys (*Alouatta pigra*) in Belize (Horwich *et al.* 1993).

17.6 Case studies

In this next section we provide two case studies of species programs for a comparison of their unique challenges for meeting population sustainability. Of note is the interplay of husbandry data from *ex situ* programs and information obtained from field studies both of which have refined methodologies for captive breeding and informed our understanding of the species' biology—ultimately impacting our effectiveness at conserving them in nature.

17.6.1 Western lowland gorilla

Few species fascinate the public like the gorilla and, as a result, it is perhaps one of the most sought after species in zoos. Their sheer size coupled with close similarity to humans make them truly fascinating to both the public and professional world alike. A century and a half of discovery of gorilla natural history has greatly impacted their management in zoos. Gorillas were first known to scientists in 1847 and in the 160 + years since many critical historical findings have led to the successful establishment of captive gorilla breeding programs in zoos.

The late 1800s was the start of the period of exploration into the heart of Africa and, as a corollary, several young gorillas were brought from Africa to European and North American zoos but survived only days, weeks, or months (Crandall 1964; Willoughby 1978). This dismal situation even provoked the then Bronx Zoo Director, William T. Hornaday, to state: "there is not the slightest reason to hope that an adult gorilla . . . will ever be seen living in a zoological park or garden" (Hornaday 1915, p. 7). A century later, zoo research that uncovered the key to successful gorilla husbandry would prove him wrong. In the 1920s several more gorillas were acquired by zoos but this time experience advanced longevity through more appropriate trans-Atlantic transportation modes and greater attention to diet, health, and temperature needs (Crandall 1964). In 1956 the first gorilla birth outside Africa occurred in North America at the Columbus Zoo (Thomas 1958). The female, Colo, and other gorillas in her age class lived well into adulthood and their descendants (F1 and F2 generations) formed the founder population of the North American gorilla population (Wharton 2000).

Although gorillas were living until adulthood and reproducing successfully, their health was in part compromised due to a high-protein diet. Obesity is another factor that impacted longevity and contributed to the high incidence of mortalities of young adult males in particular (Crandall 1964; McGuire *et al.* 1989). In 1959, the New York Zoological Society's (now Wildlife Conservation Society/WCS) Dr. George Schaller conducted the first study of gorillas in the wild, yielding important information on their diet, ecology, and social behavior (Schaller 1963). This was subsequently followed by the long-term research of Dian Fossey (1983). The information from these long-term pioneering studies in turn informed the management of captive gorillas, shifting dietary recommendations from high fat to high fiber (Calvert 1985). As a result, gorilla health and longevity were enhanced.

The next challenge in gorilla husbandry centered on unlocking the key to their social management. The pioneering work of Schaller (1963) revealed that gorillas are polygynous and gorilla troops typically consist of an adult silverback, several females, and offspring of both sexes and age classes. However, this was contrary to how most zoos managed gorillas, modeling their social structure after monogamous humans and maintaining troops with animals of the same age class. This was largely due to the fact that the first generation of gorillas from Africa were of similar age. Continued field studies of gorilla social behavior further revealed that their mating system shows preferences for unfamiliar mates (Watts 1990). Again, this is contrary to how zoos had traditionally managed gorillas, often rearing them with the same individuals that they later paired for mating, which often proved unsuccessful (Thomas 1958; Beck and Power 1988).

Applying this critical information about gorilla mating systems to the growing husbandry knowledge yielded great reproductive success. However, gorilla captive managers soon learned that gorilla parenting skills, like all social behavior, have a significant learned component as evidenced by the unsuccessful rearing of infants in the early years of the breeding program (1960s–1980s) with only 10% of gorilla infants being mother-reared (Martin 1976; Beck and Power 1988). Hand-rearing gorilla infants proved to be highly successful and was likely prolonged as a rearing strategy for its added public appeal for zoo exhibits. Nonetheless, in subsequent years the emphasis was placed on providing gorillas with the appropriate social environment to acquire important social skills, including parenting behavior, while emphasizing the impact of rearing history on an individual's future social development. This led to a shift in infant management protocols and an increase in mother-reared infants as evidenced by the 85% mother-reared infants the captive breeding program experiences today (Wharton 2010).

In the last decade and a half the management of males became the next challenge for the captive breeding program. Our understanding of male gorilla behavior unfolded in later years, partly due to the fact that the first males brought into captivity were of the same age class, and thus, captive managers had little experience and knowledge of adult male gorilla behavior. Information on male gorilla behavior in the wild also became available (Robbins 1996), noting that the young adult males can move over varying time periods from being solitary to loosely formed all-male groups. This new information, along with a captive gorilla sex ratio of approximately 50:50 and a polygynous mating system, meant that there are far too many male gorillas to manage than there are available social groups. Thus, applying management strategies for the formation and stability of all-male groups becomes a primary focus for the current gorilla breeding programs (Stoinski *et al.* 2004).

In sum, the *ex situ* program for gorillas evolved into a successful and viable program with the application of information on critical aspects of gorilla natural history and biology from field studies applied to gorilla husbandry methods. As is the case with gorillas, *ex situ* breeding programs are viable and sustainable only when the species' physical, dietary, reproductive, social, and psychological needs are met and incorporated into successful husbandry techniques for the species and applied to population management strategies.

17.6.1.1 Ex situ *contributions to conservation*

In 1980 the inception of the AZA Gorilla SSP laid the foundation for a formal cooperative breeding program for gorillas in North American zoos. And the EEP for gorillas soon followed in 1985. The early years of the Gorilla SSP reached great reproductive success through selected pairings of individuals based on mean kinship and gene diversity (Lacy 1995). As of 2010 there were 351 gorillas (168 males, 183 females) in 55 North American zoos and an impressive 856 gorillas (396 males, 460 females) in 150 institutions in the global population. Today, the achievements of the Gorilla SSP program are evident on several accounts: 76% infant survivorship with 85% of infants parent-reared, first reproduction at age 6 and extending up to age 42, a male median life expectancy of 31.1 years, a female median life expectancy of 7.4 years, and a maximum longevity record of 56 years—surpassing observed estimates for gorillas in the wild (Wharton 2010; Lucas *et al.* 2011; McCann 2012). Five generations after the first gorillas arrived in North America, the gorilla captive breeding program has shown great success. Unfortunately, their wild counterparts have not experienced the same success. According to IUCN all four subspecies of gorillas are endangered, with two of them classified as critically endangered; thus, their survival in the wild continues to be of great conservation concern. So while the concern of Bronx Zoo Director William T. Hornaday (1915) as to the unlikelihood of seeing gorillas thrive in zoological parks has passed, the concern today by all gorilla conservationists–both *in situ* and *ex situ*—has shifted to the unlikelihood of seeing thriving gorilla populations in their native habitat today and in perpetuity. Communicating this global concern and call to action has become the main conservation goal for gorilla zoo programs worldwide. And an exemplary model for this call to action and support for gorilla conservation efforts was made by the WCS's Bronx Zoo Congo Gorilla Forest exhibit. In 1999 the Bronx Zoo opened the Congo Gorilla Forest and became the first zoo to directly contribute exhibit admission fees to *in situ* conservation efforts making a visit to that exhibit an act of conservation support (Conway 1999). In its first decade of operation, more than $10 million dollars was raised for conservation in Africa. Additionally, this award-winning exhibit has provided state-of-the-art facilities to enhance the gorilla and other AZA African species breeding programs, as well as enrolling more than 100,000 students and 24,000 teachers in the zoo's conservation education programs. This indeed demonstrates the significance of captive breeding programs and the valuable role they play in providing assurance populations, communicating conservation awareness to the millions of yearly zoo visitors, and directly supporting species conservation efforts (McCann 2012).

The mounting crisis in ape conservation makes it increasingly evident that it will take all resources available from *in situ* and *ex situ* conservation efforts to conserve gorillas and other endangered ape species and their fragile habitats. In response to this growing crisis, in 2010 the AZA's Ape Taxon Advisory Group initiated a Great Ape Conservation Fund, committing more than $100,000 for *in situ* conservation support of ape conservation work by AZA zoos (<http://www.clemetzoo.com/

gorillassp/ConservationInitiative.html>). This initiative comes hot on the heels of the United Nations-backed effort to raise awareness and support for gorilla conservation resulting in the "Year of the Gorilla" campaign (YoG 2009). WAZA and the UNEP Convention on the Conservation of Migratory Species (CMS) joined forces to launch the first internationally recognized joint conservation effort for the gorilla. As we begin the 165th year after the first gorilla was discovered, their long-term preservation is the primary mission of both *in situ* and *ex situ* programs alike.

17.6.2 Aye-aye

The aye-aye (*Daubentonia madagascariensis*), a monotypic species at the family level that was first discovered in 1780, is the most unusual and distinctive primate on Earth (Petter 1977; Feistner and Sterling 1994; Mittermeier *et al.* 2010). This highly specialized prosimian is endemic to Madagascar and well adapted for its unique woodpecker-like role and nocturnal lifestyle. Its distinguishing characteristics: large bat-like ears, atypical dental formula, ever-growing, chisel-like front teeth, and an elongated, very thin and highly flexible middle digit, had taxonomists so puzzled that it was mis-classified as a rodent for almost 100 years and not correctly classified as a primate until 1859 (Petter 1977; Sterling 1994a). Similar to many primate species, the aye-aye is threatened with extinction due to habitat destruction and hunting pressures. But unlike other lemurs, aye-ayes face additional hunting pressures due to cultural taboos in some parts of Madagascar that link it to harbingers or omens of death and due to fear or disgust. Much like snake species, aye-ayes are killed on sight (Haring *et al.* 1994; Sterling and Feistner 2000). Fortunately there are other regions of Madagascar where aye-ayes, similar to several other lemur species, are highly valued and protected rather than hunted (Sterling 2003).

17.6.2.1 Captive history

Although first imported to Europe in the late 1800s, unlike gorillas the aye-aye does not have a long captive history and is still not well known or understood. As with many primate species, early records were absent or vague, survival rates were low, and reproduction was non-existent. The current International studbook records document only a handful of animals, although it acknowledges additional animals, yet to be recorded, being imported to Europe from 1862 to 1930, with most living three years or less (Wright 2011). However, Jones (1986) documented 35 animals being imported to Europe during the same timeframe, with most living less than five months but with one individual surviving 23 years from 1914 to 1937. Both sources agree on low survivorship and the absence of reproduction.

With the death of the sole surviving captive aye-aye in either 1932 or 1937, this species was absent from zoos until a second small wave of imports to Europe and North America which began in the late 1980s. There were continuous captures of just a few wild aye-ayes each year from 1986 to 1994, from 1998 to 2004, and in 2009, for a total of 36 animals (Wright 2011).

Studbook records show that three institutions: the Duke Lemur Center (formally the Duke University Primate Center), the Durrell Wildlife Conservation Trust (formally the Jersey Wildlife Preservation Trust), and Parc Zoologique de Paris imported a total of 20 wild-caught aye-ayes from Madagascar between 1986 and 1994. Twelve of these animals went to Europe and eight to North America. Eight additional animals were captured at this time and remained at in-country Malagasy zoos, Parc Botanique et Zoologique de Tsimbazaza (PBZT) and Parc Zoologique d'Ivoloina (PZI). Several captures were conducted again from 1998 through 2004, which resulted in seven aye-ayes added to PBZT. Two of these animals were exported from PBZT to Japan's Ueno Zoo, Tokyo in 2001. In 2009 one additional animal was captured and transferred to PBZT. These 36 animals are the founders for today's *ex situ* population, which totals 67 living animals worldwide (Wright 2011; Gibson 2012).

Not surprisingly, husbandry information was not recorded in the very early days of captivity. In addition, minimal data from the field was available regarding this species in the late 1980s. In spite of its poor captive history and essentially unknown requirements, these few institutions already committed to conservation in Madagascar, and lemurs in particular, devoted the necessary resources to import and attempt to establish this rare and unique species in captivity. *Ex situ* success requires institutional commitment and for a vast number of species the commitment of a single institution can make the difference between sustainability or not (Wiese and Gray 2010). This history, in contrast to the gorilla's history, is an example of how just a few institutions partnered with the government of Madagascar and made a significant difference for *ex situ* conservation of the aye-aye.

Working both together and independently, basic husbandry guidelines for maintaining aye-ayes were quickly established (Winn 1989; Carroll and Beattie 1993; Haring *et al.* 1994) and to everyone's surprise, wild-caught (WC) aye-ayes proved to be quite hardy in captivity. Twelve of the twenty aye-ayes imported in the late 1980s through early 1990s are still surviving at 20 to 30+ years of age (six of the animals are in Europe and six in North America). The first births from the imports occurred in 1992 and have continued through 2011, with the most recent births from a 25-year old female and a 30-year old male. The oldest of the first generation (F1) of captive-born aye-ayes are sexually mature and have provided an opportunity to gain critical developmental data for this species. However, as is the case with many such opportunities, as information/knowledge is gained additional questions and uncertainties present themselves.

The F1 captive born aye-ayes have been fairly consistent with developmental milestones with the exception of sexual maturation rates. As the F1 individuals were housed under different parameters, questions surrounding reproduction such as photoperiods, light intensity and wave lengths, socialization, hormone suppression, and pheromonal stimulation have been proposed and are not easily answered without additional research. In spite of several unproductive years, the North American F1 population has been the most viable and is now successfully producing the second generation (F2) population, the first of which was born in 2005. With the booming F1 population at Duke, aye-ayes became available to other institutions in North America in 2007. This was also an opportunity to gain additional reproductive information as pairs were sent out to be housed outside of the colony situation at Duke. Similar dispersals of WCx F1 pairs and single males occurred in Europe as more institutions began participating in the aye-aye program. Ten F2 offspring have been produced of which nine are still living. The majority of F2 aye-ayes have been produced in North America. Japan's Ueno Zoo produced their first F2 aye-aye in 2009 and the first for the European population was in 2011. One F2 male in the North America population marked the next major population milestone in early 2012 by breeding successfully with a F1 female. Although questions persist, reproduction appears to be trending on a positive path for this species.

In this case, unlike the gorilla population, there is much work still to be done to ensure that the *ex situ* aye-aye population achieves sustainability. Survivorship and reproduction were greatly improved with the second attempt to establish this species in captivity. Additionally, the first long-term field study on this species was completed by the early 1990s and provided valuable behavioral and ecological information that was applied to captive husbandry efforts (Sterling 1993a; Sterling *et al.* 1994). However, the basic biology of this species is still not comprehensively known. For example, life span, general physiology, development and maturation rates, reproductive parameters, and nutritional requirements still require further research. The current husbandry success with the *ex situ* population provides an opportunity for this valuable research to continue.

17.6.2.2 Population sustainability challenges

The most challenging aspects for establishing a sustainable captive population are acquiring sufficient space and maintaining institutional commitment. Aye-ayes are nocturnal primates, and unfortunately most zoos do not have nocturnal exhibits because they are considered to be architecturally challenging and less appealing to the public. As a result, nocturnal housing is in short supply and most zoos are not enthusiastic about meeting this challenge in spite of the fact that there is a vital need for *ex situ* conservation efforts for many nocturnal primate species. Lack of space is inhibiting the amount of breeding in the aye-aye *ex situ* population (Kuhar *et al.* 2011). Without additional space this population is unlikely to reach sustainability.

Neonatal mortality is also a serious concern in the *ex situ* population, especially with inexperienced females or infants that weigh less than 100 g at birth. The neonatal mortality rate for wild aye-ayes is unknown; however, studbook data shows that there have been 23 neonatal deaths (including 6 stillbirths), which represents a 30% overall infant mortality rate. Neonatal deaths can be mitigated in captivity with an assertive "hands on" approach (daily weighing and temporary nutritional support when required) as demonstrated by the high infant survivorship rate of 90% for the North American aye-aye population.

Finally, for achieving sustainability above and beyond the basic Perry *et al.* criteria, it will be necessary to maintain 90% of the genetic diversity in this population. Without the option of additional wild-caught animals, a global population management plan and cooperation between all holding institutions is the sole solution. There are currently only 16 institutions holding aye-ayes worldwide: Europe (6), Japan (1), Madagascar (2), and USA (7).

The road to *ex situ* sustainability for this species is certainly achievable with global management in place and an increase in institutional support to exhibit and/or hold this species. To initiate the global management effort, San Diego Zoo Global has recently become involved by partnering with Duke and Japan's Ueno Zoo to exchange F1 males for genetic purposes. This is an important first step toward on-going exchanges of animals as well as sharing life history and husbandry data between the four populations. International shipments are laborious and very time consuming. Global species managers must be persistent in order to maintain population viability and reproductive health. Additional field research is also required before the biology of this species can be fully understood. However, given the difficulties with nocturnal field work, especially on steep and/or uneven terrain, it is not surprising that there continues to be relatively few aye-aye field studies. This case study serves as an example when an *ex situ* population provides research opportunities that are valuable to both the captive and wild populations. Unlike gorillas, where field data was the primary source for scientific knowledge and also invaluable for *ex situ* success, with aye-ayes it is captive research, such as the ground breaking research on foraging strategy (Erickson 1991; Erickson 1994) for example, that is the principal source for gaining scientific knowledge and husbandry success.

17.7 Conclusion

Many of today's zoos function as conservation organizations by providing conservation education opportunities, population management expertise, and financial support for *in situ* conservation programs. The question of whether or not *ex situ* efforts via captive breeding of primates can play a role in assisting *in situ* primate conservation is complex and often species- and case-specific. In general, *ex situ* primate populations support *in situ* conservation efforts through varied valuable research and training opportunities as well as by serving as conservation ambassadors for *in situ* populations and their threatened habitats. Additionally, today's primate husbandry expertise has the potential to greatly benefit *in situ* captive populations and should be utilized especially for those cases where there are plausible restocking, re-introduction, or translocation opportunities.

The intensive management of populations by zoos can play a major role in a species' recovery and conservation. But in order to achieve this, populations must be genetically robust, demographically viable, and sustainable in perpetuity. Zoos have made tremendous progress breeding and maintaining primates in captive settings, yet they continue to be hindered with population management challenges. Population failures are often species-specific and include both intrinsic and extrinsic variables. Overall, there is cause for concern regarding the sustainability of many *ex situ* primate species. This review utilized the basic criteria for determining potential sustainability (survivorship and successful reproduction) to compare global *ex situ* primate populations over 38 years. Fewer primate species were found to be potentially sustainable in 2009 than in 1986. Applying more rigorous parameters (demographics and genetics) such as those used by today's population managers would likely result in even fewer species being sustainable. Global management, along with institutional cooperation, a possible solution for achieving sustainability for many populations, has yet to be utilized to its full capacity and should be initiated wherever feasible.

Acknowledgments

We wish to thank Talitha Matlin, Bob Wiese, Pat Thomas, Carmi Penny, Laurie Bingaman-Lackey, Nate Flesness, Dan Wharton, and Roby Elsner for their expert assistance and support with this chapter.

18

Primates in trade

Joshua M. Linder, Sarah C. Sawyer, and Justin S. Brashares

18.1 Introduction

The wildlife trade involves procurement, transport, and distribution of animals and their parts to supply a wide range of products such as live pets and animal exhibitions, experimental subjects in biomedical research, trophies, ornaments, ingredients for traditional medicines and ceremonies, and meat (Wyler and Sheikh 2008). Wildlife and associated products are traded locally (between neighboring communities), nationally (within the country of origin of a given species), regionally (across borders of adjacent countries), and increasingly, between continental areas (e.g., between Africa and Europe; Chaber *et al.* 2010). The wildlife trade is a burgeoning business that poses one of the greatest threats to the conservation of animal species across taxonomic groups (Sutherland *et al.* 2009). While some of the trade is legal, a substantial proportion of it is illegal, resulting in significant population declines of wild species involved in the trade, introduction of harmful invasive alien species, and reduction of ecosystem function (Broad *et al.* 2003). More than just an environmental issue, however, the wildlife trade is also significant from the perspective of human livelihoods (Brashares *et al.* 2011). It directly benefits rural (especially economically poor) people, professional traders, criminal entrepreneurs and crime syndicates, businesses, and national governments through employment, income, and, in the case of the trade in wild meat, protein intake (Broad *et al.* 2003; Golden *et al.* 2011). Additionally, the regional and transcontinental trade in wildlife poses significant risks to human health (Jones-Engel *et al.* 2005) and national security (Lin 2005).

Although the legal trade in wildlife resources (including plants, but excluding timber and fish) was estimated in the early 1990s to be worth USD 15 billion annually in export value (WWF/TRAFFIC 2002), the true extent of the trade in wildlife is unknown because much of it is illegal and carried out through informal trade networks, making it difficult to measure (Broad *et al.* 2003; Milner-Gulland

Primate Ecology and Conservation: A Handbook of Techniques. First Edition. Edited by Eleanor J. Sterling, Nora Bynum, and Mary E. Blair. © Oxford University Press 2013. Published 2013 by Oxford University Press.

et al. 2003). Attempts at quantifying the economic value of the illegal wildlife trade have suggested its annual worth to be between $5 billion and $20 billion, making it one of the world's most lucrative illicit activities (Wyler and Sheikh 2008).

18.1.1 Wild meat trade

Of all the ways in which wildlife is traded, broadly speaking, across the tropics, the hunting for wild meat trade poses one of the greatest risks to large-bodied mammals including, and especially, the primates (Milner-Gulland *et al.* 2003). Relatively large-bodied, group-living, and/or ecologically specialized primates are especially vulnerable to the growing commercial trade in wild meat and their populations are at the greatest risk of significant declines in the near future (Linder and Oates 2011). Rapidly growing human populations, a preference for the taste of wild meat, increasing per capita wealth, improved hunting technology and access to forests, and the lack of alternative protein sources and income-earning activities, have all led to the commercialization of hunting and to dramatic increases in harvest rates (Fa *et al.* 2002b; Lee *et al.* 2005; Altherr 2007). Today, the wild meat trade has become a multibillion dollar annual business and is a major source of income for millions of people worldwide, but especially those in economically developing countries (Brashares *et al.* 2011).

The growing scale of the wild meat trade has prompted studies from biological, socioeconomic, legal, and human health perspectives. Research techniques are varied, but tend to focus on enumeration of hunter harvests, records of household production, consumption, surveys of "points of sale" (including rural and urban markets, restaurants, and informal roadside trading), and official confiscation records.

These methods have been frequently used to study the wild meat trade in the west and central African forest zone, where the so-called "African bushmeat crisis" threatens populations of many game species and has long-term consequences for human food security (Milner-Gulland *et al.* 2003). In the Congo Basin forests of central Africa, for example, hunter offtake surveys suggest that over 4.9 million tons of wild mammal meat are extracted from the region annually (Fa *et al.* 2002a). To put these figures in perspective, surveys of hunter harvests in the Amazon forest, which is over twice the size of the Congo basin forest, suggest that between 67,000 and 148,171 tons of wild meat are harvested annually (Peres 2000). Even in the Neotropics, however, hunting is becoming increasingly unsustainable and trending towards unsustainable "crisis" levels (Altherr 2007). In the Brazilian Amazon alone, data from hunter offtake and household surveys indicate that rural populations consume between 9.6 and 23.5 million vertebrate animals annually (Peres 2000).

There is no comparable estimate of the rate at which people harvest or consume wild meat from such broad ecoregions in Asia. However, more localized studies indicate that hunting for and trading of wild meat can be significant. In the province of North Sulawesi, Indonesia, for example, the number of wild meat dealers increased dramatically from 1970 to 1996, increasingly selling such meats as babirusa, wild pig, and macaque (Clayton and Milner-Gulland 2000). Based on surveys conducted between 2001 and 2003 of 6 markets in North Sulawesi, 96,586 animals (most in the form of meat) were recorded for sale (Lee *et al.* 2005). In Sabah and Sarawak, the two Malaysian states in northern Borneo, subsistence hunting is being supplemented by an increasing wild meat trade estimated in the mid-1990s to be worth $3.75 million annually (Bennett *et al.* 2000). In Sarawak alone over 1000 tons of wild meat is estimated to have been traded in 1996 (Bennett and Rao 2002). Wild meat is for sale in all provinces in Vietnam, where the illegal wildlife trade (including meat and live animals) is estimated to be worth $21 million annually, with over 1 million animals traded per year (Nguyen Van Song 2008). Overhunting, for meat and other purposes, has been implicated in the extinction or near extinction of 12 species of large animals in Vietnam in the last 40 years (TRAFFIC 2008). These trends of increasing wildlife harvest are also evident in Madagascar. Increasing movement of people, exposure to modern living, recent political instability, and reduction in foreign aid to Madagascar has led to an emerging trade in lemur bushmeat (Jenkins *et al.* 2011). Prior to these developments, the extent to which lemurs were represented in household meat consumption or the trade in wild meat was limited due to traditional taboos against eating lemur meat (Randrianandrianina *et al.* 2010).

Across these tropical regions, primates have been shown to constitute a significant portion of the wild meat consumed and traded (Table 18.1). In the central and west-central African forest zone, for example, primates regularly comprise between 8 and 30% of the total animals harvested or for sale in bushmeat markets, and can represent over 30% of the total vertebrate biomass sold. Estimates for the Brazilian Amazon suggest that rural populations consume between 2.2 and 5.4 million primates annually, representing nearly 23% of the total annual vertebrate harvest and almost 14% of the total biomass harvest (Peres 2000). In Asia, although the meat of large ungulates, including pigs and deer, tends to be the most popular taxa for consumption or trade, primates can often comprise a large proportion of the biomass hunted and traded (Corlett 2007). Macaque species, for example, are frequently hunted, consumed, and traded in North Sulawesi (Lee *et al.* 2005). Similarly, in northern Myanmar, macaque and langur species are frequently hunted for trade, with the Assamese macaque comprising one of the five most traded species (for meat and other products; Rao *et al.* 2011). In general,

Table 18.1 Overview of studies examining the contribution of primates to the wildlife trade in Africa, Asia, and South America.

Region	Study site	Method	Duration of study	% of total animal harvest	% of total animal biomass
Africa	Cross-Sanaga area[1]	Urban and rural market surveys	5 months	10.2 (Nigeria) 17.1 (Cameroon)	7.3 (Nigeria) 13.8 (Cameroon)
Africa	Banyang-Mbo, Cameroon[2]	Hunter offtake surveys	34 months	12.5	10.4
Africa	Banyang-Mbo, Cameroon[3]	Rural market surveys	6 months	12.7	10.9
Africa	Korup National Park, Cameroon[4]	Rural market surveys	9 months	27	20
Africa	Bioko, Equatorial Guinea[5]	Urban market surveys	64 months	13.7	not reported
Africa	Bioko, Equatorial Guinea[6]	Hunter questionnaires Hunter follows	1 month	15.2	25.3
Africa	Bioko, Equatorial Guinea[7]	Urban market surveys	10 months in 1991 14 months in 1996/7	25.7 (1991) 5.1 (1996/7)	37.9 (1991) 11.7 (1996/7)
Africa	Bioko, Equatorial Guinea[8]	Urban market surveys	4 months	27.8	31.2
Africa	Bioko, Equatorial Guinea[9]	Urban market surveys	24.2 months	8.9	17.4
Africa	Rio Muni, Equatorial Guinea[10]	Various "points of sale" surveys Hunter offtake surveys Household surveys	5 months	19–30	not reported
Africa	Rio Muni and Bioko, Equatorial Guinea[11]	Urban market surveys	10 months	24.7 (Bioko) 21.5 (Rio Muni)	not reported
Africa	Rio Muni, Equatorial Guinea[12]	Hunter offtake surveys	16 months	10.8	17.1
Africa	Pokola logging concession Republic of Congo[13]	Household and hunter interviews	10 days	3.4	1.9
Africa	Pokola logging concession Republic of Congo[14]	Hunter offtake surveys	6 months	13.1	9.1

Africa	Ouesso, Republic of Congo[15]	Urban market surveys	4 months	22	not reported
Africa	Dzanga-Sangha Special Reserve, Central African Republic[16]	Urban market surveys	12 months	8.8	4.9
Africa	Dzanga-Sangha Special Reserve, Central African Republic[17]	Urban market surveys and other points of sale	19 months	18	not reported
Africa	Bendel State, Nigeria[18]	Rural and urban market surveys and other points of sale	4 months	16.5	not reported
Africa	Sekondi-Takoradi, Ghana[19]	Urban market surveys	2 months	0	0
Africa	Kumasi, Ghana[20]	Urban market surveys	15.5 years	0.7	not reported
Africa	Digya National Park, Ghana[21]	Rural market surveys	8 months	1.9	not reported
Africa	Haute Niger National Park, Republic of Guinea[22]	Rural market survey	12 months	8.7	5.5
South America	Loreto, Peru[23]	Urban market surveys	12 months	1.6 (urban markets)	0.7 (urban markets)
		Rural market surveys		28.6 (rural markets)	rural markets not reported
South America	Amazonia forests[24]	Hunter offtake surveys	1–4 years	22.9	13.9
South America	Yasuni National Park, Ecuador[25]	Household surveys, Rural market surveys	29 months	6.9	4.7
South America	Yasuni National Park, Ecuador[26]	Hunter offtake surveys	5 months	23.5	20
Asia	Hkakaborazi National Park, Myanmar[27]	Hunter offtake surveys	4 months	27.1	not reported
Asia	North Sulawesi, Indonesia[28]	Urban market surveys	2 years	< 1.2%	not reported

[1] Fa et al. (2006). [2] Wilcox and Nambu (2007). [3] Abugiche (2008). [4] Infield (1988). [5] Reid et al. (2005). [6] Colell et al. (1994). [7] Fa et al. (2000). [8] Albrechtsen et al. (2005). [9] Albrechtsen et al. (2007). [10] Fa et al. (2009). [11] Juste et al. (1995). [12] Fa and Yuste (2001). [13] Tieguhong and Zwolinski (2009). [14] Auzel and Wilkie (2000). [15] Hennessey and Rogers (2008). [16] Hodgkinson (2009). [17] Jost-Robinson et al. (2011). [18] Anadu et al. (1988). [19] Cowlishaw et al. (2005). [20] Crookes et al. (2005). [21] Owusu-Ansah (2010). [22] Brugiere and Magassouba (2009). [23] Bodmer et al. (2004). [24] Peres (2000). [25] Suarez et al. (2009). [26] Franzen (2006). [27] Rao et al. (2005). [28] Lee et al. (2005).

harvest rates and consumption of wild meat in many Asian (especially Southeast Asian) forests may be relatively low because unsustainable hunting levels have already caused game densities in many forest areas to decline to the point where they no longer support significant hunting (Bennett and Rao 2002; Corlett 2007).

18.1.2 Live pet and biomedical trade

Besides their importance in the wild meat trade, primates are also well represented in the live animal trade, which supplies pet markets and the exhibition business (Shepherd *et al.* 2004; Agoramoorthy and Hsu 2005; Nijman *et al.* 2011). Primates can be specifically targeted to supply the live animal trade, an activity most common in Asia and South America (Mittermeier 1987; Nekaris *et al.* 2010). In these regions the live animal trade does not discriminate against body size or conservation status (in Asia: Shepherd *et al.* 2004; Jones-Engel *et al.* 2005; Nijman 2005; Nekaris *et al.* 2010; Shepherd 2010; in South America: Duarte-Quiroga and Estrada 2003; Ceballos-Mago and Chivers 2010). In fact, every primate family from South America and Asia is represented in the pet trade, which can have a significant impact on primate populations in the wild. Surveys conducted through-out Sumatra, Kalimantan, Java, and Bali, Indonesia of wildlife markets, pet shops, private owners, public and private zoological gardens, and wildlife rescue centers indicate that the primary reason orangutans and gibbons are hunted is to obtain a young animal to keep or sell as a pet (Nijman 2005, 2009). Given that at least one adult is typically killed to obtain the young, such a trade is likely having a significant impact on orangutan and gibbon populations in Indonesia.

To our knowledge, the only published studies directly examining the primate pet trade in the African forest zone have focused on gorillas and chimpanzees (Farmer 2002; Kabasawa 2009). Infectious disease studies indicate that monkeys, as well as apes, are often kept as pets in Africa, but we failed to find any publication that provided details about an extensive African primate pet trade. This is probably because the African primate pet trade is considered an unintentional by-product of the broader wild meat trade in Africa (e.g., Cormier 2002), which has been studied extensively. These "orphans" of the wild meat trade have dramatically increased in number as hunting for meat has become more commercialized in the African forest zone. This has resulted in an increase in the number of wildlife sanctuaries (Farmer 2002). Although records kept by managers of these sanctuaries can aid in under-standing the extent of the live primate trade (e.g., Farmer 2002; Kabasawa 2009), these figures vastly underestimate the true number of primates kept as pets. Similarly, lemurs appear to be kept as pets in Madagascar (e.g., Sussman *et al.* 2003), but the extent to which there is a lemur pet trade has not been investigated.

The demand for experimental subjects in biomedical research has also driven the live primate trade (Fernandez and Luxmoore 1997). This kind of trade was especially prevalent and largely uncontrolled between the 1950s and 1970s, prior to the establishment of the Convention on International Trade in Endangered Species of Wild Fauna and Flora (CITES) and national import and export bans, when the United States and European countries regularly imported wild caught primates from source countries in Asia, Africa, and South America (Mittermeier et al. 1994; Held and Wolfle 1994; Southwick and Siddiqi 1994; Nijman et al. 2011). Despite these regulations, the trade of live primates for biomedical reasons still continues and, since 1995, this trade has been increasing, in part because of China's emergence in the biomedical trade (Nijman et al. 2011). The USA remains the single largest importer of live primates for biomedical purposes, followed by Japan and China (Nijman et al. 2011).

The most important species to the international biomedical trade are macaques (*Macaca* spp). Rhesus macaques (*Macaca mulatta*) were once so overexploited for trade that a blanket ban was imposed on their exportation from India (Southwick and Siddiqi 2001; Maldonado et al. 2009). The pressure of the biomedical market then shifted to long-tailed macaques (*Macaca fascicularis*), and the USA imported more than 24,000 individuals of the species annually between 2005 and 2007 for pharmaceutical purposes alone (Cowlishaw and Dunbar 2000). Other species of importance for the international pharmaceutical market are: squirrel monkeys (4500 annually used in the USA), marmosets and tamarins (2400 annually in the USA), African green monkeys, owl/night monkeys, and baboons (Held and Wolfle 1994; Mittermeier and Konstant 1996; Maldonado et al. 2009). Despite recent progress in captive breeding programs and trade regulations, many primate species of South and Southeast Asia remain threatened by trade for pharmaceutical research and development (Nekaris et al. 2010). Wild-caught South American primates continue to be extensively traded as well, and the undocumented cross-border trade of night monkeys for biomedical laboratories near the Colombian border represents an important livelihood strategy for many people in the region (Maldonado et al. 2009).

18.1.3 Traditional and medicinal trade

Across many cultures, non-human primates play significant roles in human myth-ologies, religious ceremonies, and traditional folk medicine, due, in part, to our close evolutionary relationship resulting in morphological, behavioral, and psycho-logical similarities (Alves et al. 2010). Primate body parts, including bones, blood, skin, brain, fingers, teeth, and penis, are used to make both clinical remedies for medical treatment, and charms for magical or religious ceremonies (Adeola 1992;

Soewu 2008; Alves *et al.* 2010). Although primate parts have been used in traditional ceremonies and as ingredients in folk medicine for millennia, the increasing commercial trade in primates for these purposes is raising conservation concerns. At least one-quarter of the approximately 390 species of primate are used for ceremonial or medicinal purposes across Africa, Madagascar, Asia, and the Neotropics (Alves *et al.* 2010). Like the pet trade, primates of all sizes are targeted in the trade for medicinal or ceremonial ingredients (e.g., Nekaris *et al.* 2010).

The trade in primates and other animals for medicines and ceremonial purposes is especially extensive and problematic in Asia, where increasing demand from China and the ubiquitous practice of traditional medicine (which relies on wild plant and animal derivatives) in South and East Asia, drive the trade (Corlett 2007; Nadler *et al.* 2007). Lorises are among the most frequently traded species for traditional medicinal practices in Asia, and Cambodia's two slow loris species are the most commonly requested animals in traditional medicine stores in the capital city Phnom Penh (Nekaris *et al.* 2010; Nijman *et al.* 2011). Gibbons and langurs are also regularly targeted for traditional medicinal purposes, mainly for use of their bones (Hilaluddin and Ghose 2005; Alves *et al.* 2010). In addition to the medicinal demand for primate body parts, the trophy and skin trade represent significant primate markets for both rare and more common purposes. Species like gorillas and baboons often fall prey to the trophy trade, while black and white colobus species and patas monkeys are targeted for the skin trade to make items like cloaks, headdresses, and rugs (Oates 1977; Adeola 1992; Mittermeier and Konstant 1996). In northeast India, the skin of capped langurs is commonly used for making sheaths for daggers (Hilaluddin and Ghose 2005).

While it is difficult to obtain data to assess the sustainability of trade in primate parts, there is anecdotal evidence to suggest that in many cases medicinal, magi-religious, trinket, and trophy markets are unsustainable. One study shows that most of the species used today for traditional medicinal and magi-religious treatments in Nigeria were once considered less-desirable substitutes for preferred species, but became staples as preferred species became more difficult to obtain (Soewu 2008). Additionally, it has been suggested that historical overhunting of black and white colobus monkeys for the skin trade may partially explain the species' patchy distribution today (Oates 1977; Oates *et al.* 1994). Further data collection on the scale and impacts of the primate part trade may therefore significantly aid conservation policy and management decision-making.

As summarized above, the illegal wildlife trade—for meat, live animals, and medicinal or ceremonial purposes—poses among the greatest threats to wild primate populations. It is critical that we apply conservation science to understand and solve these conservation challenges. Studies of the wildlife trade can be conducted on many different scales. Data can be collected from (1) hunters, who procure the wildlife; (2) families, who purchase, consume, and sell wildlife; (3) local and national wildlife traders and markets; (4) trade data gathered by CITES, customs agencies, and local law enforcement; and (5) international markets. Within each of these levels of investigation, a diverse set of techniques can be used to examine the legal and illegal trade in wild primates and other animals. We review the key methods below and, when relevant, present their advantages and disadvantages and discuss the logistical and ethical considerations involved in choosing a particular technique. Throughout, we refer to studies that have employed these techniques to study the wildlife trade. We encourage the reader to seek out the original studies to gain a deeper understanding of the connections between the questions being asked and the chosen methods of data collection and analysis (Table 18.2). This chapter aims to provide a framework for designing and implementing research projects that will, ultimately, effectively track, regulate, and mitigate impacts of trade on wild primates.

18.2 Hunter and household surveys

Hunter and household surveys are frequently used methods for assessing the primate trade and are especially useful for examining the extent and impact of the primate meat trade. Methods used to survey hunters and households include interviews and questionnaires, surveys of hunter offtake, and hunter follows. Hunter and household surveys can help answer a diverse array of questions related to the wild meat trade including:

- How frequently are different wildlife species harvested and/or consumed?
- How much of a household's income is derived from the sale of wildlife products?
- How much of household expenditure goes towards purchasing wildlife products?
- To what extent do the diets of rural and urban living people in the tropics depend on wild meat as a source of protein?
- What is the seasonality of the use of wild foods in the diet and household economies?
- What is the relationship between household wealth and wild meat consumption and sales?
- What proportion of wild meat is used for subsistence and commercial purposes?
- Which kinds of species are most frequently sold?

Table 18.2 A non-exhaustive overview of studies and methods associated with various aspects of primate utilization.

Study site	Objectives/questions	Research techniques
Bioko, Equatorial Guinea[1]	Assessed availability and consumption of animal protein in Bioko, Equatorial Guinea	Market surveys; household surveys (interviews, questionnaires; 24-h recall)
Bioko, Equatorial Guinea[2]	(i) Examined the consumption and preference patterns of wild meats	Household surveys (interviews)
	(ii) Examined whether consumption and preferences of bushmeat species obtainable on the island were related to their availability and prices in the market	Market surveys
Rio Muni, Equatorial Guinea[3]	(i) Examined the position of bushmeat hunting amongst other forms of production through interviews with hunters and household members on sources of income and hunting offtake	Hunter surveys (interviews and offtake)
	(ii) Examined both consumption of, and expenditure on, bushmeat by households compared to other meat and fish.	Household surveys (interviews, 24- recall) h
	(iii) Examined the influence of wealth, income, and other socioeconomic factors on production, consumption, and expenditure.	
Rio Muni, Equatorial Guinea[4]	Investigated consumption of meats and fish in relation to household wealth	Household surveys (questionnaires)
Rio Muni, Equatorial Guinea[5]	(i) Assessed the composition of diurnal primates in the forest	Forest surveys
	(ii) Evaluated the scale and selectivity of gun-hunting	
	(iii) Evaluated the impacts of gun-hunting on the primate community	Hunter surveys (interviews, follows, offtake)
	(iv) Evaluated policies for the conservation of primate populations in the study area	

Rio Muni, Equatorial Guinea[6]	Assessed the relationships between consumption of different types of meat and fish, income, preferences, price and availability of these products	Points of sale surveys Household surveys (interviews, 24-h recall)
Rio Muni, Equatorial Guinea[7] Southeastern Cameroon[8]	Examined the extent and impact of commercial hunting (i) Assessed the value of bushmeat from hunting to local household economies, both for home consumption and for sale (ii) Examined the role of bushmeat to household protein needs and income in two logging towns around the Lobeke National Park; (iii) Offered recommendations for bushmeat policy shifts in logging towns that can promote both development and conservation outcomes	Hunter surveys (offtake) Household surveys (questionnaires)
Banyang Mbo Wildlife Sanctuary, Cameroon[9]	Compared bushmeat hunting between two tribes to examine cultural, socioeconomic, and rural development reasons for intertribal differences and similarities	Hunter surveys (interviews, follows, offtake)
Northern Republic of Congo[10]	(i) Examined industrial logging influences human demographics in frontier forest (ii) Assessed the degree to which demographic differences among logging towns affect the supply of bushmeat and patterns of hunting (iii) Evaluated factors that determine the consumption of animal protein in logging towns (e.g., season, ethnic origin, principal economic activity, and price)	Household surveys (interviews) Market surveys
Brazzaville, Republic of Congo[11]	(i) Examined the urban bushmeat consumers' demographic profile and motivations for eating bushmeat (ii) Identify the most popular species consumed	Household surveys (questionnaires and interviews)

(continued)

Table 18.2 *Continued*

Study site	Objectives/questions	Research techniques
	(iii) Assessed consumers' perceptions in relation to the safety of bushmeat as food and their interest in breeding game animals for human consumption	
Salonga National Park, Democratic Republic of Congo[12]	(i) Examined the impact of hunting on the distribution and abundance of bonobos	Forest surveys Hunter surveys (interviews) Household surveys (interviews)
	(ii) Identified specific hunting practices and economic and ecological contexts that are likely to pose a significant threat	
Democratic Republic of Congo[13]	(i) Examined the value of wild foods in terms of both household consumption and market sales	Household surveys (24-h recall)
	(ii) Assessed the relative value of wild foods in the lean agricultural season	
	(iii) Evaluated the value of wild foods relative to economic standing in the community	
Dzanga Sangha Reserve, Central African Republic[14]	(i) Examined how hunting has changed over time in the Dzanga Sangha Reserve	Forest surveys
	(ii) Assessed how critical prey species have been affected by hunting	Hunter surveys (interviews, offtake) Market surveys
Central Gabon[15]	Investigated the relationship between hunting offtake and household wealth, gender differences in spending patterns, and the use of hunting incomes in two rural forest communities from 2003 to 2005	Hunter surveys (interviews and offtake) Household surveys (interviews)
Tai National Park, Cote d'Ivoire[16]	Assessed the impact of hunting on wildlife populations by comparing the current harvest rate with the maximal production rate.	Forest surveys Market surveys
Western Madagascar[17]	(i) Identified species traded as bushmeat	Hunter surveys

Location	Objectives	Methods
	(ii) Examined their relative importance compared to meat from domestic animals in settlements along a main road	Household surveys (interviews, 24-h recall)
Ghana, Cameroon, Tanzania, and Madagascar[18]	(i) Examined the relationship between household wealth and wildlife consumption among rural and more urban areas (ii) Assessed how proximity to harvestable wildlife populations affects prices of bushmeat relative to alternative food sources and rates of consumption at the settlement level (iii) Evaluated the extent to which the fate of harvested wildlife is determined by a hunter's access to urban bushmeat markets	Houshold surveys (interviews; annual recall; diet diaries)
Sumatra, Indonesia[19]	Assessed the trade in gibbons and orangutans in Sumatra	Household surveys (interviews) Market surveys Surveys of public and primate zoological gardens and wildlife centers
North Sulawesi, Indonesia[20]	Assessed the hunting practices of the local population by (i) identifying the kinds of animals hunted (ii) describing the hunting methods used (iii) surveying local markets for the bushmeat species sold	Hunter surveys (interviews) Market surveys
Sulawesi, Indonesia[21]	Examined primate pet ownership practices as they relate to inter-specific disease transmission	Household surveys (interviews)
Northern Borneo, Malaysia[22]	(i) Investigated the current importance of wildlife and its product to rural people, especially in their diets (ii) Examined the sustainability of hunting	Forest surveys Hunter surveys (interviews, offtake)
Hkakaborazi National Park, Myanmar[23]	(i) Examined the significance of hunting and trade for livelihoods	Household surveys (interviews) Hunter surveys (interviews, offtake)

(continued)

Table 18.2 Continued

Study site	Objectives/questions	Research techniques
Hkakaborazi National Park, Myanmar[24]	(ii) Assessed the impacts of hunting on targeted species	Official confiscation records
	Examined general patterns of hunting and wild meat consumption to assess the significance of proximity to settlements and markets for prey abundance and the influence of relative abundance and intrinsic preference on prey offtake	Forest surveys Hunter surveys (interviews) Household surveys (questionnaires)
Northeast India[25]	(i) Assessed whether the extraction of wild meat by local communities was a conservation problem in the region	Household surveys (interviews)
	(ii) Determined whether consumption of wild meat was linked to people's income	Market surveys
Northern French Guiana[26]	(i) Assessed habitat features, including anthropogenic disturbance, structuring primate communities,	Forest surveys
	(ii) Described current hunting practices	Hunter surveys (interviews, offtake)
	(iii) Evaluated the sustainability of harvests for wild meat	
Yasuni National Park, Ecuador[27]	(i) Quantified the extent of total hunting and	Hunter surveys (interviews, offtake)
	(ii) Evaluated the sustainability of current hunting patterns	
Nueva Esparta, Venezuela[28]	Characterized the knowledge and perceptions of local people keeping primates as pets towards primates	Household surveys (interviews)
Central and South America[29]	Examined how income and price influence the consumption of wildlife	Household surveys (interviews)

[1]Albrechtsen et al. (2005). [2]Fa et al. (2002za). [3]Kümpel et al. (2010). [4]Fa et al. (2009). [5]Kümpel et al. (2008). [6]East et al. (2005). [7]Fa and Garcia Yuste (2001). [8]Tieguhong and Zwolinski (2009). [9]Wilcox and Nambu (2007). [10]Poulsen et al. (2009). [11]Mbete et al. (2011). [12]Hart et al. (2008). [13]de Merode et al. (2004). [14]Jost-Robinson et al. (2011). [15]Coad et al. (2010). [16]Refisch and Koné (2005). [17]Randrianandrianina et al. (2010). [18]Brashares et al. (2011). [19]Nijman, V. (2009). [20]Onibala and Laatung (2007). [21]Jones-Engel et al. (2005). [22]Bennett et al. (2000). [23]Rao et al. (2011). [24]Rao et al. (2005). [25]Hilaluddin and Ghose (2005). [26]de Thoisy et al. (2005). [27]Franzen (2006). [28]Ceballos-Mago and Chivers (2010). [29]Wilkie and Godoy (2001).

- How has the proportion of different species in hunter harvests changed over time?
- What is the relationship between consumer preferences for and actual consumption of wild meat?

Answers to these kinds of questions can help elucidate the drivers of the wildlife trade, predict the effects of changing economic conditions on use and demand for wildlife products, assess the extent to which domestic meat can serve as a substitute for wild meat in consumer diets, and evaluate the implications of policy decisions on regulating wildlife trade for the people who use or depend on the sale or consumption of wildlife products. Furthermore, for each question, a researcher can ask to what extent *primates* are used in the diets and livelihoods of rural and urban living people in primate-range countries. With such information it may be possible to assess the degree to which prohibiting the consumption and sale of primates (especially threatened primate species) would affect nutritional intake and economic status.

Involving human subjects in studies on the trade in wild primates usually involves sensitive ethical issues and necessitates understanding and navigating different cultural landscapes. Many of these issues are not directly addressed in this piece but we encourage the reader to review Chapter 8 (Ethnoprimatology) in this volume, which discusses important points for researchers to consider prior to conducting fieldwork with human subjects.

18.2.1 Selecting households and hunters to include in a study

Hunting and wildlife trade are often (but not always) illegal activities and people may be reluctant to provide information or the information they do provide may not be reliable. To help overcome these obstacles it may be critically important to conduct trust-building activities before data collection, such as holding meetings and discussions with each village, household, or hunter involved in the study (Brashares *et al.* 2011; Rao *et al.* 2011). Additionally, in areas where much of the hunting and wildlife trading is illegal, relevant participants may be sought using a snowball sampling technique, whereby interview subjects are recruited from individuals who are already included in the study (Nielsen and Treue 2012). It may also help to hire local field assistants who are residents of the village in which the study is being conducted (e.g., Kümpel *et al.* 2008). It is critical for researchers, whether local or non-local, to ensure that sensitive information shared by hunters and their households is kept in strict confidence.

It is often not possible to interview everyone in the target population about their use of wildlife products. The need for a large sample size (to have statistically significant results and to more accurately reflect reality) is constrained by logistical constraints such as the time available to complete the field work, budget, and the number of field assistants who will be conducting the interviews (Curran *et al.* 2000). For large sample sizes, when choosing households to include in a study, it is important to select a *stratified random sample* of the population to ensure that the sample is representative of the broader population. Stratified samples separate the population into separate categories (e.g., based on ethnicity, age, sex, socio-economic status). Within each category, households are randomly surveyed to minimize bias. For example, within each village or town surveyed by Randrianan-drianina *et al.* (2010) in western Madagascar, a "zig-zag" transect walk was used, whereby researchers walked for a 1 min period (which allowed large areas to be sampled) then chose the nearest available house in which someone was available to be interviewed. Other random sampling techniques can be found in most eco-logical technical handbooks such as White and Edwards (2000). However, in areas where people are reluctant to participate in a research study it may not be possible to use a standard random sampling method. Instead, the only option may be to include in a study those people who show a willingness to cooperate (Wilcox and Nambu 2007).

18.2.2 Interviews and questionnaires

Two of the most commonly used techniques to survey hunters and households concerning their use of wildlife and wildlife products are interviews and question-naires, both of which can yield qualitative and quantitative data (Curran *et al.* 2000). Interviews and questionnaires are used to gather ethnographic, demo-graphic, and socioeconomic information from hunters and other household members and to collect data on species preferences (for trading and consumption), hunting practices, hunting catchment areas, offtake, consumption, and sales of wildlife. Interviews, which are usually informal or semistructured, are designed to stimulate conversation and maintain an open, relaxed dialogue but the interviewer attempts to seek answers from each respondent to a prepared list of questions or topics (Bernard 2011). Thus, interviews are typically guided by prepared questions but the interviewer is free to address these topics during the conversation. In fully structured interviews, typically conducted through questionnaires administered by the interviewer, study participants respond to identical questions or other stimuli (e.g., photographs or pictures). Details on preparing and conducting interviews and questionnaires to monitor hunters or households can be found in Curran *et al.* (2000).

These data collection methods often rely on the respondent being able to recall past events. For example, data on household consumption and budgets can be collected using 24 h recalls. This "recall" technique can be used to collect data on food items purchased and/or consumed the day before the interview, state of food when acquired, the quantity consumed, the total cost for that quantity if bought, and how and from where it was obtained. Because the recall technique does not measure such variables directly, relying entirely on the ability of respondents to accurately remember past events, it may result in biased, inaccurate results (Bingham 1987). Jenkins *et al.* (2011) report that the ability to remember past activities declines substantially after the third day and suggest limiting recalls to a maximum of three days prior to the interview. However, Bingham (1987) argues that even 24 h recalls may substantially underestimate food intake. Furthermore, identification of species traded or consumed may vary between local consumers and researchers, resulting in inaccurate data.

To validate recall data, it may be advisable to include an empirical observation technique for each household included in the study (or a sample of the households). For example, to examine the drivers of wildlife consumption in Ghana, Cameroon, Tanzania, and Madagascar, Brashares *et al.* (2011) estimated annual wildlife consumption through annual recalls of hunting and/or wildlife purchases and supplemented this technique by asking a subsample of households to record consumed foods and their weight in a daily diet diary. Annual wildlife consumption rates were then calculated by combining data collected from both techniques. Jenkins *et al.* (2011) supplemented household interviews in localities in eastern Madagascar with direct monitoring of wild animal carcasses consumed, for sale, and transported through the study villages. Similarly, to examine the impact of hunting on and increasing trade in wildlife among Huaorani communities in Ecuador, Franzen (2006) facilitated recall of past hunting activity by asking each household to photograph harvests over the study period.

18.2.3 Hunter follows, offtake surveys, and hunter interviews

Hunter follows, whereby a researcher accompanies a hunter on a hunting trip, are used to collect data on wildlife encounter rates, likelihood of pursuing a species upon encounter, number of animals killed or captured per encounter, and the behavior of prey species upon sighting a hunter (Kümpel *et al.* 2008). This data collection technique is less frequently used than offtake surveys and hunter interviews due to the sensitive nature of joining a hunter on a hunting trip, especially in areas where hunting is illegal. Hunters may not want to reveal catchment areas or the presence of an observer may disturb hunting activities. Hunter offtake surveys can be used to enumerate species harvested by hunters and to collect biological data of the harvested individuals (Wilcox and Nambu 2007). Hunter interviews (semistructured or questionnaires) provide information on hunter demographics, number and kinds of species hunted, catchment area, hunting techniques, number and length of hunting trips, and the proportion of harvested individuals traded and consumed locally. Empirical observation of hunter harvest through offtake surveys is often combined with hunter interviews to obtain more detailed information on hunting activities and impact. For example, in a study of hunting and the wildlife trade around Myanmar's Hkakaborazi National Park, Rao *et al.* (2011) asked hunters to recall information on the species targeted, prices of traded wildlife parts, and weapons used for hunting. Simultaneously, the authors administered questionnaires to hunters following their return from hunting activities to gather data on prey species hunted and the number of carcasses or individuals hunted per species. These questionnaires also asked the hunters about their primary occupation and the relative economic significance of different income-earning activities. Finally, official confiscation records collected by park staff were used to further assess the illegal trade in wildlife.

While surveys of hunters often provide a direct assessment of wildlife harvest, hunters are often reluctant to discuss information about catchment area and other details of their hunting and trading activity (and may not even be willing to allow carcasses to be weighed, sexed, or measured; Jost-Robinson *et al.* 2011). This may be a particular challenge for researchers in areas where hunting is illegal or where conservation activities are frequently at odds with local human populations. In Cameroon's Korup National Park, for example, local residents can be antagonistic towards park staff and researchers due to past interactions and conflicts of interest (Linder, pers. ob.). Given this sensitivity, Linder and Oates (2011) avoided using more intrusive data collection techniques such as measuring body weights, taking tissue samples, or inquiring about catchment area or destination of the meat. Instead, they reached an agreement with hunters to simply identify and record carcasses they harvested. In conjunction with forest surveys, this technique provided a powerful way to assess the primate trade and impact of hunting on primates in Korup. This example demonstrates that, depending on the research questions, less invasive data collection techniques can still result in valuable data.

18.3 Market surveys

Surveys of hunters and households often provide the most detailed assessments of the harvest, consumption, and trade of primates and other species, but to be effective, such approaches require large samples of households, repeated household visits and time-consuming validation efforts. Surveys of wildlife markets are often employed as a direct, less invasive, and less demanding approach to quantify wildlife supply and demand (e.g., Cowlishaw *et al.* 2005; Allebone-Webb *et al.*

2011). Market surveys typically fail to fully account for non-monetized components of the wildlife trade (e.g., subsistence harvest, household to household bartering), but they can provide valuable information on the scale, sustainability, economics, and politics of wildlife utilization. In fact, market surveys have become a primary tool used by researchers to assess local, regional, and even international trends in the trade of primates and other animals for meat (Cowlishaw *et al.* 2005), medicinal/spiritual uses (Malone *et al.* 2003), and for biomedical and pet markets (Duarte-Quiroga and Estrada 2003; Shepherd *et al.* 2004). This tool has proven particularly powerful when applied in conjunction with household and hunter surveys.

18.3.1 Local and regional markets

The elements of a given market survey will vary depending on the goals, region, and scale of the study; however, several features are common to both live pet and wild meat surveys. These include questions targeted towards gaining a general assessment of animals sold in markets such as: (1) species for sale and (2) amounts for sale (estimated weight and number of individuals); and also questions aimed at broader quantification of market dynamics and sustainability such as (3) price per animal or unit weight; (4) condition (health status of live individuals, preparation of meat as smoked, salted, whole, butchered, etc.); (5) animal gender; (6) animal age class; (7) days animals remain for sale (turnover); (8) animal source area (as reported by vendor); (9) mode of transport to market; (10) socioeconomic characteristics of buyers; and (11) amount and price of non-wildlife items for sale. Questions regarding knowledge of wildlife laws can also be asked of buyers and sellers. Such data can produce an informative snapshot of commodity chains, the scale and potential impact of a given market, and effectiveness of law enforcement and other conservation measures. When surveys are repeated across months or seasons, they may provide rare opportunities to assess rigorously questions of harvest sustainability while also illuminating potential socioeconomic or other avenues to enhancing conservation (e.g., Brashares *et al.* 2004, 2011; Cowlishaw *et al.* 2005; Allebone-Webb *et al.* 2011).

As with nearly all methods involving human subjects, market surveys must be conducted with great sensitivity and assurances of confidentiality for both sellers and buyers. Furthermore, similar to household and hunter surveys, identifying the most reliable and informative subjects for study (i.e., markets and individual vendors) often requires local knowledge and a bit of sleuthing. Some wildlife sellers are easily identified within large urban or suburban markets, whereas others may operate on roadsides, or out of restaurants and houses where they are difficult to detect and monitor. Different types of markets often cater to different clientele and may specialize in unique species of primates and other animals. Thus, gaining a more or less complete picture of the wildlife trade in a given area requires a truly integrative and comprehensive survey approach (Cowlishaw *et al.* 2005).

18.3.2 International markets and trade

Two primary sources of data on the international trade of primates and other wildlife include national government agencies (e.g., customs records) and annual reports for CITES, both of which can help to assess the legal trade in wildlife (Broad *et al.* 2003). Countries that have joined CITES compile annual reports on the different species and number of animals traded across their borders. The United Nations Environment Program—World Conservation Monitoring Center (UNEP-WCMC) gathers data from these reports and makes them public on the CITES Trade Database (<http://www.unep-wcmc-apps.org/citestrade/>). Researchers can then use these data to assess the scale and temporal changes of the wildlife trade and its potential impacts on wildlife populations (Mittermeier 1987; WWF/TRAFFIC 2002; Nekaris *et al.* 2010; Nijman 2009). National governments also collect wildlife trade data through customs control, but these data often fail to detail the species and number of animals involved (Broad *et al.* 2003). Not surprisingly given their politically sensitive nature, these customs records are often difficult to obtain (WWF/TRAFFIC 2002). Moreover, customs and CITES records underestimate export/import rates because they fail to fully account for illegal transport of primates and other animals (Nijman 2009). As a result, many estimates of the live trade of primates between countries and continents come in the form of incomplete extrapolations from limited data and these estimates commonly vary by orders of magnitude (WWF/TRAFFIC 2002). Thus, our understanding of the legal trade in primates remains murky and there appear to be considerable challenges to rigorous study of this trade. For researchers able to assemble reliable export/import records, slightly modified versions of the survey questions outlined above for local and regional markets can go a long way towards clarifying the scope and impact of the legal, international trade in live primates.

The complexities of quantifying the legal trade in live primates, though considerable, may pale in comparison to the challenges involved in characterizing the vast and illegal trade of primate meat and parts across international borders (WWF/TRAFFIC 2002). It is generally believed that the current unprecedented ease of human travel and globalization of commerce has resulted in dramatic increases in the international trade of wild meat and wildlife parts (Milner-Gulland *et al.* 2003; Chaber *et al.* 2010). This growing trade is increasingly driven underground as law enforcement and public health officials have increased policing efforts in response to the disease risk posed by wildlife products. This combination of scale and illegality has led some to compare the international trade in wild meat and animal products to the globalized trade of narcotics. While such a comparison ignores the cultural and religious significance of wildlife for many communities and likely overstates the human health risks of the wildlife trade, it is true that studies attempting to quantify this trade often rely on methods commonly applied in drug enforcement. Specifically, most of our current estimates of the scale and composition of the international trade in primate meat and body parts are based on data from airport and other import seizures (e.g., Chaber *et al.* 2010) or from information provided by knowledgeable and well placed informants (e.g., Milius 2005).

Official data on wildlife seizures and observations of informants provide only a limited glimpse at the international trade of primates and other wildlife, but they are a valuable starting point. Like other methods outlined in this chapter, these approaches have greater power when information is combined across larger spatial and temporal scales, and when information is gleaned through multiple, complementary forms of inference. One such approach is currently employed by the Bushmeat Monitoring Network (Milius 2005; Brashares *et al.* 2011) which

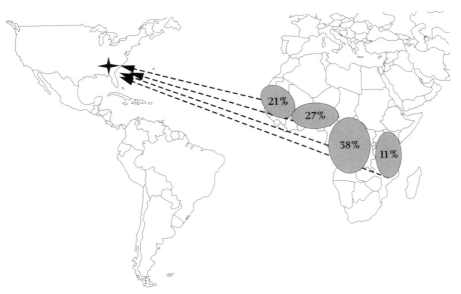

Fig. 18.1 After marking tens of thousands of African wildlife products prior to export, the Bushmeat Monitoring Network surveys wildlife markets in Europe and North America to "recapture" marked animals and thereby learn about the dynamics and scale of the international trade in wildlife. This map shows the source region for 280 African primates marked prior to export and "recaptured" in a single bushmeat market in Atlanta, USA from 2006–2012.

combines weekly surveys of 81 wildlife markets in 14 African countries with informant-based, weekly surveys of 46 wildlife markets in Europe and North America and marking and tracking of individual animals from capture to final sale to estimate the dynamics and composition of the international trade of African wildlife. Such an approach has value for illuminating the movement of wildlife among continents (Fig. 18.1), but it cannot address the likely large component of the international trade that does not take place in formal urban markets.

18.4 Conclusion

Unfortunately, there is no simple blueprint for studying the many dimensions and elements of the primate trade. However, a large and growing body of research illuminates pathways for assessing ecological, conservation, socioeconomic, and other components of primate harvest, consumption, and trade. This research has developed a range of approaches that often include structured and unstructured household surveys, hunter follows, market surveys, and field-based ecological assessments. In fact, many recent studies employ combinations of these approaches to not only triangulate and validate results but also to provide a multifaceted and comprehensive perspective on the trade of wildlife in a given area (e.g., Bennett *et al.* 2000; Jost-Robinson *et al.* 2011; Linder and Oates 2011; Brashares *et al.* 2011). The message from these and other studies is that any one approach to examining the primate or, broadly, wildlife trade, will likely reveal an incomplete picture and researchers should strive to combine approaches whenever possible.

Like researchers in other fields, those attempting to study the wildlife trade will face tough decisions and trade-offs when designing their work. Commonly faced questions include "With limited time and research funds, should I survey many households once or attempt repeated visits of a smaller number of households?," "Is it worth conducting hunter follows if I can only enlist four hunters?," or "Is there value in surveying only one wildlife market?" Of course, the answer to each of these questions is "it depends," but general rules of thumb still apply. First, in regions where wildlife harvest and trade are highly seasonal, one-off "snapshot" studies are less likely than longitudinal (i.e., repeated sampling) studies to provide a true representation of wildlife utilization and associated commerce. Second, studies that triangulate their results (i.e., validate findings using dissimilar methods) have the greatest value for evaluating the wildlife trade. When in doubt, add other forms of inference, even if the sample sizes involved (e.g., hunters or markets) are limited.

Although we have tried to highlight some common methods used to examine hunting and trading of wildlife, hopefully we have also made clear that vast gaps remain in our understanding of the scale, drivers, and options for mitigating the damaging effects of this trade. Wildlife utilization is officially "on the map" in the field of Conservation Biology and studies of wildlife harvest and trade from sites around the world regularly appear in conservation journals. Together these studies help us paint a broad picture of the wildlife trade, but this is possible only insofar as these studies employ comparable frameworks, methods, and metrics. We do not promote a one-size-fits-all approach to studying the wildlife trade, but we strongly encourage researchers to design research questions and methods that will allow others to consider their results within the context of the larger body of relevant work. Last, we emphasize the need for researchers to publish their findings such that their results may inform our current understanding of the wildlife trade and also mark the path for future research. Results from these studies should be shared and discussed with conservation practitioners to help reduce the impact of hunting and trade on wild animal populations.

Acknowledgments

We thank the editors and the anonymous reviewers for their comments on this manuscript. We are grateful to the editors for inviting us to contribute to this important book.

19

Conclusion: the future of studying primates in a changing world

Eleanor J. Sterling, Nora Bynum, and Mary E. Blair

The rapid advancement in techniques used to study non-human primate ecology and conservation over the past several decades has helped to expand the horizons of primatologists, particularly in terms of the increasingly sophisticated questions that can be asked and addressed. The chapters in this book point to exciting opportunities for continued innovation. Assimilating ideas from across the chapters, we note several emerging patterns in primate studies that we group into five categories, including the importance of learning from other, related fields; the increased prevalence of multidisciplinary efforts; the tantalizing vistas afforded by larger quantities of and higher quality data; the role of modeling in helping to understand primates; and the decreasing schism between laboratory and field studies.

Primatologists have benefited from creatively drawing from fields well beyond primatology or biological anthropology. Much of the progress in primatology to date stems from approaches and technologies borrowed from other fields, for instance in analysis of movement data, inter-group interactions, or survey techniques for rare species. Both theoretical as well as empirical research on other long-lived or group-living species such as social carnivores, ungulates, birds, and invertebrates can continue to serve as inspiration for primate studies. For example, as we continue to draw from these other fields, we facilitate the generation of more robust, comparative studies about relative uniqueness or non-uniqueness of primate socioecology. Such studies will enable the development of more comprehensive theoretical frameworks, moving towards the explanation of variation in socioecological patterns across space, time, and phylogeny. Other techniques developed in other fields, such as next-generation, whole-genome sequencing and rapid genome sequencing for large samples of individuals, will allow us to explore the adaptive genetic diversity of complex individual behaviors like temperament, personality, or cognitive ability. This adopted technology will also

Primate Ecology and Conservation: A Handbook of Techniques. First Edition. Edited by Eleanor J. Sterling, Nora Bynum, and Mary E. Blair. © Oxford University Press 2013. Published 2013 by Oxford University Press.

facilitate investigations of ecological genomics, such as the phenotypic plasticity of life history traits, longevity, and aging.

Moving beyond learning from other single individual fields, use of multidisciplinary research methods can enable the investigation of research questions across perspectives and scales, which in turn may facilitate improved understanding of primate ecology and behavior, as well as effectiveness and sustainability of conservation actions. As illustrated by studies in ethnoprimatology and disease transmission, among others, multiple methods can be drawn from a variety of disciplines, including cultural anthropology, ecology, ethology, genetics, chemistry, cognitive sciences, agricultural sciences, and conservation biology, to understand non-human primates, their future prospects, and their relationships to humans. This is increasingly important in a world where non-human primate habitats are almost universally affected by human activities.

Advanced technology has enhanced our ability to collect and analyze ever-larger quantities of both quantitative and qualitative data. Examples include our ability to detect primate habitats from space via remotely sensed data, to track individual and group movements with telemetry, to recognize individuals and their relatives with genetic techniques such as PCR, and to track reproductive status using hormones found in feces and urine. As remotely-sensed data continues on the trend towards finer-scales (for instance data from drones, near infrared reflectance spectroscopy, etc.), we should be able to characterize more accurately primate habitat at a variety of scales, with elements including availability and timing of food resources as well as sleeping sites and other critical habitat elements. These approaches are increasingly important to improve our understanding of how temporal and spatial variability in resources and habitat affect primate socioecology, dispersal patterns, genetic structure, and population viability.

Devices for field digital data collection are already facilitating the collection of vast amounts of behavioral and ecological data, and digital databasing allows for a more efficient and accurate pipeline to bring data from the collection phase into data analysis and interpretation. Data collection of this type is also greatly accelerating the analysis pipeline for qualitative data from surveys and interviews of human communities, through automatic transcription and coding of survey responses through digital voice recording or the use of smart pens. More sophisticated data storage and analyses as well as new statistical techniques have allowed for, in some cases, fuller use of data gathered in the field, and for more robust inferences to be drawn.

While the management, storage, and maintenance of these large data sets is daunting, the questions they can illuminate take us several steps farther than any we could address previously. For instance, long-term data sets that are now

Box 19.1 Modeling the effects of climate change on species' distributions and extinction risk

Climate change is expected to result in shifts in species' distributions, causing some parts of a species' current range to become less suitable, other parts to become more suitable, and some parts to move into novel areas (Huntley *et al.* 2008; Thomas 2010), increasing the likelihood of local and global extinctions (Parmesan *et al.* 1999; Pounds *et al.* 1999). To date, several studies have documented observed impacts of climate change on species' distributions (e.g., Parmesan and Yohe 2003; Root *et al.* 2003; Rosenzweig *et al.* 2008). Thus there is a push to develop practical methods for incorporating climate change within conservation planning (Mawdsley *et al.* 2009; Ackerly *et al.* 2010; Boutin 2010). For example, the IUCN is developing methods to incorporate species susceptibility to climate change into its estimates of extinction risk (Foden *et al.* 2008).

Although we might not expect climate change to be very important for primates compared to other non-climatic threats such as hunting and habitat disturbance, assessing likely future exposure across a species' range to threats such as climate change can make a difference in conservation prioritization processes and planning for the long-term survival of a species. Also, a few recent studies provide evidence for effects of climate change on primate population persistence: Wiederholt and Post (2010) documented the negative influence of the intensification of El Niño southern oscillation events on ateline primate resource levels as well as population dynamics across Central and South America.

In general, climate regimes are expected to shift upward in elevation and poleward in latitude (Parmesan 2006). However, models and empirical analyses of range shifts suggest that species show individualistic responses to changing climates (Peterson *et al.* 2011). Thus, species-specific considerations of the likely effects of climate change will be important, and they can be forecast using species niche modeling (see Chapter 6), which can identify the parts of a species' range that are expected to be more exposed to changes in temperature and precipitation (e.g., Thomas *et al.* 2004; Araújo *et al.* 2006; Dawson *et al.* 2011).

A typical method to capture the uncertainty of modeling into the future is *ensemble forecasting*, which involves the calibration of ensembles or sets of species niche models using various algorithms (see Chapter 6) projected to a suite of future climate scenarios and exploration of the resulting range of uncertainties (Araújo and New 2007). Future climate projection data from the Intergovernmental Panel on Climate Change (IPCC)'s Fourth Assessment are available for at least three emissions scenarios (A1, A2, B2), seven global circulation models, and seven time periods (averages for every decade from the 2020s to the 2080s; <http://www.ccafs-climate.org/>). Ensemble forecasting has been theoretically and empirically

Box 19.1 *Continued*

proven to outperform forecasts by individual models in predictive ability (Araújo and New 2007).

There are several issues to consider when using niche models to predict the effects of climate change, and especially when interpreting model predictions. For example, a careful and complete estimation of the full dimensions of the ecological niche of a species is required in order to calibrate a model that will be representative of a species' potential response (Peterson *et al.* 2011). Thus, niche models calibrated with very few occurrence points, or with an insufficient set of points that may not characterize a species' full extent of occurrence, may result in inaccurate predictions of a species' response to climate change. It may also be important to couple niche models with models of dispersal as well as interactions among species in order to more accurately characterize the niche (Peterson *et al.* 2011).

Also, correlative niche models may not be adequate to estimate the extinction risk of small populations, because processes leading to their extinction may be more stochastic than deterministic, or more demographic than geographic (Lande *et al.* 2003). It is possible to couple niche models with spatially explicit stochastic population models to explore the interactions of mechanisms causing population decline (e.g., Keith *et al.* 2008).

emerging from field sites established decades ago will allow us to better understand longevity and reproduction given the long lifetimes of primates, and to monitor the effects of changing conditions, including those due to climate change or habitat modification over time on primate socioecology and population status.

With better data sets we can also use advanced modeling techniques to help us characterize and test hypotheses about primate ecology, behavior, and evolution, from the individual to species levels. Modeling techniques have helped us to focus survey efforts for endangered populations and to estimate probability of population persistence, with the goal of allowing more effective allocation of scarce resources. Application of underutilized techniques such as agent-based modeling or matrix models may afford us better predictions of primate behavior. Computer models may help us to understand changes in primate body shape, dietary choices, social structures, and cognitive evolution with changing environments. Niche models will help conservationists to better predict the effects of climate change on primate habitats and focus conservation investment (Box 19.1).

Advanced modeling will also enable the generation of robust hypotheses that can be tested in the field. Indeed, researchers have increasingly integrated field and laboratory approaches, for instance by using in the field tools such as portable PCR machines or metabolic chambers, as well as tracking devices that measure core and subcutaneous body temperatures, heart rate, activity levels, and three-dimensional distance traveled. The ability to bring such techniques that were formerly lab-based to the field will afford rapid results that may better reflect actual conditions. Combining techniques and results from field and lab efforts will undoubtedly afford us a more comprehensive and holistic understanding of primate ecology, behavior, and conservation, including with regard to public and wildlife health issues. As humans accelerate the pace of modifying terrestrial environments, they will increasingly come in contact with non-human primates whose populations were previously isolated, changing infection profiles for human and non-human primates alike. Techniques that span the field and laboratory may further elucidate the role of pathogens in primate sociality and social behavior, including dominance relationships and other intra-groups interactions, as well as relationships among groups, including dispersal of individuals.

The recent and rapid expansion of techniques used by primatologists is exciting and fruitful. Building on the progress we have made, we face continued opportunity for innovation in our toolbox. The increasing integration of ecological studies with conservation science will enhance our understanding of the ecology of non-human primate communities and will be critical to effective and sustainable conservation actions. These approaches will help us understand how evolutionary and ecological processes in primate populations are changing as humans continue to transform the habitats in which non-human primates live, and more importantly, will help us predict how primate populations may respond in the future to continued change. We sincerely hope that this increased understanding will improve the chances for conserving primate species in the long term.

Bibliography

Abernethy, K. A., White, L. J. T., and Wickings, E. J. (2002). Hordes of mandrills (*Mandrillus sphinx*): extreme group size and seasonal male presence. *Journal of Zoology*, 258, 131–7.

Abbott, D. H., Keverne, E. B., Bercovitch, F. B., et al. (2003). Are subordinates always stressed? A comparative analysis of rank differences in cortisol levels among primates. *Hormones and Behavior*, 43, 67–82.

Aber, J. D. (1979). Foliage-height profiles and succession in northern hardwood forests. *Ecology*, 60, 18–23.

Abugiche, S. A. (2008). Impact of Hunting and Bushmeat Trade on Biodiversity Loss in Cameroon: A Case Study of the Banyang-Mbo Wildlife Sanctuary (Ph.D. Dissertation), Brandenburg University of Technology, Cottbus, Germany.

Ackerly, D. D., Loarie, S. R., Cornwell, W. K., et al. (2010). The geography of climate change: implications for conservation biogeography. *Diversity and Distributions*, 16, 476–87.

Adeola, M. O. (1992). Importance of wild animals and their parts in the culture, religious festivals, and traditional medicine of Nigeria. *Environmental Conservation*, 19, 125–34.

Adkins-Regan, E. (2005). *Hormones and Animal Social Behavior*. Princeton University Press, Princeton, NJ.

Agoramoorthy, G., and Hsu, M. J. (2005). Use of nonhuman primates in entertainment in Southeast Asia. *Journal of Applied Animal Welfare Science*, 8, 141–9.

Agrawal, A., and Gibson, C. C. (1999). Enchantment and disenchantment: The role of community in natural resource conservation. *World Development*, 27, 629–49.

Aguiar, L. M., Ludwig, G., Svoboda, W. K., et al. (2007). Use of traps to capture black and gold howlers (*Alouatta caraya*) on the Islands of the upper paraná river, Southern Brazil. *American Journal of Primatology*, 69, 241–7.

Akçakaya, H. R. (1997). RAMAS Metapop: viability analysis for stage-structured metapopulations. Version 2. Applied Biomathematics, Setauket, NY.

Albert, E. M., San Mauro, D., Garcia-Paris, M., et al. (2009). Effect of taxon sampling on recovering the phylogeny of squamate reptiles based on complete mitochondrial genome and nuclear gene sequence data. *Gene*, 441, 12–21.

Alberts, S. C., and Altmann, J. (2002). Matrix models for primate life history analysis. In P. M. Kappeler, and M. E. Pereira, eds. *Primate life histories and socioecology*, pp. 66–102. The University of Chicago Press, Chicago.

Alberts, S. C., Buchan, J.C., and Altmann, J. (2006). Sexual selection in wild baboons: from mating opportunities to paternity success. *Animal Behaviour*, 72, 1177–96.

Albrechtsen, L., Fa, J. E., Barry, B., and Macdonald, D. W. (2005). Contrasts in availability and consumption of animal protein in Bioko Island, West Africa: the role of bushmeat. *Environmental Conservation*, 32, 340–8.

Albrechtsen, L., Macdonald, D. W., Johnson, P. J., Castelo, R., and Fa, J. E. (2007). Faunal loss from bushmeat hunting: empirical evidence and policy implications in Bioko Island. *Environmental Science and Policy*, 10, 654–67.

Allebone-Webb, S. M., Kümpel, N. F., Rist, J., Cowlishaw, G., Rowcliffe, J. M., and Milner-Gulland, E. J. (2011). Use of market data to assess bushmeat hunting sustainability in Equatorial Guinea. *Conservation Biology*, 25, 597–606.

Altherr, S. (2007). *Going to pot—The Neotropical bushmeat crisis and its impact on primate populations.* Care for the Wild & Pro Wildlife, West Sussex, UK and Munich, Germany.

Altizer, S., Nunn, C., Thrall, P., *et al.* (2003). Social organization and parasite risk in mammals: integrating theory and empirical studies. *Annual Review of Ecology, Evolution and Systematics*, 34, 517–47.

Altmann, J. (1974). Observational Study of Behavior: Sampling Methods. *Behavior*, 49, 227–67.

Altmann, J. (1980). *Baboon Mothers and Infants.* Harvard University Press, Cambridge, MA.

Altmann, J., Gesquiere, L., Galbany, J., Onyango, P. O., and Alberts, S. C. (2010). Life history context of reproductive aging in a wild primate model. *Annals of the New York Academy of Sciences*, 1204, 127–38.

Altmann, S. A., and Altmann, J. (2003). The transformation of behaviour field studies. *Animal Behaviour*, 65, 413–23.

Alves, R. R. N., Souto, W. M. S., and Barboza, R. R. D. (2010). Primates in traditional folk medicine: a world overview. *Mammal Review*, 40, 155–80.

Ambrose, S. H. (1990). Preparation and characterization of bone and tooth collagen for isotopic analysis. *Journal of Archaeological Science*, 17, 431–51.

Ambrose, S. H. (1991). Effects of diet, climate, and physiology on nitrogen isotope abundances in terrestrial foodwebs. *Journal of Archaeological Science*, 18, 293–317.

Ambrose, S. H. (1993). Isotopic analysis of paleodiets: methodological and interpretive considerations. In M. K. Sanford, ed. *Investigations of Ancient Human Tissue*, pp. 59–130. Gordon and Breach Science Publishers, Langhorne, PA.

Ambrose, S. H., and DeNiro, M. J. (1986). The isotopic ecology of east African mammals. *Oecologia*, 69, 395–406.

Ambrose, S. H., and Norr, L. (1993). Experimental evidence for the relationship of the carbon isotope ratios of whole diet and dietary protein to those of bone collagen and carbonate. In J. B. Lambert, and G. Grupe, eds. *Prehistoric Human Bone: Archaeology at the Molecular Level*, pp. 1–37. Springer, Berlin.

Anadu, P. A., Elamah, P. O., and Oates, J. F. (1988). The bushmeat trade in southwestern Nigeria: a case study. *Human Ecology*, 16, 199–208.

Ancrenaz, M., Lackman-Ancrenaz, I., and Mundy, N. (1994). Field observations of aye-ayes (*Daubentonia madagascariensis*) in Madagascar. *Folia Primatologica*, 62, 22–36.

Ancrenaz, M., Giminez, O., Ambu, L. *et al.* (2005). Aerial surveys give new estimates for orangutans in Sabah, Malaysia. *PLoS Biology*, 3, e3.

Anderson, C. O., and Mason, W. A. (1974). Early experience and complexity of social organization in groups of young rhesus monkeys (*Macaca mulatta*). *Journal of Comparative and Physiological Psychology*, 87, 681–90.

Anderson, H. B., Thompson, M. E., Knott, C. D., and Perkins, L. (2008). Fertility and mortality patterns of captive Bornean and Sumatran orangutans: is there a species difference in life history? *Journal of Human Evolution*, 54, 34–42.

Anderson, R. M., and May, R. M. (1992). *Infectious Diseases of Humans: Dynamics and Control.* Oxford University Press, Oxford.

Anderson, R. P., Gomez-Laverde, M., and Peterson, A. T. (2002). Geographical distributions of spiny pocket mice in South America: insights from predictive models. *Global Ecology and Biogeography,* **11**, 131–41.

Anderson, R. P., Lew, D., and Peterson, A. T. (2003). Evaluating predictive models of species' distributions: criteria for selecting optimal models. *Ecological Modeling,* **162**, 211–32.

Andriamasimanana, M. (1994). Ecoethological study of free-ranging aye-ayes (*Daubentonia madagascariensis)* in Madagascar. *Folia Primatologica,* **62**, 37–45.

AOAC. (1990). *Official methods of analysis.* Association of Official Analytical Chemists, Arlington, VA.

Arandjelovic, M., Guschanski, K., Schubert, G., *et al.* (2009). Two-step PCR improves the speed and accuracy of genotyping using DNA from noninvasive and museum samples. *Molecular Ecology Resources,* **9**, 28–36.

Arandjelovic, M., Head, J., Rabanal, L. I., *et al.* (2011). Non-Invasive Genetic Monitoring of Wild Central Chimpanzees. *PLoS ONE,* **6**, e14761.

Archer, J. (2006). Testosterone and human aggression: an evaluation of the challenge hypothesis. *Neuroscience and Biobehavioral Reviews,* **30**, 319–45.

Araújo, M. B., and New, M. (2007). Ensemble forecasting of species distributions. *Trends in Ecology & Evolution,* **22**, 42–7.

Araújo, M. B., Thuiller, W., and Pearson, R. G. (2006). Climate warming and the decline of amphibians and reptiles in Europe. *Journal of Biogeography,* **33**, 1712–28.

Asa, C., and Porton, I., eds. (2009). *Wildlife Contraception: Issues, Methods and Applications.* John Hopkins University Press, Baltimore, pp. 119–48.

Asensio, N., Korstjens, A., Schaffner, C., and Aureli, F. (2008). Intragroup aggression, fission-fusion dynamics and feeding competition in spider monkeys. *Behaviour,* **145**, 983–1001.

Ash, L., and Orihel, T. (1991). *Parasites: A Guide to Laboratory Procedures and Identification.* ASCP Press, Chicago.

Ashton, P. S. (1969). Speciation among tropical forest trees: some deductions in the light of recent evidence. *Biological Journal of the Linnean Society,* **1**, 155–96.

Audet, D., and Thomas, D. W. (1996). Evaluation of the accuracy of body temperature measurement using external transmitters. *Canadian Journal of Zoology,* **74**, 1778–81.

August, P. V. (1983). The role of habitat complexity and heterogeneity in structuring tropical mammal communities. *Ecology,* **64**, 1495–507.

Aureli, F., and de Waal, F. B. M., eds. (2000). *Natural Conflict Resolution.* University of California Press, Berkeley.

Auzel, P., and Wilkie, D. S. (2000). Wildlife use in northern Congo: Hunting in a commercial logging concession. In J. G. Robinson and E. L. Bennett, eds. *Hunting for Sustainability in Tropical Forests,* pp. 413–26. Columbia University Press, New York.

Avise, J. C. (2000). *Phylogeography: The History and Formation of Species.* Harvard University Press, Cambridge, MA.

Avise, J. C. (2004). *Molecular Markers, Natural History and Evolution,* 2nd edition. Sinauer, Sunderland, MA.

AZA (Association of Zoos and Aquariums). (2011). *About us*. AZA, Silver Spring, MD. Available at: http://www.aza.org/about-aza/(Accessed 9/7/2011).

Baillie, J., and Groombridge, B. (1996). *1996 IUCN Red List of threatened animals*. IUCN, Gland, Switzerland.

Baker, L. R., ed. (2002). Guidelines for Nonhuman Primate Re-introductions. *IUCN/SSC Re-introduction Specialist Group Newsletter*, 21.

Baker, A. M., Lacy, R. C., Leus, K., and Traylor-Holzer, K. (2011). Intensive management of populations for conservation. *WAZA Magazine*, 12, 40–3.

Baker, A. J., and Woods, F. (1992). Reproduction of the emperor tamarin (*Saguinus imperator*) in captivity, with comparisons to cotton-top and golden lion tamarins. *American Journal of Primatology*, 26, 1–10.

Bales, K. L., French, J. A., Hostetler, C. M., and Dietz, J. M. (2005). Social and reproductive factors affecting cortisol levels in wild female golden lion tamarins (*Leontopithecus rosalia*). *American Journal of Primatology*, 67, 25–35.

Balkenhol, N., Waits, L. P., and Dezzani, R. J. (2009). Statistical approaches in landscape genetics: an evaluation of methods for linking landscape and genetic data. *Ecography*, 32, 818–30.

Ballou, J., Lees, C., Faust, L., *et al.* (2010). Demographic and genetic management of captive populations for conservation. In D. G. Kleiman, K. V. Thompson, and C. Kirk Baer, eds. *Wild Mammals in Captivity*, 2nd edition, pp. 219–52. University of Chicago Press, Chicago.

Barrett, L., Henzi, P., Weingrill, T., *et al.* (2000). Female baboons do not raise the stakes but they give as good as they get. *Animal Behaviour*, 59, 763–70.

Bass, M. S., Finer, M., Jenkins, C. N., *et al.* (2010). Global conservation significance of Ecuador's Yasuní National Park. *PLoS ONE*, 5, e8767.

Baverstock, P. R., and Craig, M. (1996). Project Design. In D. M. Hillis, C. Moritz, and B. K. Mable, eds. *Molecular Systematics*, 2nd edition, pp. 17–27. Sinauer, Sunderland, MA.

Bearder, S. (1999). Physical and social diversity among nocturnal primates: a new view based on long term research. *Primates*, 40, 267–82.

Bearder, S. K., Nekaris, K. A. I., and Curtis, D. J. (2006). A Re-Evaluation of the role of vision in the activity and communication of nocturnal primates. *Folia Primatologica*, 77, 50–71.

Beaumont, M. A., and Rannala, B. (2004). The Bayesian revolution in genetics. *Nature Reviews Genetics*, 5, 251–61.

Beck, B. B., Castro, M. I., Stoinski, T. S., and Ballou, J. D. (2002). The effects of prerelease environments and postrelease management on survivorship in reintroduced golden lion tamarins. In D. G. Kleiman, and A. B. Rylands, eds. *Lion Tamarins Biology and Conservation*, pp. 283–300. Smithsonian Institute Press, Washington, D.C.

Beck, B. B., and Power, M. (1988). Correlates of sexual and maternal competence in captive gorillas. *Zoo Biology*, 2, 253–7.

Beck, B. B., Rapaport, L. G., Price, M. R. S., and Wilson, A. C. (1994). Reintroduction of captive born animals. In P. J. S. Olney, G. M. Mace, and A. T. C. Feistner, eds. *Creative Conservation: Interactive Management of Wild and Captive Animals*, pp. 265–86. Chapman & Hall, London.

Beck, B. B., Walkup, K., Rodrigues, M., *et al.* (2007). *Best Practice Guidelines for the Reintroduction of Great Apes.* SSC Primate Specialist Group of the World Conservation Union, Gland Switzerland.

Becker, J. B., Breedlove, S. M., Crews, D., and McCarthy, M. M., eds. (2002). *Behavioral Endocrinology*, 2nd edition. MIT Press, Cambridge, MA.

Beehner, J. C., and McCann, C. (2008). Cortisol levels in captive and wild geladas (*Theropithecus gelada*) *Physiology and Behavior*, **95**, 508–14.

Beehner, J. C., and Whitten, P. L. (2004). Modifications of a field method for fecal steroid analysis in baboons. *Physiology and Behavior*, **82**, 269–77.

Beehner, J. C., Nguyen, N., Wango, E. O., Alberts, S. C., & Altmann, J. (2006). The endocrinology of pregnancy and fetal loss in wild baboons. *Hormones and Behavior*, 49, 688–699.

Begley, M. R., and Mackin, T. J. (2004). Spherical indentation of freestanding circular thin films in the membrane regime. *Journal of the Mechanics and Physics of Solids*, **52**, 2005–23.

Bellows, A. S., Pagels, J. F., and Mitchell, J. C. (2001). Macrohabitat and microhabitat affinities of small mammals in a fragmented landscape on the upper coastal plain of Virginia. *The American Midland Naturalist*, **146**, 345–60.

Belzung, C., and Anderson, J. R. (1986). Social rank and responses to feeding competition in Rhesus monkeys. *Behavioral Processes*, **12**, 307–16.

Bennett, E. L., Nyaoi, A. J., and Sompud, J. (2000). Saving Borneo's bacon: the sustainability of hunting in Sarawak and Sabah. In J. G. Robinson and E. L. Bennett, eds. *Hunting for Sustainability in Tropical Forests*, pp. 305–24. Columbia University Press, New York.

Bennett, E. L., and Rao, M. (2002). Wild meat consumption in Asian tropical forest countries: Is this a glimpse of the future for Africa? In *Proceedings of the Conference On Links Between Biodiversity Conservation, Livelihoods And Food Security: The Sustainable Use Of Wild Meat 2001*. World Conservation Union, Gland, Switzerland.

Bensasson, D., Zhang, D. X., Hartl, D. L., and Hewitt, G. M. (2001). Mitochondrial pseudogenes: evolution's misplaced witnesses. *Trends in Ecology & Evolution*, **16**, 314–21.

Bentley-Condit, V. K., and Hare, T. S. (2007). Using Geographic Information Systems and spatial statistics to examine the spatial dimension of animal social behavior: a baboon example. *Journal of Mathematics, Statistics and Allied Fields* **1**, 1028. (Available at http://www.scientificjournals.org/journals2007/articles/1028.htm).

Bercovitch, F. B., and Strum, S. C. (1993). Dominance rank, resource availability, and reproductive maturation in female savanna baboons. *Behavioral Ecology and Sociobiology*, **33**, 313–18.

Bergey, C. M. (2011). AluHunter: a database of potentially polymorphic Alu insertions for use in primate phylogeny and population genetics. *Bioinformatics*, **27**, 2924–5.

Bernard, H. R. (2006). *Research Methods in Anthropology: Qualitative and Quantitative Approaches*, 4th edition. Altamira Press, Walnut Creek, MD.

Bernard, H. R. (2011). *Research Methods in Anthropology: Qualitative and Quantitative Approaches*, 5th edition. Altamira Press, Walnut Creek, MD.

Bernstein, I. S. (1991). An empirical comparison of focal and ad libitum scoring with commentary on instantaneous scans, all occurrence and one-zero techniques. *Animal Behaviour*, **42**, 721–8.

Bicca-Marques, J. C., and Garber, P. A. (2005). Use of social and ecological information in tamarin foraging decisions. *International Journal of Primatology*, **26**, 1321–44.

Biebouw, K., Bearder, S., and Nekaris, A. (2009). Tree hole utilisation by the hairy-eared dwarf lemur (*Allocebus trichotis*) in Analamazaotra Special Reserve. *Folia Primatologica*, **80**, 89–103.

Bingham, S. A. (1987). The dietary assessment of individuals: methods, accuracy, new techniques and recommendations. *Nutrition Abstracts and Reviews, Series A*, 57, 705–42.

Blair, M. E. (2011). Habitat Modification and Gene Flow in *Saimiri oerstedii*: Landscape Genetics, Intraspecific Molecular Phylogenetics, and Conservation (Ph.D. Dissertation), Columbia University, New York, NY.

Blair, M. E., and Melnick, D. J. Scale-dependent effects of a heterogeneous landscape on genetic differentiation in the central american squirrel monkey (*Saimiri oerstedii*). *PLoS ONE*, 7, e43027.

Blair, M. E., Sterling, E. J., and Hurley, M. M. (2011). Taxonomy and conservation of Vietnam's primates: A review. *American Journal of Primatology*, 73, 1093–106.

Blas, J., Bortolotti, G. R., Tella, J. L., *et al.* (2007). Stress response during development predicts fitness in a wild, long-lived vertebrate. *Proceedings of the National Academy of Sciences USA*, 104, 8880–4.

Blumenthal, S. A., Chritz, K. L., and Rothman, J. M. (In Press). Stable carbon isotope composition of wild mountain gorilla feces record intra-annual dietary variability. *Proceedings of the National Academy of Sciences USA*.

Bodmer, R. E., Pezo Lozana, E., and Fang, T. G. (2004). Economic analysis of wildlife use in the Peruvian Amazon. In K. M. Silvius, R. E. Bodmer, and M. V. Fragoso, eds. *People in Nature: Wildlife Conservation in South and Central America*, pp. 191–207. Columbia University Press, New York.

Bolter, D. R. (2011). A comparative study of growth patterns in Crested Langurs and Vervet Monkeys. *Anatomy Research International*, 2011, Article ID 948671.

Bonnor, G. M. (1967). Estimation of ground canopy density from ground measurements. *Journal of Forestry*, 65, 544–7.

Borchers, D. L., Laake, J. L., Southwell, C., and Paxton, C. G. M. (2006). Accommodating unmodeled heterogeneity in double-observer distance sampling surveys. *Biometrics*, 62, 372–8.

Borgatti, S. P. (2002). *Netdraw Network Visualization*. Analytic Technologies, Natick, MA.

Borgatti, S. P. (2004). *ANTHROPAC 4.983*. Analytic Technologies, Natick, MA.

Borgerhoff Mulder, M., and Caro, T. M. (1985). The use of quantitative observational techniques in anthropology. *Current Anthropology*, 26, 323–30.

Boubli, J. P. (1999). Feeding ecology of black-headed uacaris in Pico de Neblina National Park, Brazil. *International Journal of Primatology*, 20, 719–49.

Boughton, R. K., Joop, G., and Armitage, S. A. O. (2011). Outdoor immunology: Methodological considerations for ecologists. *Functional Ecology*, 25, 81–100.

Boutin, S. (2010). Conservation planning within emerging global climate and economic realities. *Biological Conservation*, 143, 1569–70.

Bowlby, J. (1969). *Attachment and loss. Vol. 1. Attachment*. Basic Books, New York.

Boyer, D., and Walsh, P. D. (2010). Modelling the mobility of living organisms in heterogeneous landscapes: does memory improve foraging success? *Philosophical Transactions of the Royal Society a-Mathematical Physical and Engineering Sciences*, 368, 5645–59.

Bradley, B. J., Doran-Sheehy, D. M., and Vigilant, L. (2008). Genetic identification of elusive animals: re-evaluating tracking and nesting data for wild western gorillas. *Journal of Zoology*, 275, 333–40.

Brashares, J. S., Arcese, P, Sam, M. K., *et al.* (2004). Bushmeat hunting, wildlife declines, and fish supply in West Africa. *Science*, 306, 1180–3.

Brashares, J. S., Golden, C. D., Weinbaum, K. Z., *et al.* (2011). Economic and geographic drivers of wildlife consumption in rural Africa. *Proceedings of the National Academy of Sciences USA,* **108**, 13931–6.

Brauer, F., and Castillo-Chavez, C. (2001). *Mathematical Models in Population Biology and Epidemiology.* Spring-Verlag, New York.

Brent, L. J. N., Lehmann, J., and Ramos- Fernandez, G. (2011). Social network analysis in the study of nonhuman primates: A historical perspective. *American Journal of Primatology,* **73**, 720–30.

Brett, F. L., Turner, T. R., Jolly, C. J., *et al.* (1982). Trapping baboons and vervet monkeys from wild, free-ranging populations. *Journal of Wildlife Management,* **46**, 164–74.

Breuer, T., Robbins A. M., Olejniczak, C. *et al.* (2010). Variance in the male reproductive success of western gorillas: acquiring females is just the beginning. *Behavioral Ecology and Sociobiology,* **64**, 515–28.

Britt, A., Lambana, R. R., Welch, C. R., and Katz, A. S. (2002). Project Betampona: Re-stocking of *Varecia variegata variegata* into the Betampona Reserve. In S. Goodman, and J. Benstead, eds. *The Natural History of Madagascar,* pp. 1545–51. University of Chicago Press, Chicago.

Britt, A., Welch, C., Katz, A., *et al.* (2004). The re-stocking of captive-bred ruffed lemurs (*Varecia variegata variegata*) into the Betampona Reserve, Madagascar: methodology and recommendations. *Biodiversity and Conservation,* **13**, 635–57.

Broad, S., Mulliken, T., and Roe, D. (2003). The nature and extent of legal and illegal trade in wildlife. In S. Oldfield, ed., *The Trade in Wildlife: Regulation for Conservation,* pp. 3–22. Earthscan Publications, Ltd., London, UK.

Brockelman, W. Y., and Srikosamatara, S. (1993). Estimating density of gibbon groups by use of the loud songs. *American Journal of Primatology,* **29**, 93–108.

Brockman, D. K., Whitten, P. L., Russell, E., *et al.* (1995). Application of fecal steroid techniques to the reproductive endocrinology of female Verreaux's Sifaka (*Propithecus verreauxi*). *American Journal of Primatology,* **36**, 313–25.

Brook, B. W., Burgman, M. A., and Frankham, R. (2000). Differences and congruencies between PVA packages: the importance of sex ratio for predictions of extinction risk. *Conservation Ecology,* **4**, 6.

Bronikowski, A. M., and Flatt, T. (2010). Aging and its demographic measurement. *Nature Education Knowledge,* **1**, 3.

Brosnan, S. F. (2010). What do capuchin monkeys tell us about cooperation? In D. R. Forsyth, and C. L. Hoyt, eds. *For the Greater Good of All: Perspectives on Individualism, Society, and Leadership,* pp. 11–28. Palgrave Macmillan Publishers, New York.

Brosnan, S. F., and de Waal, F. B. M. (2002). A proximate perspective on reciprocal altruism. *Human Nature,* **13**, 129–52.

Brosnan, S. F., Grady, M., Lambeth, S., *et al.* (2008). Chimpanzee autarky. *PLoS ONE,* **3**, e1518.

Brosnan, S. F., Parrish, A. R., Beran, M. J., *et al.* (2011). Responses to the Assurance game in monkeys, apes, and humans using equivalent procedures. *Proceedings of the National Academy of Sciences USA,* **108**, 3442–7.

Brown, A. K., and Brown, P. (2005). *Ultra low-power GPS recorder (TrackTag®)*. Paper presented at the Institute of Navigation 61st Annual Meeting, Cambridge, MA.

Brown, J. H. (1964). The evolution of diversity in avian territorial systems. *Wilson Bulletin*, 76, 160–9.

Brown, M. (2011). Intergroup Encounters in Grey-Cheeked Mangabeys (*Lophocebus albigena*) and Redtail Monkeys (*Cercopithecus ascanius*): Form and Function (Ph.D. Dissertation), Columbia University, New York.

Brügger, R., Dobbertin, M., and Kräuchi, N. (2003). Phenological variation in forest trees. In M. Schwartz, ed. *Phenology: An Integrative Environmental Science*, pp. 255–67. Kluwer Academic Publishers, Dordercht.

Brugiere, D., and Magassouba, B. (2009). Pattern and sustainability of the bushmeat trade in the Haut Niger National Park, Republic of Guinea. *African Journal of Ecology*, 47, 630–9.

Bryson, J. J., Yasushi, A., and Lehmann, H. (2007). Agent-based modelling as a scientific methodology: a case study analyzing primate social behaviour. *Philosophical Transactions of the Royal Society*, 362, 1686–98.

Buckingham, F., and Shanee, S. (2009). Conservation priorities for the Peruvian yellow-tailed woolly monkey (*Oreonax flavicauda*): a GIS risk assessment and gap analysis. *Primate Conservation*, 24, 65–71.

Buckland, S. T., Anderson, D. R., Burnham, K. P., *et al.* (2001). *Introduction to Distance Sampling: Estimating Abundance Of Biological Populations*. Oxford University Press, Oxford.

Buckland, S. T., Summers, R. W., Borchers, D. L., and Thomas, L. (2006). Point transect sampling with traps or lures. *Journal of Applied Ecology*, 43, 377–84.

Buckland, S. T., Plumptre, A. J., Thomas, L., and Rexstad, E. (2010a). Line transect surveys of primates: can animal-to-observer distance methods work? *International Journal of Primatology*, 31, 485–99.

Buckland, S. T., Plumptre, A. J., Thomas, L., and Rexstad, E. (2010b). Design and analysis of line transect surveys for primates. *International Journal of Primatology*, 31, 833–47.

Buckley, C., Nekaris, K. A. I., and Husson, S. J. (2006). Survey of *Hylobates agilis albibarbis* in a logged peat-swamp forest: Sabangau catchment, Central Kalimantan. *Primates*, 47, 327–35.

Buij, R., Wich, S. A., Lubis, A. H., and Sterck, E. H. M. (2002). Seasonal movements in the Sumatran orangutan (*Pongo pygmaeus abelii*) and consequences for conservation. *Biological Conservation*, 107, 83–7.

Bullock, J. (1996). Plants. In W. J. Sutherland, ed. *Ecological Census Techniques: A Handbook*, pp. 111–38. Cambridge University Press, Cambridge.

Burgos-Rodriquez, A. G. (2011). Zoonotic diseases of primates. *Veterinary Clinics of North America: Exotic Animal Practice*, 14, 557–75.

Burnham, K. P., Anderson, D. R., and Laake, J. L. (1980). Estimation of density from line transect sampling of biological populations. *Wildlife Monographs*, 72, 1–202.

Butler, P. J., Green, J. A., Boyd, I. L., and Speakman, J. R. (2004). Measuring metabolic rate in the field: the pros and cons of the doubly labelled water and heart rate methods. *Functional Ecology*, 18, 168–83.

Butler, P. J., Woakes, A. J., Boyd, I. L., and Kanatous, S. (1992). Relationship between heart rate and oxygen consumption during steady-state swimming in California sea lions. *Journal of Experimental Biology*, **170**, 35–42.

Butynski, T. M., and Kalina, J. (1998). Gorilla tourism: A critical look. In E. J. Millner-Gulland and R. Mace (Eds.), *Conservation of Biological Resources*, pp. 294–313. Blackwell Science, Oxford.

Bynum, D. Z. (1999). Assessment and monitoring of anthropogenic disturbance in Lore Lindu National Park, Central Sulawesi, Indonesia. *Tropical Biodiversity*, **6**, 43–57.

Bynum, E. L., Bynum, D. Z., and Supriatna, J. (1997). Confirmation and location of the hybrid zone between wild populations of *Macaca tonkeana* and *Macaca hecki* in Central Sulawesi, Indonesia. *American Journal of Primatology*, **43**, 181–209.

Bynum, N. (2002). Morphological variation within a macaque hybrid zone. *American Journal of Physical Anthropology*, **118**, 45–9.

Byrne, R. W., Noser, R., Bates, L. A., and Jupp, P. E. (2009). How did they get here from there? Detecting changes of direction in terrestrial ranging. *Animal Behaviour*, 77, 619–31.

Caillaud, D., Crofoot, M. C., Scarpino, S. V., *et al.* (2010). Modeling the spatial distribution and fruiting pattern of a key tree species in a neotropical forest: methodology and potential applications. *PLoS ONE*, **5**, e15002.

Calenge, C. (2006). The package "adehabitat" for the R software: a tool for the analysis of space and habitat use by animals. *Ecological Modelling*, **197**, 516–19.

Calvert, J. J. (1985). Food Selection By Western Gorillas In Relation To Food Chemistry And Selective Logging In Cameroon, West Africa (Ph.D. Dissertation), University of California, Los Angeles.

Campbell, A. F. (1994). Patterns of home range use by *Ateles geoffroyj* and *Cebus capucinus* at La Selva Biological Station, Northeast Costa Rica. *American Journal of Primatology*, **33**, 199–200.

Campbell, C. J., Fuentes, A., MacKinnon, K. C., *et al.* (2011a). *Primates in Perspective*, 2nd edition. Oxford University Press, Oxford.

Campbell, G., Kuehl, H., Diarrassouba, A., *et al.* (2011b). Long-term research sites as refugia for threatened and over-harvested species. *Biology Letters*, 7, 723–6.

Cannon, C. H., Curran, L. M., Marshall, A. J., and Leighton, M. (2007a). Beyond mast-fruiting events: Community asynchrony and individual dormancy dominate woody plant reproductive behavior across seven Bornean forest types. *Current Science*, **93**, 1558–66.

Cannon, C. H., Curran, L. M., Marshall, A. J., and Leighton, M. (2007b). Long-term reproductive behavior of woody plants across seven Bornean forest types in the Gunung Palung National Park (Indonesia): suprannual synchrony, temporal productivity, and fruiting diversity diversity. *Ecology Letters*, **10**, 956–69.

Carpenter, C. R. (1935). Behavior of Red Spider Monkeys in Panama. *Journal of Mammalogy*, **16**, 171–80.

Carroll, J. B., and Beattie, J. C. (1993). Maintenance and breeding of the aye-aye *Daubentonia madagascariensis* at the Jersey Wildlife Preservation Trust. *Dodo J. Wildlife Preservation Trusts*, **29**, 45–54.

Carter, M. R., and Gregorich, E. G. (2008). *Soil Sampling and Methods of Analysis*. CRC Press, Boca Raton, FL.

Carter, S. (2005). Don't Fear the Monkey: a review of the lion-tailed (*Macaca silenus*), Japanese (*Macaca fuscata*) and Sulawesi macaque (*Macaca nigra*) populations in AZA. American Zoo and Aquarium Association Annual Conference Symposium Presentation, Chicago.

Carter, S. (2010). The North American Studbook for the Lion-tail macaque (Macaca silenus). AZA and the Detroit Zoological Society, Silver Spring, MD.

Carthew, S. M., and Goldingay, R. L. (1997). Non-flying mammals as pollinators. *Trends in Ecology & Evolution,* **12,** 104–8.

Casagrande, D. G. (2004). Conceptions of primary forest in a Tzeltal Maya community: Implications for conservation. *Human Organization,* **63,** 180–202.

Caswell, H. (2001). *Matrix Population Models: Construction, Analysis and Interpretation.* 2nd edition. Sinauer, Sunderland, MA.

Catchpole, C. K., and Slater, P. J. B. (2008). *Bird song: Biological Themes and Variations.* Cambridge University Press, Cambridge.

Ceballos-Mago, N., and Chivers, D. J. (2010). Local knowledge and perceptions of pet primates and wild Margarita capuchins on Isla de Margarita and Isla de Coche in Venezuela. *Endangered Species Research,* **13,** 63–72.

Cerling, T. E., Hart, J., and Hart, T. (2004). Stable isotope ecology in the Ituri Forest. *Oecologia,* **138,** 5–12.

Chaber, A.-L., Allebone-Webb, S., Lignereux, Y., *et al.* (2010). The scale of illegal meat importation from Africa to Europe via Paris. *Conservation Letters,* **3,** 317–23.

Chakraborty, D., Meagher, T., and Smouse, P. E. (1988). Parentage analysis with genetic markers in natural populations. I. The expected proportion of offspring with unambiguous paternity. *Genetics,* **118,** 527–36.

Chancellor, R. L., Rundus, A. S., and Nyandwi, S. (2012). The influence of seasonal variation on chimpanzee (*Pan troglodytes schweinfurthii*) fallback food consumption, nest group size, and habitat use in gishwati, a montane rain forest fragment in Rwanda. *International Journal of Primatology,* **33,** 115–33.

Chapman, C. (1988). Patch use and patch depletion by the Spider and Howling Monkeys of Santa Rosa National Park, Costa Rica. *Behaviour,* **105,** 99–116.

Chapman, C. A., and Chapman, L. J. (1996). Frugivory and the fate of dispersed and non-dispersed seeds of six African tree species. *Journal of Tropical Ecology,* **12,** 491–504.

Chapman, C. A., Chapman, L. J., Wrangham, R. W. *et al.* (1992). Estimators of fruit abundance of tropical trees. *Biotropica,* **24,** 527–31.

Chapman, C. A., and Chapman, L. J. (2000a). Determinants of group size in primates: the importance of travel costs. In S. Boinski, and P. A. Garber, eds. *On the Move: How and Why Animals Travel in Groups.* University of Chicago Press, Chicago.

Chapman, C. A., and Chapman, L. J. (2000b). Interdemic variation in mixed-species association patterns: common diurnal primates of Kibale National Park, Uganda. *Behavioral Ecology and Sociobiology,* **47,** 129–39.

Chapman, C. A., Chapman, L. J., Cords, M., *et al.* (2002). Variation in the diets of *Cercopithecus* species: differences within forests, among forests, and across species. In M. E. Glenn, and M. Cords, eds. *The Guenons: Diversity and Adaptation in African Monkeys,* pp. 325–50. Kluwer Academic/Plenum Publishers, New York.

Chapman, C. A., Chapman, L. J., Rode, K. D., *et al.* (2003). Variation in the nutritional value of primate foods: among trees, time periods, and areas. *International Journal of Primatology,* 24, 317–33.

Chapman, C. A., Wrangham, R. W., and Chapman, L. J. (1994). Indices of habitat-wide fruit abundance in tropical forests. *Biotropica,* 26, 160–71.

Chapman, C. A., Wrangham, R. W., and Chapman, L. J. (1995). Ecological constraints on group size: an analysis of spider monkey and chimpanzee groups. *Behavioural Ecology and Sociobiology,* 36, 59–60.

Chapman, C. A., Chapman, L. J., Struhsaker, T., *et al.* (2005a). A long-term evaluation of fruiting phenology: importance of climate change. *Journal of Tropical Ecology,* 21, 1–14.

Chapman, C. A., Gillespie, T. R., and Goldberg, T. L. (2005b). Primates and the ecology of their infectious diseases: how will anthropogenic change affect host-parasite interactions? *Evolutionary Anthropology: Issues, News, and Reviews,* 14, 134–44.

Charles-Dominique, P. (1978). Solitary and gregarious prosimians: evolution of social structures in primates. In D. J. Chivers, and K. A. Joysey, eds. *Recent advances in primatology.* Academic Press, London.

Chave, J., Navarrete, D., Almeida, S., *et al.* (2010). Regional and seasonal patterns of litterfall in tropical South America. *Biogeosciences,* 7, 43–55.

Cheesbrough, M. (2005). *District Laboratory Practice in Tropical Countries, Part 1,* 2nd edition. Cambridge University Press, Cambridge.

Chen, C., Durand, E., Forbes, F., and Francois, O. (2007). Bayesian clustering algorithms ascertaining spatial population structure: A new computer program and a comparison study. *Molecular Ecology Notes,* 7, 747–56.

Cheney, D. L. (1981). Intergroup encounters among free-ranging vervet monkeys. *Folia Primatologica,* 35, 124–46.

Cheney, D. L., and Seyfarth, R. M. (1987). The influence of intergroup competition on the survival and reproduction of female vervet monkeys. *Behavioral Ecology and Sociobiology,* 21, 375–86.

Cheney, D. L., and Seyfarth, R. M. (1988). Assessment of meaning and the detection of unreliable signals by vervet monkeys. *Animal Behaviour,* 36, 477–86.

Chomel, B. B., Belotto, A., and Meslin, F. X. (2007). Wildlife, exotic pets, and emerging zoonoses. *Emerging Infectious Diseases,* 13, 6–11.

Chua, W. K., and Oyen, M. L. (2009). Viscoelastic properties of membrates measured by spherical indentation. *Cellular and Molecular Bioengineering,* 2, 49–56.

Cincotta, R. P., Wisnewski, J., Engelman, R., *et al.* (2000). Human population in the biodiversity hotspots. *Nature,* 404, 990–2.

Clayton, L., and Milner-Gulland, E. J. (2000). The trade in wildlife in North Sulawesi, Indonesia. In J. G. Robinson and E. L. Bennett, eds. *Hunting for Sustainability in Tropical Forests,* pp. 473–96. Columbia University Press, New York.

Cleveland, S., Laurenson, M. K., and Taylor, L. H. (2001). Diseases of humans and their domestic mammals: Pathogen characteristics, host range and the risk of emergence. *Philosophical Transactions of the Royal Society of London Series B,* 356, 991–9.

Coad, L., Abernethy, K., Balmford, A., *et al.* (2010). Distribution and use of income from bushmeat in a rural village, central Gabon. *Conservation Biology,* 24, 1510–18.

Coatney, G. R. (1971). The simian malarias: Zoonoses, anthroponoses, or both? *American Journal of Tropical Medicine and Hygiene,* **20**, 795–803.

Coatney, G. R., Collins, W. E., Warren, M., and Contacos, P. G. (1971). *The Primate Malarias.* National Institutes of Health, Bethesda, MD.

Codron, D., Lee-Thorp, J. A., and Sponheimer, M. (2006). Inter- and intrahabitat dietary variability of Chacma baboons (*Papio ursinus*) in South African savannas based on fecal δ13C, δ15N, and %N. *American Journal of Physical Anthropology,* **129**, 204–14.

Colell, M., Mate, C., and Fa, J. E. (1994). Hunting among Moka Bubis in Bioko: dynamics of faunal exploitation at the village level. *Biodiversity and Conservation,* **3**, 939–50.

Combes, C. (2004). *Parasitism: The Ecology and Evolution of Intimate Interactions.* University of Chicago Press, Chicago.

Conde, D. A., Flesness, N., Colchero, F., *et al.* (2011). An emerging role of zoos to conserve biodiversity. *Science,* **331**, 1390–1.

Conklin-Brittain, N. L., Dierenfeld, E. S., Wrangham, R. W., *et al.* (1999). Chemical protein analysis: a comparison of Kjeldahl crude protein and total ninhydrin protein from wild, tropical vegetation. *Journal of Chemical Ecology,* **25**, 2601–22.

Conklin-Brittain, N. L., Knott, C. D., and Wrangham, R. W. (2006). Energy intake by wild chimpanzees and orangutans: methodological considerations and a preliminary comparison. In G. Hohmann, M. M. Robbins, and C. Boesch, eds. *Feeding Ecology in Apes and Other Primates,* pp. 445–571. Cambridge University Press, Cambridge.

Conklin-Brittain, N. L., Wrangham, R. W., and Hunt, K. (1998). Dietary response of chimpanzees and cercopithecines to seasonal variation in fruit abundance II Macronutrients. *International Journal of Primatology,* **19**, 971–97.

Conradt, L., and Roper, T. J. (2005). Consensus decision making in animals. *Trends in Ecology & Evolution,* **20**, 449–56.

Conway, W. G. (1999). Congo: A zoo experiment in participatory conservation. In *American Zoo and Aquarium Association Annual Conference Proceedings,* pp. 101–3. Minnesota Zoo, Minneapolis.

Conway, W. G. (2011). Buying time for wild animals in zoos. *Zoo Biology,* **30**, 1–8.

Corbin, G. D., and Schmid, J. (1995). Insect secretions determine habitat use patterns by a female lesser mouse lemur (*Microcebus murinus*). *American Journal of Primatology,* **37**, 317–24.

Cords, M. (2007). Variable participation in the defense of communal feeding territories by blue monkeys in the Kakamega Forest, Kenya. *Behaviour,* **144**, 1537–50.

Corlett, R. T. (2007). The impact of hunting on the mammalian fauna of tropical Asian forests. *Biotropica,* **39**, 292–303.

Corlett, R. T., and Lucas, P. W. (1990). Alternative seed-handling strategies in primates: seed-spitting by long-tailed macaques (*Macaca fascicularis*). *Oecologia,* **82**, 166–71.

Cormier, L. A. (2002). Monkey as food, monkey as child: Guaja symbolic cannibalism. In A. Fuentes and L. D. Wolfe, eds. *Primates Face to Face: The Conservation Implications of Human-Nonhuman Primate Interconnections,* pp. 63–84. Cambridge University Press, Cambridge UK.

Cormier, L. A. (2003). *Kinship with Monkeys.* Columbia University Press, New York.

Costa, D. P., and Sinervo, B. (2004). Field physiology: Physiological insights from animals in nature. *Annual Review of Physiology,* **66**, 209–38.

Cowlishaw, G., and Dunbar, R. (2000). *Primate Conservation Biology*. University of Chicago Press, Chicago.

Cowlishaw, G., Mendelson, S., and Rowcliffe, J. M. (2005). Evidence for post-depletion sustainability in a mature bushmeat market. *Journal of Applied Ecology*, 42, 460–8.

Cox-Singh, J., Davis, T. M. E., Lee, K. S., *et al.* (2008). *Plasmodium knowlesi* malaria in humans is widely distributed and potentially life-threatening. *Clinical Infectious Diseases*, 46, 165–71.

Cracraft, J. (1983). Species concepts and speciation analysis. *Current Ornithology*, 1, 159–87.

Cramer, A. E., and Gallistel, C. R. (1997). Vervet monkeys as travelling salesmen. *Nature*, 387, 464–464.

Crandall, K. A., Bininda-Emonds, O. R. P., Mace, G. M., and Wayne, R. K. (2000). Considering evolutionary processes in conservation biology. *Trends in Ecology & Evolution*, 15, 290–5.

Crandall, L. S. (1964). *The Management of Wild Mammals in Captivity*, University of Chicago Press, Chicago.

Cranfield, M. R. (2008). Mountain gorilla research: The risk of disease transmission relative to the benefit from the perspective of ecosystem health. *American Journal of Primatology*, 70, 751–4.

Crews, D. (2011). Epigenetic modifications of brain and behavior: Theory and practice. *Hormones and Behavior*, 59, 393–8.

Crews, D., and McLachlan, J. A. (2006). Epigenetics, evolution, endocrine disruption, health, and disease. *Endocrinology*, 147, s4–s10.

Croat, T. B. (1979). The sexuality of the Barrio Colorado Island flora (Panama). *Phytologia*, 42, 319–48.

Crofoot, M. C., Gilby, I. C., Wikelski, M. C., and Kays, R. W. (2008). Interaction location outweighs the competitive advantage of numerical superiority in *Cebus capucinus* inter-group contests. *Proceedings of the National Academy of Sciences USA*, 105, 577–81.

Crofoot, M. C., Lambert, T. D., Kays, R., and Wikelski, M. C. (2010). Does watching a monkey change its behaviour? Quantifying observer effects in habituated wild primates using automated radiotelemetry. *Animal Behaviour*, 80, 475–80.

Croft, D. P., James, R., and Krause, J. (2008). *Exploring animal social networks*. Princeton University Press, Princeton, NJ.

Crookes, D. J., Ankudey, N., and Milner-Gulland, E. J. (2005). The value of a long-term bushmeat market dataset as an indicator of system dynamics. *Environmental Conservation*, 32, 333–9.

Crowley, B. E. (2010). Stable carbon and nitrogen isotope enrichment in primate tissues. *Oecologia*, 164, 611–26.

Csillery, K., Johnson, T., Beraldi, D., *et al.* (2006). Performance of marker-based relatedness estimators in natural populations of outbred vertebrates. *Genetics*, 173, 2091–101.

Culik, B. M., and Wilson, R. P. (1991). Swimming energetics and performance of instrumented Adelie penguins (*Pygoscelis adeliae*). *Journal of Experimental Biology*, 158, 355–68.

Cuozzo, F. P., and Sauther, M. L. (2006). Severe wear and tooth loss in wild ring-tailed lemurs (*Lemur catta*): A function of feeding ecology, dental structure, and individual life history. *Journal of Human Evolution*, 51, 490–505.

Curran, L. M., and Leighton, M. (2000). Vertebrate responses to spatio-temporal variation in seed production of mast-fruiting Dipterocarpaceae. *Ecological Monographs, 70,* 101–28.

Curran, B., Wilkie, D., and Tshombe, R. (2000). Socio-economic data and their relevance to protected area management. In L. White and A. Edwards, eds. *Conservation Research in the African Rain Forests: A Technical Handbook,* pp. 331–54. Wildlife Conservation Society, New York.

Cushman, S. A., and Landguth, E. L. (2010). Spurious correlations and inference in landscape genetics. *Molecular Ecology, 19,* 3592–602.

Dallas, J. F., Coxon, K. E., Sykes, P. R., *et al.* (2003). Similar estimates of population genetic composition and sex ration derived from carcasses and faeces of Eurasian otter *Lutra lutra. Molecular Ecology, 12,* 275–82.

Dammhahn, M., and Kappeler, P. (2010). Scramble or contest competition over food in solitarily foraging mouse lemurs (*Microcebus* spp.): new insights from stable isotopes. *American Journal of Physical Anthropology, 141,* 181–9.

Darvell, B. W., Lee, P. K. D., Yuen, T. D. B., and Lucas, P. W. (1996). A portable fracture toughness tester for biological materials. *Measurement Science and Technology, 7,* 954–62.

Daugherty, C. H., Cree, A., Hay, J. M., and Thompson, M. B. (1990). Neglected taxonomy and continuing extinctions of tuatara (*Sphenodon*). *Nature, 347,* 376–89.

Dausmann, K. H. (2005). Measuring body temperature in the field—evaluation of external vs implanted transmitters in a small mammal. *Journal of Thermal Biology, 30,* 195–202.

Dausmann, K. H., Glos, J., Ganzhorn, J. U., and Heldmaier, G. (2004). Hibernation in a tropical primate. *Nature, 429,* 825–6.

Dausmann, K. H., Glos, J., Ganzhorn, J. U., and Heldmaier, G. (2005). Hibernation in the tropics: lessons from a primate. *Journal of Comparative Physiology B, 175,* 147–155.

Davies, N. B., and Houston, A. I. (1984). Territory economics. In J. R. Krebs, and N. B. Davies, eds. *Behavioural Ecology: An Evolutionary Approach,* 2nd edition, pp. 148–69. Blackwell Scientific Publications, Oxford.

Davis, J. I., and Nixon, K. C. (1992). Populations, genetic variation, and the delimitation of phylogenetic species. *Systematic Biology, 41,* 421–35.

Davison, A. C., and Hinkley, D. V. (1997). *Bootstrap methods and their applications.* Cambridge University Press, New York.

Dawson, T. P., Jackson, S. T., House, J. I., Prentice, I. C., and Mace, G. M. (2011). Beyond Predictions: Biodiversity Conservation in a Changing Climate. *Science, 332,* 53–8.

de Kroon, H., Plaisier, A., van Groenendael, J., and Caswell, H. (1986). Elasticity: the relative contribution of demographic parameters to population growth rate. *Ecology, 67,* 1427–31.

de Merode, E., Homewood, K., and Cowlishaw, G. (2004). The value of bushmeat and other wild foods to rural households living in extreme poverty in Democratic Republic of Congo. *Biological Conservation, 118,* 573–81.

de Solla, S. R., Bondurianky, R., and Brooks, R. J. (1999). Eliminating autocorrelation reduces biological relevance of home range estimates. *Journal of Animal Ecology, 68,* 221–34.

de Thoisy, B., Renoux, F., and Julliot, C. (2005). Hunting in northern French Guiana and its impact on primate communities. *Oryx, 39,* 149–57.

de Thoisy, B., Vogel, I., Reynes, J.-M., *et al.* (2001). Health evaluation of translocated free-ranging primates in French Guiana. *American Journal of Primatology, 54,* 1–16.

de Waal, F. B. M. (1986). Class structure in a rhesus monkey group: the interplay between dominance and tolerance. *Animal Behaviour* **34**, 1033–40.

de Waal, F. B. M. (1997). Food transfers through mesh in brown capuchins. *Journal of Comparative Psychology,* **111**, 370–8.

de Waal, F. B. M., and Davis, J. M. (2002). Capuchin cognitive ecology: cooperation based on projected returns. *Neuropsychologia,* **41**, 221–8.

DeGabriel, J. L., Wallis, I. R., Moore, B. D., and Foley, W. J. (2008). A simple, integrative assay to quantify nutritional quality of browses for herbivores. *Oecologia,* **156**, 107–16.

Demas, G. E., Zysling, D. A., Beechler, B. R., *et al.* (2011). Beyond phytohaemagglutinin: Assessing vertebrate immune function across ecological contexts. *Journal of Animal Ecology,* **80**, 710–30.

DeNiro, M. J., and Epstein, S. (1978). Influence of diet on the distribution of carbon isotopes in animals. *Geochimica et Cosmochimica Acta,* **42**, 495–506.

DeSalle, R., and Amato, G. (2004). The expansion of conservation genetics. *Nature Reviews Genetics,* **5**, 702–12.

Devore, I., editor. (1965). *Primate Behavior: Field Studies of Monkeys and Apes*. Holt, Rinehart, and Winston, New York.

Di Bitetti, M. S. (2001). Food associated calls in the tufted capuchin monkeys (*Cebus apella*) (Ph.D. Dissertation), SUNY Stony Brook, Stony Brook, NY.

Di Bitetti, M. S. (2003). Food-Associated Calls of Tufted Capuchin Nonkeys (*Cebus apella nigritus*) are functionally referential signals. *Behaviour,* **140**, 565–92.

Di Bitetti, M. S. (2005). Food-associated calls and audience effects in tufted capuchin monkeys, *Cebus apella nigritus*. *Animal Behaviour,* **69**, 911–19.

Di Fiore, A. (2003). Molecular genetic approaches to the study of primate behavior, social organization, and reproduction. *Yearbook of Physical Anthropology,* **46**, 62–99.

Di Fiore, A. (2005). A rapid genetic method for sex assignment in non-human primates. *Conservation Genetics,* **6**, 1053–8.

Di Fiore, A. (2009). Genetic approaches to the study of dispersal and kinship in New World primates. In P. A. Garber, A. Estrada, J. C. Bicca-Marques, E. W. Heymann, and K. B. Strier, eds. *South American Primates*, pp. 211–50. Springer, New York.

Di Fiore, A., Link, A., and Campbell, C. (2011). The Atelines: Behavioral and socio-ecological diversity in a new world monkey radiation. In C.J. Campbell, A. Fuentes, K. C. Mackinnon, S. K. Bearder, and R. M. Stumpf, eds. *Primates in perspective*, 2nd edition. Oxford University Press, Oxford, UK.

Dindo, M., Whiten, A., and De Waal, F. B. M. (2009). In-group conformity sustains different foraging traditions in capuchin monkeys (*Cebus apella*). *PLoS ONE,* **4**, e7858.

Donovan, T. M., and Hines, J. (2007). *Exercises in occupancy modeling and estimation*. The University of Vermont, Burlington, VT. http://www.uvm.edu/rsenr/vtcfwru/spreadsheets/occupancy/occupancy.htm. (accessed October 2012).

Doran-Sheehy, D. M., Derby, A. M., Greer, D., and Mongo, P. (2007). Habituation of western gorillas: the process and factors that influence it. *American Journal of Primatology,* **69**, 1–16.

Drummond, A. J., Ho, S. Y., Phillips, M. J., and Rambaut, A. (2006). Relaxed phylogenetics and dating with confidence. *PLoS Biology,* **4**, e88.

Drummond, A. J., and Rambaut, A. (2007). BEAST: Bayesian evolutionary analysis by sampling trees. *BMC Evolutionary Biology*, 7, 214.

Duarte-Quiroga, A., and Estrada, A. (2003). Primates as pets in Mexico City: An assessment of the species involved, source of origin, and general aspects of treatment. *American Journal of Primatology*, 61, 53–60.

Duckworth, J. W. (1993). Feeding damage left in bamboos, probably by aye-ayes *(Daubentonia madagascariensis)*. *International Journal of Primatology*, 14, 927–31.

Duckworth, J. W. (1998). The difficulty of estimating population densities of nocturnal forest mammals from transect counts of animals. *Journal of Zoology*, 246, 466–8.

Dunbar, R. I. M. (1980). Demographic and life history variables of a population of gelada baboons (*Theropithecus gelada*). *Journal of Animal Ecology*, 49, 485–506.

Dunbar, R. I. M. (1988). *Primate social systems*. Cornell University Press, Ithaca, NY.

Dunbar, R. I. M. (1991). Functional Significance of Social Grooming in Primates. *Folia Primatologica*, 57, 121–31.

Dunbar, R. I. M. (1992). Neocortex size as a constraint on group size in primates. *Journal of Human Evolution*, 20, 469–93.

East, T., Kümpel N. F., Milner-Gulland, E. J., and Rowcliffe, J. M. (2005). Determinants of urban bushmeat consumption in Río Muni, Equatorial Guinea. *Biological Conservation*, 126, 206–15.

Eaton, M. J., Meyers, G. L., Kolokotronis, S.-O., *et al.* (2010). Barcoding bushmeat: molecular identification of Central African and South American harvested vertebrates. *Conservation Genetics*, 11, 1389–404.

Edgar, R. C. (2004). MUSCLE: a multiple sequence alignment method with reduced time and space complexity. *BMC Bioinformatics*, 5, 113.

Edwards, M. S., and Ullrey, D. E. (1999). Effect of dietary fiber concentration on apparent digestibility and digesta passage in non-human primates. II. Hindgut and foregut fermenting folivores. *Zoo Biology*, 18, 537–49.

Eggert, L. S., Eggert, J. A., and Woodruff, D. S. (2003). Estimating population sizes of elusive animals: the forest elephants of Kakum National Park, Ghana. *Molecular Ecology*, 12, 1398–402.

Elgart-Berry, A. (2004). Fracture toughness of mountain gorilla (*Gorilla gorilla beringei*) food plants. *American Journal of Primatology*, 62, 275–85.

Elith, J., Graham, C. H., Anderson, R. P., *et al.* (2006). Novel methods improve prediction of species' distributions from occurrence data. *Ecography*, 29, 129–51.

Ellison, P. T. (2009). Social relationships and reproductive ecology. In P. T. Ellison, and P. B. Gray, eds. *Endocrinology of social relationships*, pp. 54–73. Harvard University Press, Cambridge, MA.

Ellison, P. T., and Gray, P. B. (2009). *Endocrinology of social relationships*. Harvard University Press, Cambridge, MA.

Emery Thompson, M., and Knott, C. D. (2008). Urinary c-peptide of insulin as a non-invasive marker of energy balance in wild orangutans. *Hormones and Behavior*, 53, 526–35.

Emlen, S. T., and Oring, L. W. (1977). Ecology, sexual selection, and the evolution of mating systems. *Science*, 197, 215–23.

Engh, A. L., Beehner, J. C., Bergman, T. J., *et al.* (2006). Behavioural and hormonal responses to predation in female chacma baboons (*Papio hamadryas ursinus*). *Proceedings of the Royal Society B-Biological Sciences*, **273**, 707–12.

Enstam, K. L., and Isbell, L. A. (2004). Microhabitat preference and vertical use of space by Patas monkeys (*Erythrocebus patas*) in relation to predation risk and habitat structure. *Folia Primatologica*, **75**, 70–84.

Epps, C. W., Wehausen, J. D., Bleich, V. C., *et al.* (2007). Optimizing dispersal and corridor models using landscape genetics. *Journal of Applied Ecology*, **44**, 714–24.

Erdfelder, E., Faul, F., and Buchner, A. (1996). GPOWER: A general power analysis program. *Behavior Research Methods, Instruments, & Computers*, **28**, 1–11.

Erickson, C. J. (1991). Percussive foraging in the aye-aye (*Daubentonia madagascariensis*). *Animal Behavior*, **41**, 793–801.

Erickson, C. J. (1994). Tap-Scanning and Extractive Foraging in Aye-Ayes, *Daubentonia madagascariensis. Folia Primatologica*, **62**, 125–35.

Escalante, A. A., Freeland, D. E., Collins, W. E., and Lal, A. A. (1998). The evolution of primate malaria parasites based on the gene encoding cytochrome b from the linear mitochondrial genome. *Proceedings of the National Academy of Sciences USA*, **95**, 8124–9.

Etherington, T. R. (2010). Python based GIS tools for landscape genetics: visualising genetic relatedness and measuring landscape connectivity. *Methods in Ecology and Evolution*, **2**, 52–5.

Eudey, A. A. (1987). *Action Plan for Asian primate conservation: 1987–1991*. IUCN, Gland, Switzerland.

Evans, B. J., Supriatna, J., Andayani, N., *et al.* (2003). Monkeys and toads define areas of endemism on Sulawesi. *Evolution*, **57**, 1436–43.

Evans, T. A., and Westergaard, G. C. (2006). Self-control and tool use in tufted capuchin monkeys (*Cebus apella*). *Journal of Comparative Psychology*, **120**, 163–6.

Excoffier, L., and Heckel, G. (2006). Computer programs for population genetics data analysis: a survival guide. *Nature Reviews Genetics*, **7**, 745–58.

Excoffier, L., Laval, G., and Schneider, S. (2005). Arlequin ver. 3.0: An integrated software package for population genetics data analysis. *Evolutionary Bioinformatics Online*, **1**, 47–50.

Ezenwa, V. (2004). Host social behavior and parasitic infection: A multifactorial approach. *Behavioral Ecology*, **15**, 446–54.

Fa, J. E., Albrechtsen, L., Johnson, P. J., and Macdonald, D. W. (2009). Linkages between household wealth, bushmeat and other animal protein consumption are not invariant: evidence from Rio Muni, Equatorial Guinea. *Animal Conservation*, **12**, 599–610.

Fa, J. E., Garcia Yuste, J. E., and Castelo, R. (2000). Bushmeat markets on Bioko as a measure of hunting pressure. *Conservation Biology*, **14**, 1602–13.

Fa, J. E., and Garcia Yuste, J. E. (2001). Commercial bushmeat hunting in the Monte Mitra forests, Equatorial Guinea: extent and impact. *Animal Biodiversity and Conservation*, **24.1**, 31–52.

Fa, J. E., Juste, J., Burn, R. W., and Broad, G. (2002a). Bushmeat consumption and preferences of two ethnic groups in Bioko Island, West Africa. *Human Ecology*, **30**, 397–416.

Fa, J. E., Peres, C. A., and Meeuwig, J. (2002b). Bushmeat exploitation in tropical forests: an intercontinental comparison. *Conservation Biology*, **16**, 232–7.

Fa, J. E., Seymour, S., Dupain, J., *et al.* (2006). Getting to grips with the magnitude of exploitation: Bushmeat in the Cross–Sanaga rivers region, Nigeria and Cameroon. *Biological Conservation,* **129,** 497–510.

Falush, D., Stephens, M., and Pritchard, J. K. (2003). Inference of population structure using multilocus genotype data: Linked loci and correlated allele frequencies. *Genetics,* **164,** 1567–87.

Farmer, K. H. (2002). Pan-African Sanctuary Alliance: Status and range of activities for great ape conservation. *American Journal of Primatology,* **58,** 117–32.

Farnert, A., Arez, A., Correia, A., *et al.* (1999). Sampling and storage of blood and the detection of malaria parasites by polymerase chain reaction. *Transactions of the Royal Society of Tropical Medicine and Hygiene,* **93,** 50–3.

Fashing, P. J. (2001). Male and female strategies during intergroup encounters in guerezas (*Colobus guereza*): evidence for resource defense mediated through males and a comparison with other primates. *Behavioral Ecology and Sociobiology,* **50,** 219–30.

Fashing, P. J., Dierenfeld, E., and Mowry, C. B. (2007a). Influence of plant and soil chemistry on food selection, ranging patterns, and biomass of *Colobus guereza* in Kakamega Forest, Kenya. *International Journal of Primatology,* **28,** 673–703.

Fashing, P. J., Mulindahabi, F., Gakima, J. B., *et al.* (2007b). Activity and Ranging Patterns of *Colobus angolensis ruwenzorii* in Nyungwe Forest, Rwanda: Possible Costs of Large Group Size. *International Journal of Primatology,* **28,** 529–50.

Fashing, P. J., and Nguyen, N. (2011). Behavior toward the dying, diseased, or disabled among animals and its relevance to paleopathology. *International Journal of Paleopathology,* **1,** 128–9.

Faubet, P., Waples, R. S., and Gaggiotti, O. E. (2007). Evaluating the performance of a multilocus Bayesian method for the estimation of migration rates. *Molecular Ecology,* **16,** 1149–66.

Faul, F., Erdfelder, E., Buchner, A., and Lang, A.-G. (2009). Statistical power analyses using G*Power 3.1: tests for correlation and regression analyses. *Behavior Research Methods,* **41,** 1149–60.

Fedigan, L. M. (2010). Ethical issues faced by field primatologists: Asking the relevant questions. *American Journal of Primatology,* **72,** 754–71.

Feistner, A. T. C., and Sterling, E. J. (1994). Aye-ayes: Madagascar's most puzzling primate. *Folia Primatologica,* **62,** 6–7.

Felsenstein, J. (2004). *Inferring Phylogenies.* Sinauer, Sunderland, MA.

Feret, J.-B., and Asner, G. P. (2011). Spectroscopic classification of tropical forest species using radiative transfer modeling. *Remote Sensing of Environment,* **115,** 2415–22.

Fernandez, C., and Luxmoore, R. (1997). *The Value of the Wildlife Trade. Industrial Reliance on Biodiversity—WCMC Biodiversity Series 7.* World Conservation Press, Cambridge, UK.

Fernandez-Duque, E. (2012). Owl monkeys *Aotus* spp. in the wild and in captivity. *International Zoo Yearbook,* **46,** 80–94.

Fieberg, J., and Kochanny, C. O. (2005). Quantifying home-range overlap: the importance of the utilization distribution. *Journal of Wildlife Management,* **69,** 1346–59.

Fietz, J., Klose, S. M., and Kalko, E. K. V. (2010). Behavioural and physiological consequences of male reproduction trade-offs in edible dormice (*Glis glis*). *Naturwissenschaften*, 97, 883–90.

Finer, M., Moncel, R., and Jenkins, C. N. (2010). Leaving the oil under the Amazon: Ecuador's Yasuní-ITT Initiative. *Biotropica*, 42, 63–6.

Fitch-Snyder, H., Schulze, H., and Streicher, U. (2008). Enclosure design for captive slow and pygmy lorises. In M. Shekelle, I. Maryanto, C. Groves, H. Schulze, and H. Fitch-Snyder, eds. *Primates of the Oriental Night*, pp. 123–35. LIPI Press, Bogor.

Flesness, N. R. (1986). Captive status and genetic considerations in primates. In K. Benirschke, ed. *Primates, the Road to Self-Sustaining Populations*, pp. 847–56. Springer-Verlag, New York.

Flombaum, J. I., and Santos, L. R. (2005). Rhesus monkeys attribute perceptions to others. *Current Biology*, 15, 447–52.

Foden, W., Mace, G. M., Vie, J.-C., *et al.* (2008). Species susceptibility to climate change impacts. In J.-C. Vie, C. Hilton-Taylor, and S. Stuart, eds. *The 2008 Review of the IUCN Red List of Threatened Species*, pp. 77–88. IUCN, Gland, Switzerland.

Forthman, D. L., Strum, S. C., and Muchemi, G. M. (2005). Applied conditioned taste aversion and the management and conservation of crop-raiding primates. In J. Paterson, and J. Wallis, eds. *Commensalism and Conflict: The Human-Primate Interface*, pp. 420–43. American Society of Primatologists, Norman, OK.

Fossey, D. (1983). *Gorillas in the Mist*. Houghton Mifflin Company, Boston, MA.

Fourie, N. H., and Bernstein, R. M. (2011). Hair cortisol levels track phylogenetic and age related differences in hypothalamic-pituitary-adrenal (HPA) axis activity in non-human primates. *General and Comparative Endocrinology*, 174, 150–5.

Fournier, L. A., and Charpantier, C. (1975). El tamaño de la muestra y la frecuencia de las observaciones en el estudio de las característícas fenolóígicas de los árboles tropicales. *Turrialba*, 25, 45–48.

Fragaszy, D. M., Boinski, S., and Whipple, J. (1992). Behavioral sampling in the field: Comparison of individual and group sampling methods. *American Journal of Primatology*, 26, 259–75.

Fragaszy, D., Johnson-Pynn, J., Hirsh, E., and Brakke, K. (2003). Strategic navigation of two-dimensional alley mazes: comparing capuchin monkeys and chimpanzees. *Animal Cognition*, 6, 149–60.

Fragaszy, D. M., Izar, P., Visalberghi, E., *et al.* (2004). Wild capuchin monkeys (*Cebus libidinosus*) use anvils and stone pounding tools. *American Journal of Primatology*, 64, 359–66.

Frankham, R., Ballou, J. D., and Briscoe, D. A. (2002). *Introduction to Conservation Genetics*. Cambridge University Press, Cambridge, UK.

Franklin, J. (2009). *Mapping species distributions: spatial inference and prediction*. Cambridge University Press, Cambridge.

Franzen, M. (2006). Evaluating the sustainability of hunting: a comparison of harvest profiles across three Huaorani communities. *Environmental Conservation*, 33, 36–45.

Freedman, D. A. (1999). *Ecological Inference and the Ecological Fallacy.* University of California, Berkeley. Prepared for the International Encyclopedia of the Social & Behavioral Sciences Technical Report No. 549. Available online: http://www.stanford.edu/class/ ed260/freedman549.pdf (accessed 7 June 2012).

Freeland, W. (1976). Pathogens and evolution of primate sociality. *Biotropica,* **8**, 12–24.

Freeman, L. (2004). *The Development of Social Network Analysis.* Empirical Press, Vancouver, BC.

Friis, R. H. (2009). *Epidemiology 101.* Jones & Bartlett, Sudbury, MA.

Fruteau, C., Lemoine, S., Hellard, E., *et al.* (2011). When females trade grooming for grooming: testing partner control and partner choice models of cooperation in two primate species. *Animal Behaviour,* **81**, 1223–30.

Fuentes, A. (2006). Human culture and monkey behavior: Assessing the contexts of potential pathogen transmission between macaques and humans. *American Journal of Primatology,* **68**, 880–96.

Fuentes, A. (2010). Natural cultural encounters in Bali: Monkeys, temples, tourists, and ethnoprimatology. *Cultural Anthropology,* **25**, 620–4.

Fuentes, A., and Gamerl, S. (2005). Disproportionate participation by age/sex classes in aggressive interactions between long-tailed macaques (*Macaca fascicularis*) and human tourists at Padangtegal Monkey Forest, Bali, Indonesia. *American Journal of Primatology,* **66**, 197–204.

Fuentes, A., and Hockings, K. J. (2010). The ethnoprimatological approach in primatology. *American Journal of Primatology,* **72**, 841–7.

Fuentes, A., Kalchik, S., Gettler, L., Kwiatt, A., Konecki, M., and Jones-Engel, L. (2008). Characterizing human-macaque interactions in Singapore. *American Journal of Primatology,* **70**, 879–83.

Fuentes, A., and Wolfe, L. D., eds. (2002). *Primates Face to Face: The Conservation Implications of Human-Nonhuman Primate Interconnections.* Cambridge University Press, Cambridge.

Fusani, L., Canoine, V., Goymann, W., *et al.* (2005). Difficulties and special issues associated with field research in behavioral neuroendocrinology. *Hormones and Behavior,* **48**, 484–91.

Gadberry, M. D., Malcomber, S. T., Doust, A. N., and Kellogg, E. A. (2005). Primaclade—a flexible tool to find conserved PCR primers across multiple species. *Bioinformatics,* **21**, 1263–4.

Gage, T. B. (1998). The comparative demography of primates, with some comments on the evolution of life histories. *Annual Review of Anthropology,* **27**, 197–221.

Galat, G., Galat-Luong, A., and Nizinski, G. (2008). Our cousins chimpanzees and baboons face global warming by digging wells to filtrate drinking water. International Water Resource Congress, 13th World Water Congress. Montpellier, France.

Ganas, J., Nkurunungi, J. B., and Robbins, M. M. (2009). A preliminary study of the temporal and spatial biomass patterns of herbaceous vegetation consumed by mountain gorillas in an Afromontane rain forest. *Biotropica,* **41**, 37–46.

Ganzhorn, J. U. (1989). Niche separation of seven lemur species in the eastern rainforest of Madagascar. *Oecologia,* **79**, 279–86.

Ganzhorn, J. U. (2003). Habitat description and phenology. In J. M. Setchell, and D. J. Curtis, eds. *Field and laboratory methods in primatology*, pp. 40–56. Cambridge University Press, Cambridge.

Ganzhorn, J. U., and Schmid, J. (1998). Different population dynamics of *Microcebus murinus* in primary and secondary deciduous dry forests of Madagascar. *International Journal of Primatology*, **19**, 785–96.

Gao, F., Bailes, E., Robertson, D. L., *et al.* (1999). Origin of HIV-1 in the chimpanzee *Pan troglodytes troglodytes*. *Nature*, **397**, 436–41.

Garber, P. A. (1987). Foraging strategies among living primates. *Annual Review of Anthropology*, **16**, 339–64.

Garber, P. A., Moya, L., and Malaga, C. (1984). A preliminary field study of the moustached tamarin monkey (*Saguinus mystax*) in Northeastern Peru: questions concerned with the evolution of a communal breeding system. *Folia Primatologica*, **42**, 17–32.

Garber, P. A., Encarnacion, F., Moya, L., *et al.* (1993). Demographic and reproductive patterns in moustached tamarin monkeys (*Saguinus mystax*): implications for reconstructing platyrrhine mating systems. *American Journal of Primatology*, **29**, 235–54.

Garcia, L. (2007). *Diagnostic Medical Parasitology*, 5th edition. ASM Press, Washington, DC.

Garcia, C., Huffman, M. A., Shimizu, K., and Speakman, J. R. (2011). Energetic Consequences of Seasonal Breeding in Female Japanese Macaques (*Macaca fuscata*). *American Journal of Physical Anthropology*, **146**, 161–70.

Garnham, P. C. C. (1966). *Malaria Parasites and Other Haemosporidia*. Blackwell Scientific Publications, Oxford.

Garton, E. O., Wisdom, M. J., Leban, F. A., and Johnson, B. K. (2001). Experimental design for radiotelemetry studies. In J. J. Millspaugh, and J. M. Marzluff, eds. *Radio Tracking and Animal Populations*, pp. 16–41. Academic Press, San Diego, CA.

Gaubert, P., Papes, M. and Peterson, A. T. (2006). Natural history collections and the conservation of poorly known taxa: ecological niche modeling in central African rainforest genets (*Genetta* spp.). *Biological Conservation*, **130**, 106–17.

Gautier-Hion, A., and Maisels, F. (1994). Mutualism between a leguminous tree and large African monkeys as pollinators. *Behavioral Ecology and Sociobiology*, **34**, 203–10.

Geissmann, T., and Orgeldinger, M. (2000). The relationship between duetsongs and pair bonds in siamangs, (*Hylobates syndactylus*). *Animal Behaviour*, **60**, 805–9.

Gerwing, J. J., Schnitzer, S. A., Burnham, R. J., *et al.* (2006). A standard protocol for liana censuses. *Biotropica*, **38**, 256–61.

Gesquiere, L. R., Wango, E. O., Alberts, S. C., and Altmann, J. (2007). Mechanisms of sexual selection: Sexual swellings and estrogen concentrations as fertility indicators and cues for male consort decisions in wild baboons. *Hormones and Behavior*, **51**, 114–25.

Gessaman, J. A., and Nagy, K. A. (1988). Energy metabolism: errors in gas exchange conversion factors. *Physiological Zoology*, **61**, 507–13.

Gholz, H. L. (1982). Environmental limits on above ground net primary production, leaf area, and biomass in vegetation zones of the Pacific Northwest. *Ecology*, **63**, 469–81.

Gibbons, E.F. (1995). Conservation of Primates in Captivity. In E. F. Gibbons, B. S. Durrant, and J. Demarest, eds. *Conservation of Endangered Species in Captivity*, pp. 485–501. State University of New York Press, Albany, N.Y.

Gibson, D. (2012). Aye-aye (*Daubentonia madagascariensis*) North American Regional Stud-book, 4th edition. AZA and San Diego Zoo Global, Silver Spring, MD and San Diego, CA.

Gicquel, C., and Le Bouc, Y. (2006). Hormonal regulation of fetal growth. *Hormone Research,* 65, 28–33.

Gilbert, K. (1997). Red howling monkey use of specific defacation sites as a parasite avoidance strategy. *Animal Behaviour,* 54, 451–5.

Gilby, I. C., Pokempner, A. A., and Wrangham, R. W. (2010). A direct comparison of scan and focal sampling methods for measuring wild chimpanzee feeding behaviour. *Folia Primatologica,* 81, 254–64.

Gill, S. J., Biging, G. S., and Murphy, E. C. (2000). Modeling conifer tree crown radius and estimating canopy cover. *Forest Ecology and Management,* 126, 405–16.

Gillespie, T. R. (2006). Noninvasive assessment of gastrointestinal parasite infections in free-ranging primates. *International Journal of Primatology,* 27, 1129–43.

Gillespie, T. R., and Chapman, C. A. (2006). Prediction of parasite infection dynamics in primate metapopulations based on attributes of forest fragmentation. *Conservation Biology,* 20, 441–8.

Gillespie, T. R., and Chapman, C. A. (2008). Forest fragmentation, the decline of an endangered primate, and changes in host-parasite interactions relative to an unfragmented forest. *American Journal of Primatology,* 70, 222–30.

Gillespie, T. R., Lonsdorf, E. V., Canfield, E. P., *et al.* (2010). Demographic and ecological effects on patterns of parasitism in Eastern Chimpanzees (*Pan troglodytes schweinfurthii*) in Gombe National Park, Tanzania. *American Journal of Physical Anthropology,* 143, 534–44.

Gillespie, T. R., Nunn, C. L., and Leendertz, F. H. (2008). Integrative approaches to the study of primate infectious disease: Implications for biodiversity conservation and global health. *Yearbook of Physical Anthropology,* 51, 53–69.

Glander, K. E. (1982). The impact of plant secondary compounds on primate feeding behavior. *Yearbook of Physical Anthropology,* 25, 1–18.

Glander, K. E., Fedigan, L. M., Fedigan, L., and Chapman, C. (1991). Capture techniques and measurements of three monkey species in Costa Rica. *Folia Primatologica,* 57, 70–82.

Glander, K. E., Wright, P. C., Daniels, P. S., *et al.* (1992). Morphometrics and testicle size of rainforest lemur species from southeastern Madagascar. *Journal of Human Evolution,* 22, 1–17.

Glander, K. E. (2006). Average body weight for mantled howling monkeys (*Alouatta pallilata*): an assessment of average values and variability. In A. Estrada, P. A. Garber, M. Pavelka, and L. Luecke, eds. *New Perspectives In The Study Of Mesoamerican Primates,* pp. 247–63. Springer, New York.

Glenn, T. C. (2011). Field guide to next-generation DNA sequencers. *Molecular Ecology Resources,* 11, 759–69.

Glucksmann, A. (1974). Sexual dimorphism in mammals. *Biological Reviews,* 49, 423–75.

Goldberg, T. L. (2003). Application of phylogeny reconstruction and character-evolution analysis to inferring patterns of directional microbial transmission. *Preventive Veterinary Medicine,* 61, 59–70.

Goldberg, T. L., Gillespie, T. R., and Rwego, I. B., *et al.* (2008). Forest fragmentation as cause of bacterial transmission among nonhuman primates, humans, and livestock, Uganda. *Emerging Infectious Diseases,* 14, 1375–82.

Goldberg, T. L., Gillespie, T. R., and Rwego, I. B., *et al.* (2007). Patterns of gastrointestinal bacterial exchange between chimpanzees and humans involved in research and tourism in western Uganda. *Biological Conservation,* 135, 511–17.

Goldberg, T. L., Gillespie, T. R., and Singer, R. S. (2006). Optimization of analytical parameters for inferring relationships among *Escherichia coli* isolates from repetitive-element PCR by maximizing correspondence with multilocus sequence typing data. *Applied Environmental Microbiology,* 72, 6049–52.

Goldberg, T. L., Sintasath, D. M., Chapman, C. A., *et al.* (2009). Coinfection of Ugandan red colobus (*Procolobus [Piliocolobus] rufomitratus tephrosceles*) with novel, divergent Delta-, Lenti-, and Spumaretroviruses. *Journal of Virology,* 83, 11318–29.

Golden, C. D., Fernald, C. H., Brashares, J. S., *et al.* (2011). Benefits of wildlife consumption to child nutrition in a biodiversity hotspot. *Proceedings of the National Academy of Sciences USA,* 108, 19653–6.

Goldizen, A. W. (2006). Tamarins and marmosets: communal care of offspring. In N. B. Smuts, D. L. Cheney, R. M. Seyfarth, R. W. Wrangham, and T. T. Struhsaker, eds. *Primate Societies.* The University of Chicago Press, Chicago, IL.

Goldizen, A. W., Terborgh, J., Cornejo, F., *et al.* (1988). Seasonal food shortage, weight loss, and the timing of births in saddle-back tamarins (*Saguinus fuscicollis*). *Journal of Animal Ecology,* 57, 893–901.

Goldstein, D. L. (1988). Estimates of daily energy expenditure in birds: the time-energy budget as an integrator of laboratory and field studies. *American Zoologist,* 28, 829–44.

Goodall, J. (1973). The behaviour of the chimpanzee in their natural habitat. *American Journal of Psychiatry,* 130, 1–12.

Goodman, S., and Sterling, E. J. (1996). The utlization of *Canarium* (Burseraceae) seeds by vertebrates in the RNI d'Andringitra, Madagascar. *Fieldana: Zoology,* 85, 83–92.

Goodnight, K. F., and Queller, D. C. (1999). Computer software for performing likelihood tests of pedigree relationship using genetic markers. *Molecular Ecology,* 8, 1231–4.

Goossens, B., Chikhi, L., Jalil, M. F., *et al.* (2005). Patterns of genetic diversity and migration in increasingly fragmented and declining orang-utan (*Pongo pygmaeus*) populations from Sabah, Malaysia. *Molecular Ecology,* 14, 441–56.

Gore, A. C. (2008). Developmental programming and endocrine disruptor effects on reproductive neuroendocrine systems. *Frontiers in Neuroendocrinology,* 29, 358–74.

Goudet, J., Perrin, N., and Waser, P. (2002). Tests for sex-biased dispersal using bi-parentally inherited genetic markers. *Molecular Ecology,* 11, 1103–14.

Graczyk, T. K., Nizeyi, J. B., Ssebide, B., *et al.* (2002). Anthropozoonotic *Giardia duodenalis* genotype (assemblage) A infections in habitats of free-ranging human-habituated gorillas, Uganda. *Journal of Parasitology,* 88, 905–9.

Grant, P. R., and Grant, B. R. (2000). Non-random fitness variation in two populations of Darwin's finches. *Proceeding of the Royal Society of London Series B,* 267, 131–8.

Gray, P. B., and Ellison, P. T. (2009). Introduction. In P. T. Ellison, and P. B. Gray, eds. *Endocrinology of social relationships,* pp. 1–9. Harvard University Press, Cambridge, MA.

Gray, M., McNeilage, A., Fawcett, K., *et al.* (2009). Censusing the mountain gorillas in the Virunga Volcanoes: complete sweep method versus monitoring. *African Journal of Ecology*, 48, 488–599.

Greenwood, J. J. D. (1996). Basic techniques. In W. J. Sutherland, ed. *Ecological census techniques: a handbook*, pp. 11–110. Cambridge University Press, Cambridge.

Greenwood, J. J. D., and Robinson, R. A. (2006). Principles of sampling. In W. J. Sutherland, ed. *Ecological Census Techniques: A Handbook*, 2nd edition, pp. 11–86. Cambridge University Press, Cambridge.

Greiner, E. C., and McIntosh, A. (2009). Collection methods and diagnostic procedures for primate parasitology. In M. A. Huffman, and C. A. Chapman, eds. *Primate Parasite Ecology: The Dynamics of Host-Parasite Relationships*, pp. 3–27. Cambridge University Press, Cambridge.

Grimm, V., and Railsback, S. (2005). *Individual-Based Modeling and Ecology*. Princeton University Press, Princeton, NJ.

Grinnell, J. (1917). Field tests of theories concerning distributional control. *American Naturalist*, 51, 115–28.

Gross-Camp, N., and Kaplin, B. A. (2011). Differential seed handling by two African primates affects: seed fate and establishment of large-seeded trees. *Acta Oecologica*, 37, 578–86.

Grossmann, F., Hart, J. A., Vosper, A., and Ilambu, O. (2008). Range occupation and population estimates of Bonobos in the Salonga National Park. In T. Furuichi, and J. Thompson, eds. *The Bonobos*, pp. 189–216. Springer, New York.

Groves, C. (1993). Order Primates. In D. E. Wilson, and D. M. Reeder, eds. *Mammal species of the world: A taxonomic and geographic reference*, pp. 243–77. Smithsonian Institution Press, Washington, DC.

Guillot, G., Mortier, F., and Estoup, A. (2005). Geneland: A program for landscape genetics. *Molecular Ecology Notes*, 5, 712–15.

Guindon, S., and Gascuel, O. (2003). A simple, fast, and accurate algorithm to estimate large phylogenies by maximum likelihood. *Systematic Biology*, 52, 696–704.

Gursky, S. (2003). Lunar philia in a nocturnal primate. *International Journal of Primatology*, 24, 351–67.

Guschanski, K., Vigilant, L., McNeilage, A., *et al.* (2009). Counting elusive animals: Comparing field and genetic census of the entire mountain gorilla population of Bwindi Impenetrable National Park, Uganda. *Biological Conservation*, 142, 290–300.

Hagerman, A. (2011). The Tannin Handbook. Online resource available at: <http://www.users.muohio.edu/hangermae/> (accessed May 2012).

Hall, B. G. (2011). *Phylogenetic Trees Made Easy: A How To Manual*, 4th Edition. Sinauer, Sunderland, MA.

Hall, M. B. (2009). Determination of starch, including maltooligosaccharides, in animal feeds: comparison of methods and a method recommended for AOAC collaborative study. *Journal of AOAC International*, 92, 42–9.

Hall, M. B., Hoover, W. H., Jennings, J. P., and Webster, T. K. M. (1999). A method for partitioning neutral detergent-soluble carbohydrates. *Journal of the Science of Food and Agriculture*, 79, 2079–86.

Hampe, A. (2004). Bioclimate envelope models: what they detect and what they hide. *Global Ecology and Biogeography*, 13, 469–71.

Hanya, G., Yoshihiro, S., Zamma, K., *et al*. (2003). New method to census primate groups: Estimating group density of Japanese macaques by point census. *American Journal of Primatology,* **60**, 43–56.

Haring, D., Hess, W., Coffman, B., *et al*. (1994). Natural history and captive management of the Aye-aye *(Daubentonia madagascariensis)* at the Duke University Primate Center, Durham. *Internatonal Zoo Yearbook,* **33**, 201–19.

Harris, T. R. (2006). Between-group contest competition for food in a highly folivorous population of black and white colobus monkeys (*Colobus guereza*). *Behavioral Ecology and Sociobiology,* **61**, 317–29.

Harris, T. R. (2007). Testing mate, resource and infant defence functions of intergroup aggression in non-human primates: issues and methodology. *Behaviour,* **144**, 1521–35.

Harris, T. R. (2010). Multiple resource values and fighting ability measures influence intergroup conflict in guerezas (*Colobus guereza*). *Animal Behaviour,* **79**, 89–98.

Harrison, M. E., and Marshall, A. J. (2011). Strategies for the use of fallback foods in apes. *International Journal of Primatology,* **32**, 531–65.

Hart, J. A., Grossmann, F., Vosper, A., and Ilanga, J. (2008). Human hunting and its impact on bonobos in the Salonga National Park, Democratic Republic of Congo. In T. Furuichi and J. Thompson, eds. *The Bonobos: Behavior, Ecology, and Conservation*, pp. 245–72. Springer, New York.

Hartley, S., and Kunin, W. E. (2003). Scale dependency of rarity, extinction risk, and conservation priority. *Conservation Biology,* **17**, 1559–70.

Hausfater, G., and Meade, B. (1982). Alternation of sleeping groves by yellow baboons (*Papio cynocephalus*) as a strategy for parasite avoidance. *Primates,* **23**, 287–97.

Hawkes, K., Hill, K., and O'Connell, J. F. (1982). Why hunters gather: optimal foraging and the Aché of eastern Paraguay. *American Ethnologist,* **9**, 379–98.

Hayek, L., and Buzas, M. A. (1997). *Surveying Natural Populations.* Columbia University Press, New York.

Hedley, S. H., and Buckland, S. T. (2004). Spatial models for line transect sampling. *Journal of Agricultural, Biological, and Environmental Statistics*, **9**, 181–99.

Heintz, M. R., Santymire, R. M., Parr, L. A., and Lonsdorf, E. V. (2011). Validation of a cortisol enzyme immunoassay and characterization of salivary cortisol circadian rhythm in chimpanzees (*Pan troglodytes*). *American Journal of Primatology,* **73**, 903–8.

Held, J. R., and Wolfle, T. L. (1994). Imports: Current Trends and Usage. *American Journal of Primatology*, **34**, 85–96.

Helversen, O. V., and Reyer, H. U. (1984). Nectar intake and energy expenditure in a flower visiting bat. *Oecologia,* **63**, 178–84.

Helversen, O. V., Volleth, M., and Núñez, J. (1986). A new method for obtaining blood from a small mammal without injuring the animal: use of Triatomid bugs. *Experientia,* **42**, 809–10.

Hemelrijk, C. K. (2000). Self-reinforcing dominance interactions between virtual males and females: hypothesis generation for primate studies. *Adaptive Behavior,* **8**, 13–26.

Hemingway, C. A., and Bynum, N. (2005). The influence of seasonality on primate diet and ranging. In D. K. Brockman, and C. P. van Schaik, eds. *Seasonality in Primates: Studies of Living and Extinct Human and Non-Human Primates*, pp. 57–104. Cambridge University Press, Cambridge.

Hemingway, C. A., and Overdorff, D. J. (1999). Sampling effects on food availability estimates: phenological method, sample size, and species composition. *Biotropica,* **31**, 354–64.

Henderson, J. (2005). Ernest Starling and "hormones": an historical commentary. *Journal of Endocrinology,* **184**, 5–10.

Hennessey, A. B., and Rogers, J. (2008). A study of the bushmeat trade in Ouesso, Republic of Congo. *Conservation and Society,* **6**, 179–84.

Henry, R., and Winkler, L. (2001). Foraging feeding and defecation site selection as a parasite avoidance strategy of *Alouatta palliata* in a dry tropical forest. *American Journal of Physical Anthropology,* **32**, 79.

Hernandez, P. A., Graham, C. H., Master, L. L. and Albert D. L. (2006). The effect of sample size and species characteristics on performance of different species distribution modeling methods. *Ecography,* **29**, 773–85.

Herrnstein, R. J. (1961). Relative and absolute strength of responses as a function of frequency of reinforcement. *Journal of the Experimental Analysis of Behavior,* **4**, 267–72.

Hey, J. (2010). Isolation with migration models for more than two populations. *Molecular Biology and Evolution,* **27**, 905–20.

Heymann, D. L. (2008). *Control of Communicable Diseases Manual,* 19th edition. American Public Health Association, Washington DC.

Hickerson, M. J., Carstens, B. C., Cavender-Bares, J., *et al.* (2010). Phylogeography's past, present, and future: 10 years after Avise, 2000. *Molecular Phylogenetics and Evolution,* **54**, 291–301.

Hickerson, M. J., and Meyer, C. P. (2008). Testing comparative phylogeographic models of marine vicariance and dispersal using a hierarchical Bayesian approach. *BMC Evolutionary Biology,* **8**, 322.

Higham, J. P., Ross, C., Warren, Y., *et al.* (2007). Reduced reproductive function in wild baboons (*Papio hamadryas anubis*) related to natural consumption of the African black plum (*Vitex doniana*). *Hormones and Behavior,* **52**, 384–90.

Hilaluddin, R. K., and Ghose, D. (2005). Conservation implications of wild animal biomass extractions in Northeast India. *Animal Biodiversity and Conservation,* **28**, 169–79.

Hill, C. M. (2000). Conflict of interest between people and baboons: crop raiding in Uganda. *International Journal of Primatology,* **21**, 299–315.

Hill, D., Fasham, M., Tucker, G., *et al.* (2005). *Handbook of Biodiversity Methods: Survey, Evaluation and Monitoring.* Cambridge University Press, Cambridge.

Hinde, R. A., and Atkinson, S. (1970). Assessing the roles of social partners in maintaining mutual proximity, as exemplified by mother-infant relations in rhesus monkeys. *Animal Behaviour,* **18**, 169–76.

Hird, S. M., Brumfield, R. T., and Carstens, B. C. (2011). PRGmatic: an efficient pipeline for collating genome-enriched second generation sequencing data using a "provisional-reference genome." *Molecular Ecology Resources,* **11**, 743–8.

Hirzel, A. H., Hausser, J., Chessel, D., and Perrin, N. (2002). Ecological-niche factor analysis: how to compute habitat-suitability maps without absence data? *Ecology,* **83**, 2027–36.

Hockings, K. J., Anderson, J. R., and Matsuzawa, T. (2009). Use of wild and cultivated foods by chimpanzees at Bossou, Republic of Guinea: Feeding dynamics in a human-influenced environment. *American Journal of Primatology,* **71**, 636–46.

Hockings, K. J., and Humle, T. (2009). Best Practice Guidelines for the Prevention and Mitigation of Conflict between Human and Great Apes. IUCN/SSC Primate Specialist Group (PSG), Gland, Switzerland.

Hodgen, G. D., Niemann, W. H., Turner, C. K., and Chen, H. C. (1976). Diagnosis of pregnancy in chimpanzees using the nonhuman primate pregnancy test kit. *Journal of Medical Primatology*, 5, 247–52.

Hodgkinson, C. (2009). Tourists, gorillas and guns: Integrating Conservation and Development in the Central African Republic (Ph.D. Dissertation), University College London, London.

Hoffmann, M., Hilton-Taylor, C., Angulo, A., *et al.* (2010). The impact of conservation on the status of the world's vertebrates. *Science*, 330, 1503–9.

Hoffmann, M., and O'Riain, M. J. (2012). Landscape requirements of a primate population in a human-dominated environment. *Frontiers in Zoology*, 9, 1–17.

Holst, B., Medici, E. P., Marino-Filho, O. J., *et al.*, eds. (2006). *Lion Tamarin Population and Habitat Viability Assessment Workshop 2005, final report*. IUCN/SSC Conservation Breeding Specialist Group, Apple Valley, MN.

Hopper, L. M., and Whiten, A. (2012). The evolutionary and comparative psychology of social learning and culture. In J. Vonk, and T. K. Shackelford, eds. *The Oxford Handbook of Comparative Evolutionary Psychology*. Oxford University Press, New York.

Hornaday, W. T. (1915). Gorillas past and present. *Bulletin of the New York Zoological Society*, 18, 1181–5.

Horning, N., Robinson, J., Sterling, E. J., Turner, W., and Spector, S. (2010). *Remote Sensing for Ecology and Conservation: A Handbook for Techniques*. Oxford University Press, Oxford.

Horwich, R. H., Koontz, F. W., Saqui, E., *et al.* (1993). A re-introduction program for the conservation of black howler monkeys in Belize. *Endangered Species*, 10, 1–6.

Hosey, G., Melfi, V., and Pankhurst, S. (2009). *Zoo animals: behaviour, management, and welfare*. Oxford University Press, Oxford.

Houle, A., Chapman, C. A., and Vickery, W. L. (2004). Tree climbing strategies for primate ecological studies. *International Journal of Primatology*, 25, 237–60.

Houwen, B. (2000). Blood film preparation and staining procedures. *Laboratory Hematology*, 6, 1–7.

Hsu, M. J., Kao, C., and Agoramoorthy, G. (2009). Interactions between visitors and Formosan macaques (*Macaca cyclopis*) at Shou-Shan Nature Park, Taiwan. *American Journal of Primatology*, 71, 214–22.

Hudson, I. L., and Keatley, M. R. (2010). *Phenological Research: Methods for Environmental and Climate Change Analysis*. Springer, Dordrecht.

Huff, J. L., and Barry, P. A. (2003). B-virus (*Cercopithecine herpesvirus* 1) infection in humans and macaques: potential for zoonotic disease. *Emerging Infectious Diseases*, 9, 246–50.

Huffman, M. A., Gotoh, S., Turner, L. A., *et al.* (1997). Seasonal trends in intestinal nematode infection and medicinal plant use among chimpanzees in the Mahale Mountains, Tanzania. *International Journal of Primatology*, 38, 111–25.

Hughes, K. (2003). The global positioning system, geographical information systems and remote sensing. In J. M. Setchell and D. J. Curtis, eds. *Field and Laboratory Methods in Primatology. A Practical Guide*, pp. 57–73. Cambridge University Press, Cambridge.

Humphrey, L. T. (2010). Weaning behaviour in human evolution. *Seminars in Cell and Developmental Biology*, **21**, 453–61.

Huntley, B., Collingham, Y. C., Willis, S. G., and Green, R. E. (2008). Potential impacts of climate change on European breeding birds. *PLoS ONE*, **1**, e1439.

Hutchins, M., and Conway, W. G. (1995). Beyond Noah's Ark: The Evolving Role of Modern Zoological Parks and Aquariums in Field Conservation. *International Zoo Yearbook*, **34**, 117–30.

Hutchinson, G. E. (1957). Concluding remarks. *Cold Spring Harbor Symposia on Quantitative Biology*, **22**, 415–27.

Hutchinson, J. M. C., and Waser, P. M. (2007). Use, misuse and extensions of "ideal gas" models of animal encounter. *Biological Reviews*, **82**, 335–59.

Infield, M. (1988). *Hunting, trapping and fishing in villages within and on the periphery of the Korup National Park*. Report to the World Wildlife Fund, Gland, Switzerland.

Ingicco, T., Moigne, A. M., and Gommery, D. (2011). A deciduous and permanent dental wear stage system for assessing the age of *Trachypithecus* sp. specimens (Colobinae, Primates). *Journal of Archaeological Science*, **39**, 421–7.

IPCC (Intergovernmental Panel on Climate Change). (2007) *Climate change 2007: impacts, adaptation, and vulnerability*. IPCC Secretariat, Geneva, Switzerland.

Irwin, M. T. (2007). Living in forest fragments reduces group cohesion in Diademed Sifakas (*Propithecis diadema*) in eastern Madagascar by reducing food patch size. *American Journal of Primatology*, **69**, 434–47.

Irwin, M. T. (2008). Feeding Ecology of *Propithecus diadema* in forest fragments and continuous forest. *International Journal of Primatology*, **29**, 95–115.

Irwin, M. T., Johnson, S. E., and Wright, P. C. (2005). The state of lemur conservation in south-eastern Madagascar: population and habitat assessments for diurnal and cathemeral lemurs using surveys, satellite imagery and GIS. *Oryx*, **39**, 204–18.

Isaac, N. J. B., Mallet, J., and Mace, G. M. (2004). Taxonomic inaction: its influence on macroecology and conservation. *Trends in Ecology & Evolution*, **19**, 464–9.

Isbell, L. A. (1991). Contest and scramble competition: patterns of female aggression and ranging behavior among primates. *Behavioral Ecology*, **2**, 143–55.

IUCN. (1998). *IUCN Guidelines for Re-introductions*. IUCN, Gland, Switzerland.

IUCN. (2001). *IUCN Red List Categories and Criteria: Version 3.1*. IUCN, Gland, Switzerland.

IUCN. (2003). *Guidelines for Application of IUCN Red List Criteria at Regional Levels: Version 3.0*. IUCN, Gland, Switzerland.

IUCN. (2005). Resolution 3.013: The Uses of the IUCN Red List of Threatened Species. In IUCN, ed. *Resolutions and Recommendations: World Conservation Congress, Bangkok, Thailand, 17–25 November 2004*, pp. 14–16. Gland, Switzerland: IUCN.

IUCN. (2011). *Guidelines for Using the IUCN Red List Categories and Criteria: Version 9.0*. IUCN, Gland, Switzerland.

IUCN. (2012). *IUCN Red List of Threatened Species. Version 2012.1*. IUCN, Gland, Switzerland. Available at <http://www.iucnredlist.org> (accessed October 2012).

Iwamoto, T. (1982). Food and nutritional condition of free ranging Japanese monkeys on Koshima Islet during winter. *Primates*, **23**, 153–70.

Jacobs, A., and Petit, O. (2011). Social network modeling: A powerful tool for the study of group scale phenomena in primates. *American Journal of Primatology,* **73**, 741–7.

Jansen, P. A., Bohlman, S. A., Garzon-Lopez, C. X., *et al.* (2008). Large-scale spatial variation in palm fruit abundance across a tropical moist forest estimated from high-resolution aerial photographs. *Ecography,* **31**, 33–42.

Janson, C. H. (1985). Aggressive competition and individual food consumption in wild brown capuchin monkeys (*Cebus apella*). *Behavioral Ecology and Sociobiology,* **18**, 128–38.

Janson, C. H. (1988). Food Competition in Brown Capuchin Monkeys (*Cebus apella*): Quantitative Effects of Group Size and Tree Productivity. *Behaviour,* **105**, 53–76.

Janson, C. H. (1994). Naturalistic environments in captivity: a methodological bridge between field and laboratory studies of primates. In E. Gibbons, E. Water, E. Wyers, and E. Menzel, eds. *Naturalistic Environments in Captivity for Animal Behavior Research,* pp. 271–9. SUNY Press, Albany.

Janson, C. H. (1996). Toward an Experimental Socioecology of Primates: examples from Argentine brown capuchin monkeys (*Cebus apella nigritus*). In M. A. Norconk, and W. Kinzey, eds. *Adaptive Radiations of Neotropical Primates,* pp. 309–25. Plenum Press, New York.

Janson, C. H. (1998). Experimental evidence for spatial memory in foraging wild capuchin monkeys, *Cebus apella. Animal Behaviour,* **55**, 1229–43.

Janson, C. H. (2000). Primate socio-ecology: The end of a golden age. *Evolutionary Anthropology: Issues, News, and Reviews,* **9**, 73–86.

Janson, C. H. (2007). Experimental evidence for route integration and strategic planning in wild capuchin monkeys. *Animal Cognition,* **10**, 341–56.

Janson, C. H. (2012). Reconciling Rigor and Range: Observations, Experiments, and Quasi-experiments in Field Primatology. *International Journal of Primatology,* **33**, 520–41.

Janson, C. H., Baldovino, M. C., and Di Bitetti, M. S. (2012). The group life cycle and demography of Brown Capuchin monkeys (*Cebus [apella] nigritus*) in Iguazú National Park, Argentina. In P. M. Kappeler, and D. P. Watts, eds. *Long-Term Field Studies of Primates.* Springer, Berlin.

Janson, C. H., and Goldsmith, M. L. (1995). Predicting group size in primates: foraging costs and predation risks. *Behavioral Ecology and Sociobiology,* **36**, 326–36.

Janson, C. H., and Verdolin, J. (2005). Seasonality of primate births in relation to climate. In D. K. Brockman, and C. P. van Schaik, eds. *Seasonality in Primates: Studies of Living and Extinct Human and Non-Human Primates,* pp. 307–50. Cambridge University Press, Cambridge.

Janzen, D. (1983). Physiological ecology of fruits and their seeds. In O. L. Lange, P. S. Nobel, C. B. Osmond, and H. Ziegler, eds. *Physiological plant ecology,* pp. 625–55. Springer-Verlag, Berlin.

Jenkins, R. K. B., Keane1, A., Rakotoarivelo, A. R., *et al.* (2011). Analysis of Patterns of Bushmeat Consumption Reveals Extensive Exploitation of Protected Species in Eastern Madagascar. *PLoS ONE,* **6**, 1–12.

Jennings, S. B., Brown, N. D., and Sheil, D. (1999). Assessing forest canopies and understorey illumination: canopy closure, canopy cover and other measures. *Forestry,* **72**, 59–74.

Jens, W., Mager-Melicharek, C. A. X., and Rietkerk, F. E. (2012). Free-ranging New World primates in zoos: cebids at Apenheul. *International Zoo Yearbook, 46,* 137–49.

Jiang, Z., Sugita, M., Kitahara, M., Takatsuki, S., Goto, T., and Yoshida, Y. (2008). Effects of habitat feature, antenna position, movement, and fix interval on GPS radio collar performance in Mount Fuji, central Japan. *Ecological Research, 23,* 581–8.

Johnson, A., and Sackett, R. (1998). Direct systematic observation of behavior. In H. R. Bernard, ed. *Handbook of Methods in Cultural Anthropology*, pp. 301–31. Alta Mira Press, Lanham.

Johnston, A. R., Gillespie, T. R., Rwego, I. B., *et al.* (2010). Molecular epidemiology of cross-species *Giardia duodenalis* transmission in Western Uganda. *PLoS Neglected Tropical Diseases, 4,* e683.

Jolly, C. J. (1998). A simple and inexpensive pole syringe for injecting caged primates. *Laboratory Primate Newsletter, 37,* 1–2.

Jolly, C. J., Phillips-Conroy, J. E., and Muller, A. E. (2011). Trapping primates. In J. M. Setchell, and D. J. Curtis, eds. *Field and Laboratory Methods in Primatology,* 2nd edition, pp. 110–21. Cambridge University Press, Cambridge.

Jones, J. H. (2007). demogR: A Package for the Construction and Analysis of Age-structured Demographic Models in R. *Journal of Statistical Software, 22,* 1–28.

Jones, J. H. (2011). Primates and the evolution of long slow life histories. *Current Biology, 21,* R708–17.

Jones, M. (1986). Successes and failures of captive breeding. In K. Benirschke, ed. *Primates: The Road to Self-Sustaining Populations,* pp. 251–60. Springer-Verlag, New York.

Jones-Engel, L., Engel, G. A., and Fuentes, A. (2011). An ethnoprimatological appraoch to interactions between human and non-human primates. In J. Setchell, and D. J. Curtis, eds. *Field and Lab Methods in Primatology: A Practical Guide*, pp. 21–32. Cambridge University Press, Cambridge.

Jones-Engel, L., Schillaci, M., Engel, G., *et al.* (2005). Characterizing primate pet ownership in Sulawesi: Implications for disease transmission. In J. Paterson, and J. Wallis, eds. *Commensalism and conflict: The primate-human interface*, pp. 197–221. American Society of Primatologists, Norman, OK.

Joseph, G. (2005). *Fundamentals of Remote Sensing.* Universities Press, Hyderabad, India.

Jost-Robinson, C. A., Daspit, L. L., and Remis, M. J. (2011). Multi-faceted approaches to understanding changes in wildlife and livelihoods in a protected area: a conservation case study from the Central African Republic. *Environmental Conservation, 38,* 247–55.

Juste, J., Fa, J. E., Perez del Val, J., and Castroviejo, J. (1995). Market Dynamics of Bushmeat Species in Equatorial Guinea. *Journal of Applied Ecology, 32,* 454–67.

Kabasawa, A. (2009). Current state of the chimpanzee pet trade in Sierra Leone. *African Study Monographs, 30,* 37–54.

Kacelnik, A., and Abreu, F. B. E. (1998). Risky choice and Weber's law. *Journal of Theoretical Biology, 194,* 289–98.

Kagan, R., and Vesey, J. (2010). Challenges of zoo animal welfare. In D. G. Kleiman, K. V. Thompson, and C. Kirk Baer, eds. *Wild Mammals in Captivity,* 2nd edition, pp. 11–21. University of Chicago Press, Chicago.

Kaplin, B. A., and Moermond, T. C. (1998). Variation in seed handling by two species of forest monkeys in Rwanda. *American Journal of Primatology, 45,* 83–101.

Kapos, V. (1989). Effects of isolation on the water status of forest patches in the Brazilian Amazon. *Journal of Tropical Ecology,* **5**, 173–85.

Kappeler, P. M. (2000). Primate Males: History and Theory. In P. M. Kappeler, ed. *Primate Males: Causes and Consequences of Variation in Group Composition.* Cambridge University Press, Cambridge.

Kappeler, P. M., and van Schaik, C. P. (2002). Evolution of Primate Social Systems. *International Journal of Primatology,* **23**, 707–40.

Karpanty, S. M. (2006). Direct and indirect impacts of raptor predation on lemurs in Southeastern Madagascar. *International Journal of Primatology,* **27**, 239–61.

Kaur, T., and Singh, J. (2009). Primate-parasite zoonoses and anthropozoonoses: A literature review. In M. A. Huffman, and C. A. Chapman, eds. *Primate Parasite Ecology: The Dynamics of Host-Parasite Relationships,* pp. 199–230. Cambridge University Press, Cambridge.

Kaur, T., Singh, J., Tong, S., *et al.* (2008). Descriptive epidemiology of fatal respiratory outbreaks and detection of a human-related metapneumovirus in wild chimpanzees (*Pan troglodytes*) at Mahale Mountains National Park, Western Tanzania. *American Journal of Primatology,* **70**, 755–65.

Kay, R. F. (1981). The nut-crackers—a new theory of the adaptations of the Ramapithecinae. *American Journal of Physical Anthropology,* **55**, 141–51.

Keane, A., Hobinjatovo, T., Razafimanahaka, H. J., *et al.* (2012). The potential of occupancy modeling as a tool for monitoring wild primate populations. *Animal Conservation,* **15**, 457–465.

Keith, D. A., Akcakaya, H. R, Thuiller, W., *et al.* (2008). Predicting extinction risks under climate change: Coupling stochastic population models with dynamic bioclimatic habitat models. *Biology Letters,* **4**, 560–3.

Kelley, J., and Schwartz, G. T. (2010). Dental development and life history in living African and Asian apes. *Proceedings of the National Academy of Sciences USA,* **107**, 1035–40.

Kennedy, M. (2009). *Introducing Geographic Information Systems with ArcGIS.* John Wiley & Sons, Inc., Hoboken, NJ.

Kent, M., and Coker, P. (1994). *Vegetation description and analysis. A practical approach.* John Wiley & Sons, Chichester.

Kenward, R. W., and Hodder, K. H. (1996). *Ranges V: An Analysis System for Biological Location Data.* Institute for Terrestrial Ecology, Wareham, UK.

Kernohan, B. J., Gitzen, R. A., and Millspaugh, J. J. (2001). Analysis of animal space use and movements. In J. J. Millspaugh, and J. M. Marzluff, eds. *Radio Tracking and Animal Populations,* pp. 125–66. Academic Press, San Diego, CA.

Kery, M., Gardner, B., Stoeckle, T., *et al.* (2011). Use of spatial capture-recapture modeling and DNA data to estimate densities of elusive animals. *Conservation Biology,* **25**, 356–4.

Khan, I. H., Mendoza, S., Yee, J., *et al.* (2006). Simultaneous detection of antibodies to six nonhuman-primate viruses by multiplex microbead immunoassay. *Clinical Vaccine Immunology,* **13**, 45–52.

Kierulff, M. C. M., Procopio De Oliveria, P., *et al.* (2002). Reintroduction and Translocations as Conservation Tools for Golden Lion Tamarins. In D. G. Kleiman, and A. B. Rylands, eds. *Lion Tamarins Biology and Conservation,* pp. 271–300. Smithsonian Institution Press, Washington, DC.

King, A. J., and Cowlishaw, G. (2009). All together now: behavioural synchrony in baboons. *Animal Behaviour*, **78**, 1381–7.

King, A. J., Clark, F. E., and Cowlishaw, G. (2011a). The dining etiquette of desert baboons: the roles of social bonds, kinship, and dominance in co-feeding networks. *American Journal of Primatology*, **73**, 768–74.

King, A. J., Sueur, C., Huchard, E., and Cowlishaw, G. (2011b). A rule-of-thumb based on social affiliation explains collective movements in desert baboons. *Animal Behaviour*, **82**, 1337–45.

Kinnaird, M. F. (1992). Competition for a forest palm: Use of *Phoenix reclinata* by human and nonhuman primates. *Conservation Biology*, **6**, 101–7.

Kinzey, W. G., and Norconk, M. A. (1990). Hardness as a basis of fruit choice in two sympatric primates. *American Journal of Physical Anthropology*, **81**, 5–15.

Kirksey, S. E., and Helmreich, S. (2010). The emergence of multispecies ethnography. *Cultural Anthropology*, **25**, 545–76.

Kitchen, D. M., and Beehner, J. C. (2007). Factors affecting individual participation in group-level aggression among non-human primates. *Behaviour*, **144**, 1551–81.

Kitchen, D. M., Cheney, D. L., and Seyfarth, R. M. (2004a). Factors mediating inter-group encounters in savannah baboons (*Papio cynocephalus ursinus*). *Behaviour*, **141**, 197–218.

Kitchen, D. M., Horwich, R. H., and James, R. A. (2004b). Subordinate male black howler monkey (*Alouatta pigra*) responses to loud calls: experimental evidence for the effects of intra-group male relationships and age. *Behaviour*, **141**, 703–23.

Kitchen, D. M., Seyfarth, R. M., Fischer, J., and Cheney, D. L. (2003). Loud calls as indicators of dominance in male baboons (*Papio cynocephalus ursinus*). *Behavioral Ecology and Sociobiology*, **53**, 374–84.

Klüver, H. (1933). *Behavior Mechanisms in Monkeys*. Univ. of Chicago Press, Chicago.

Knott, C. D. (1998). Changes in orangutan diet, caloric intake, and ketones in response to fluctuating fruit availability. *International Journal of Primatology*, **19**, 1061–79.

Knott, C. D. (2005). Radioimmunoassay of estrone conjugates from urine dried on filter paper. *American Journal of Primatology*, **67**, 121–35.

Knott, C. D., Beaudrot, L., Snaith, T., *et al.* (2008). Female-Female Competition in Bornean Orangutans. *International Journal of Primatology*, **29**, 975–97.

Kobbe, S., and Dausmann, K. H. (2009). Hibernation in Malagasy mouse lemurs as a strategy to counter environmental challenge. *Naturwissenschaften*, **10**, 1221–7.

Kobbe, S., Ganzhorn, J. U., and Dausmann, K. H. (2010). Extreme individual flexibility of heterothermy in free-ranging Malagasy mouse lemurs (*Microcebus griseorufus*). *Journal of Comparative Biochemistry and Physiology B*, **181**, 165–73.

Koch, P., Tuross, N., and Fogel, M. (1997). The effects of sample treatment and diagenesis on the isotopic integrity of carbonate in biogenic hydroxylapatite. *Journal of Archaeological Science*, **24**, 417–30.

Koenig, A. (2002). Competition for Resources and its Behavioral Consequences Among Female Primates. *International Journal of Primatology*, **23**, 759–83.

Koenig, A., and Borries, C. (2006). The predictive power of socioecological models: a reconsideration of resource characteristics, agonism, and dominance hierarchies. In G. Hohmann, M. M. Robbins, and C. Boesch, eds. *Feeding Ecology in Apes and Other*

Primates. Ecological, Physical and Behavioral Aspects, pp. 263–84. Cambridge University Press, Cambridge.

Koenig, A., Borries, C., Doran-Sheehy, D. M., and Janson, C. H. (2006). How important are affiliation and cooperation? A reply to Sussman *et al. American Journal of Physical Anthropology,* **131**, 522–3.

Koh, L. P., and Wich, S. A. (2012). Dawn of drone ecology: low-cost autonomous aerial vehicles for conservation. *Tropical Conservation Science,* **5**, 121–132.

Kohn, M., York, E. C., Kamradt, D. A., *et al.* (1999). Estimating population size by genotyping faeces. *Proceedings of the Royal Society of London Series B,* **266**, 657–63.

Köndgen, S., Kühl, H., N'Goran, P. K., *et al.* (2008). Pandemic human viruses cause decline of endangered great apes. *Current Biology,* **18**, 1–5.

Kowalewski, M. M., Salzer, J. S., Deutsch, J. C., *et al.* (2011). Black and gold howler monkeys (*Alouatta caraya*) as sentinels of ecosystem health: Patterns of zoonotic protozoa infection relative to degree of human-primate contact. *American Journal of Primatology,* **73**, 75–83.

Krakauer, K., Lemelin, P., and Schmitt, D. (2002). Hand and body position during locomotor behavior in the aye-aye (*Daubentonia madagascariensis*). *American Journal of Primatology,* **57**, 105–18.

Kralik, J. D., and Sampson, W. W. L. (2011). A fruit in hand is worth many more in the bush: Steep spatial discounting by free-ranging rhesus macaques (*Macaca mulatta*). *Behavioral Processes,* **89**, 197–202.

Krebs, C. J. (1999). *Ecological Methodology.* Addison-Wesley Longman, Menlo Park, California.

Krebs, C. J. (2001). *Ecology: the experimental Analysis of Distribution and Abundance*, 5th edition. Benjamin Cummings, London.

Krebs, J. R., and Davies, N. B. (1993). *An Introduction to Behavioural Ecology.* Blackwell, Oxford.

Kress, W. J., Schatz, G. E., Andrianifahanana, M., and Morland, H. S. (1994). Pollination of *Ravenala madagascariensis* (Strelitziaceae) by Lemurs in Madagascar: Evidence for an Archaic Coevolutionary System? *American Journal of Botany,* **81**, 542–51.

Kroodsma, D. E., Byers, B. E., Goodale, E., *et al.* (2001). Pseudoreplication in playback experiments, revisited a decade later. *Animal Behaviour,* **61**, 1029–33.

Kuhar, C., Gibson, D., and Marti, K. (2011) Population Analysis & Breeding and Transfer Plan for Nocturnal Prosimian Primates, Section 8, Aye-aye (*Daubentonia madagascariensis*). AZA Species Survival Plan®, Chicago, IL: AZA Population Management Center at Lincoln Park Zoo.

Kuhlmeier, V. A., and Boysen, S. T. (2001). The effect of response contingencies on scale model task performance by chimpanzees (*Pan troglodytes*). *Journal of Comparative Psychology,* **115**, 300–6.

Kummer, H. (1968). Two variations in the social organization of baboons. In P. C. Jay, ed. *Primates: Studies in Adaptation and Variability.* Rinehart and Winston, New York.

Kummer, H. (1995). *In Quest of the Sacred Baboon.* Princeton University Press, Princeton, NJ.

Kummer, H., and Kurt, F. (1963). Social unites of a free-living population of hamadryas baboons. *Folia Primatologica,* **1**, 4–19.

Kümpel N. F., Milner-Gulland, E. J., Cowlishaw, G., and Rowcliffe, J. M. (2010). Incentives for hunting: the role of bushmeat in the household economy in rural Equatorial Guinea. *Human Ecology,* **38**, 251–64.

Kümpel, N. F., Milner-Gulland, E. J., Rowcliffe, J. M., and Cowlishaw, G. (2008). Impact of gun-hunting on diurnal primates in continental Equatorial Guinea. *International Journal of Primatology*, **29**, 1065–82.

Kunin, W. E., and Gaston, K. J. (1993). The biology of rarity: Patterns, causes and consequences. *Trends in Ecology & Evolution*, **8**, 298–301.

Kuster, M., López de Alda, M., and Barceló, D. (2005). Estrogens and Progestogens in Wastewater, Sludge, Sediments, and Soil Water Pollution. In D. Barceló, ed. *The Handbook of Environmental Chemistry* Vol. 2, pp. 227–9. Springer, Berlin.

Lacy, R. C. (1995). Clarification of genetic terms and their use in the management of captive populations. *Zoo Biology*, **14**, 565–77.

Lacy, R. C. (2010). Re-thinking *ex-situ* vs. *in-situ* Species Conservation. In *World Association of Zoos and Aquariums (WAZA) Proceedings of 65th Annual Conference*, pp. 25–9. WAZA, Gland, Switzerland.

Lacy, R. C., Hughes, K. A., and Miller, P. S. (1995). *VORTEX: a stochastic simulation of the extinction process. Version 7 user's manual*. IUCN/SSC Conservation Breeding Specialist Group, Apple Valley, MN.

Lambert, J. E. (1998). Primate digestion: interactions among anatomy, physiology, and feeding ecology. *Evolutionary Anthropology*, **7**, 8–20.

Lambert, J. E. (2007). Seasonality, fallback strategies, and natural selection: A chimpanzee and cercopithecoid model for interpreting the evolution of the hominin diet. In P. S. Ungar, ed. *Evolution of the human diet: The known, the unknown, and the unknowable*. Oxford University Press, Oxford, UK.

Lambert, J. E., Chapman, C. A., Wrangham, R. W., and Conklin-Brittain, N. L. (2004). Hardness of cercopithecine foods: Implications for the critical function of enamel thickness in exploiting fallback foods. *American Journal of Physical Anthropology*, **125**, 363–8.

Lande, R., Engen, S., and Saether, B.-E. (2003). *Stochastic Population Dynamics in Ecology and Conservation*. Oxford University Press, New York.

Landguth, E. L., and Cushman, S. A. (2010). CDPOP: A spatially explicit cost distance population genetics program. *Molecular Ecology Resources*, **10**, 156–61.

Laska, M., Sanchez, E. C., Rivera, J. A. R., and Luna, E. R. (1996). Gustatory thresholds for food-associated sugars in the spider monkey (*Ateles geoffroyi*). *American Journal of Primatology*, **39**, 189–93.

Latch, E. K., Dharmarajan, G., Glaubitz, J. C., and Rhodes, O. E. (2006). Relative performance of Bayesian clustering software for inferring population substructure and individual assignment at low levels of population differentiation. *Conservation Genetics*, **7**, 295–302.

Lawler, R. R., Caswell, H. Richard, A. F., *et al.* (2009). Demography of Verreaux's sifaka in a stochastic rainfall environment. *Oecologia*, **161**, 491–504.

Lee, R. J., Gorog, A. J., Dwiyahreni, A., *et al.* (2005). Wildlife trade and implications for law enforcement in Indonesia: a case study from North Sulawesi. *Biological Conservation*, **123**, 477–88.

Lee-Thorp, J., and van der Merwe, N. J. (1987). Carbon isotope analysis of fossil bone apatite. *South African Journal of Science*, **83**, 712–15.

Leendertz, F. H., Pauli, G., Maetz-Rensing, K., *et al.* (2006). Pathogens as drivers of population declines: The importance of systematic monitoring in great apes and other threatened mammals. *Biological Conservation,* 131, 325–37.

Lees, C. M., and Wilcken, J. (2009). Sustaining the Ark: The challenges faced by zoos in maintaining viable populations. *International Zoo Yearbook,* 43, 6–18.

Lehmann, J., Korstjens, A. H., and Dunbar, R. I. M. (2007a). Fission–fusion social systems as a strategy for coping with ecological constraints: a primate case. *Evolutionary Ecology,* 21, 613–34.

Lehmann, J., Korstjens, A. H., and Dunbar, R. I. M. (2007b). Group size, grooming and social cohesion in primates. *Animal Behaviour,* 74, 1617–29.

Lehman, S. M., Rajaonson, A., and Day, S. (2006). Edge effects and their influence on lemur density and distribution in Southeast Madagascar. *American Journal of Physical Anthropology,* 129, 232–41.

Leighton, M. (1993). Modeling diet selectivity by Bornean orangutans: Evidence for integration of multiple criteria for fruit selection. *International Journal Primatology,* 14, 257–313.

Leighton, M., and Leighton, D. (1983). Vertebrate responses to fruiting seasonality within a Bornean rain forest. In S. L. Sutton, T. C. Whitmore, and A. C. Chadwick, eds. *Tropical Rain Forest: Ecology and Management,* pp. 181–96. Blackwell Scientific Publications, Boston.

Leith, H. (1974). *Phenology and Seasonality Modeling.* Springer, New York.

Lemmon, A. R., and Lemmon, E. M. (2008). A likelihood framework for estimating phylogeographic history on a continuous landscape. *Systematic Biology,* 57, 544–61.

Lemos-Espinal, J. A., Rojas-González, R. I., and Zúñiga-Vega, J. J. (2005). *Técnicas para el estudio de poblaciones de fauna silvestre.* Universidad Nacional Autónoma de México and Comisión Nacional para el Conocimiento y Uso de la Biodiversidad, Distrito Federal, México.

Lerney, P., Rambaut, A., Drummond, A. J., and Suchard, M. A. (2009). Bayesian phylogeography finds its roots. *PLoS Computational Biology,* 5, e10000520.

Leroy, E. M., Rouquet, P., Formenty, P., *et al.* (2004). Multiple Ebola virus transmission events and rapid decline of Central African wildlife. *Science,* 303, 387–90.

Li, L., Yu, S., Ren, B., Li, M., Wu, R., and Long, Y. (2009). A study on the carrying capacity of the available habitat for the *Rhinopithecus bieti* population at Mt. Laojun in Yunnan, China. *Environmental Science and Pollution Research,* 16, 474–8.

Lifson, N., and McClintock, R. (1966). Theory of use of the turnover rates of body water for measuring energy and material balance. *Journal of Theoretical Biology,* 12, 46–74.

Lin, J. (2005). Tackling Southeast Asia's Illegal Wildlife Trade. *Singapore Year Book of International Law,* 200s, 191–208.

Linder, J. M., and Oates, J. F. (2011). Differential impact of bushmeat hunting on monkey species and implications for primate conservation in Korup National Park, Cameroon. *Biological Conservation,* 144, 738–45.

Linkie, M., Dinata, Y., Nofrianto, A., and Leader-Williams, N. (2007). Patterns and perceptions of wildlife crop raiding in and around Kerinci Seblat National Park, Sumatra. *Animal Conservation,* 10, 127–35.

Liu, Z., Ren, B., Wu, R., *et al.* (2009). The effect of landscape features on population genetic structure in Yunnan snub-nosed monkeys (*Rhinopithecus bieti*) implies an anthropogenic genetic discontinuity. *Molecular Ecology,* **18**, 3831–46.

Loiselle, B. A., Howelle, C. A., Graham, C. H., *et al.* (2003). Avoiding pitfalls of using species distribution models in conservation planning. *Conservation Biology,* **17**, 1591–600.

Loudon, J. E., Howells, M. E., and Fuentes, A. (2006). The importance of integrative anthropology: A preliminary investigation employing primatological and cultural anthropological data collection methods in assessing human-monkey coexistence in Bali, Indonesia. *Ecological and Environmental Anthropology,* **2**, 1–12.

Loudon, J. E., Sauther, M. L., Fish, K. D., *et al.* (2006). One reserve, three primates: Applying a holistic approach to understand the interconnections among ring-tailed lemurs (*Lemur catta*), Verreaux's sifaka (*Propithecus verreauxi*), and humans (*Homo sapiens*) at Beza Mahafaly Special Reserve, Madagasar. *Ecological and Environmental Anthropology,* **2**, 54–74.

Lucas, K., Elser, R., Long, S., and Broome, C. (2011). Population Analysis & Breeding and Transfer Plan Western Lowland Gorilla (*Gorilla gorilla gorilla*). AZA Species Survival Plan®, Chicago, IL: AZA Population Management Center at Lincoln Park Zoo.

Lucas, P. W. (2004). *Dental functional morphology: How teeth work.* Cambridge University Press, Cambridge.

Lucas, P. W., Beta, T., Darvell, B. W., *et al.* (2001). Field kit to characterize physical, chemical and spatial aspects of potential primate foods. *Folia Primatologica,* **72**, 11–25.

Lucas, P. W., Osorio, D., Yamashita, N., *et al.* (2003). Dietary analysis 1: Food physics. In J. M. Setchell, and D. J. Curtis, eds. *Field and Laboratory Methods in Primatology,* pp. 184–98. Cambridge Univesity Press, Cambridge.

Lucas, P. W., and Pereira, B. (1990). Estimation of the fracture toughness of leaves. *Functional Ecology,* **4**, 819–22.

Lucas, P. W., Peters, C. R., and Arrandale, S. R. (1994). Seed-breaking forces exerted by orangutans with their teeth in captivity and a new technique for estimating forces produced in the wild. *American Journal of Physical Anthropology,* **94**, 365–78.

Lucas, P. W., Turner, I. M., Dominy, N. J., and Yamashita, N. (2000). Mechanical defences to herbivory. *Annals of Botany,* **86**, 913–20.

Lynch, M. (1988). Estimation of relatedness by DNA fingerprinting. *Molecular Biology and Evolution,* **5**, 584–99.

MacDonald, S. E., and Wilkie, D. M. (1990). Yellow-nosed monkeys' (*Ceropithecus ascanius whitesidei*) spatial memory in a simulated foraging environment. *Journal of Comparative Psychology,* **104**, 382–7.

Mace, G. M. (2004). The role of taxonomy in species conservation. *Proceedings of the Royal Society of London Series B—Biological Sciences,* **359**, 711–19.

Macfie, E. J., and Williamson, E. A. (2010). *Best Practice Guidelines for Great Ape Tourism.* Occasional Paper of the IUCN Species Survival Commission (38). IUCN, Gland, Switzerland.

MacKenzie, D. I., Nichols, J. D., Royle, J. A., *et al.* (2006). *Occupancy Estimation and Modeling Inferring Patterns and Dynamics of Species Occurrence.* Elsevier Inc, New York.

MacKinnon, K. C., and Riley, E. P. (2010). Field primatology of today: Current ethical issues. *American Journal of Primatology,* **72**, 749–53.

Maddison, W. P., and Maddison, D. R. (2011). Mesquite: a modular system for evolutionary analysis. Version 2.75. <http://mesquiteproject.org.> (accessed October 2012)

Magurran, A. E. (2004). *Measuring Biological Diversity*. Blackwell, Oxford.

Maher, C. R., and Lott, D. F. (1995). Definitions of territoriality used in the study of variation in vertebrate spacing systems. *Animal Behaviour*, **49**, 1581–97.

Maho, Y. L., Goffart, M., Rochas, A., *et al.* (1981). Thermoregulation in the only nocturnal simian: The night monkey *Aotus trivirgatus. American Journal of Physiology: Regulatory, Integrative and Comparative Physiology*, **240**, 156–65.

Malcolm, J. R. (1994). Edge effects in Central Amazonian forest fragments. *Ecology*, **75**, 2438–45.

Maldonado, A. M., Nijman, V., and Bearder, S. K. (2009). Trade in night monkeys *Aotus* spp. in the Brazil-Colombia-Peru tri-border area: international wildlife trade regulations are ineffectively enforced. *Endangered Species Research*, **9**, 143–9.

Malhi, R. S., Satkoski-Trask, J., Shattuck, M., *et al.* (2011). Genotyping single nucleotide polymorphisms (SNPs) across species in Old World Monkeys. *American Journal of Primatology*, **73**, 1031–40.

Malone, N. M., Fuentes, A., Purnama, A., and Adi Putra, I. M. W. (2003). Displaced hylobatids: biological, cultural, and economic aspects of the primate trade in Jawa and Bali, Indonesia. *Tropical Biodiversity*, **8**, 41–9.

Mamanova, L., Coffey, A. J., Scott, C. E., *et al.* (2010). Target-enrichment strategies for next-generation sequencing. *Nature Methods*, 7, 111–18.

Manansang, J., Traylor-Holzer, K., Reed, D., and Leus, K., eds. (2005). *Indonesian Proboscis Monkey Population and Habitat Viability Assessment: Final Report*. IUCN/SSC Conservation Breeding Specialist Group, Apple Valley, MN.

Manel, S., Schwarts, M. K., Luikart, G., and Taberlet, P. (2003). Landscape genetics: Combining landscape ecology and population genetics. *Trends in Ecology & Evolution*, **18**, 189–97.

Manson, J. H., Perry, S., and Stahl, D. (2005). Reconciliation in Wild White-Faced Capuchins (*Cebus capucinus*). *American Journal of Primatology*, **65**, 205–19.

Maricec, T., Whitten, M., and Paabo, S. (2010). Multiplexed DNA sequence capture of mitochondrial genomes using PCR products. *PLoS ONE*, 5, e14004.

Markham, C. A., and Altmann, J. (2008). Remote monitoring of primates using automated GPS technology in open habitats. *American Journal of Primatology*, **70**, 1–5.

Marler, P. (2005). Ethology and the origins of behavioral endocrinology. *Hormones and Behavior*, **47**, 493–502.

Marshall, A. J. (2004). The population ecology of gibbons and leaf monkeys across a gradient of Bornean forest types (Ph.D. Dissertation), Harvard University, Cambridge, MA.

Marshall, A. J. (2009). Are montane forests demographic sinks for Bornean white-bearded gibbons (*Hylobates albibarbis*)? *Biotropica*, **41**, 257–67.

Marshall, A. J. (2010). Effect of habitat quality on primate populations in Kalimantan: gibbons and leaf monkeys as case studies. In J. Supriatna, and S. L. Gursky, eds. *Indonesian Primates*, pp. 157–77. Springer, New York.

Marshall, A. J., Ancrenaz, M., Brearley, F. Q., *et al.* (2009a). The effects of habitat quality, phenology, and floristics on populations of Bornean and Sumatran orangutans: are Sumatran forests more productive than Bornean forests? In S. A. Wich, S. S. Utami Atmoko,

T. Mitra Setia, and C. P. van Schaik, eds. *Orangutans: Ecology, Evolution, Behaviour and Conservation*, pp. 97–117. Oxford University Press, Oxford.

Marshall, A. J., Boyko, C. M., Feilen, K. L., *et al.* (2009b). Defining fallback foods and assessing their importance in primate ecology and evolution. *American Journal of Physical Anthropology*, **140**, 603–14.

Marshall, A. J., Cannon, C. H., and Leighton, M. (2009c). Competition and niche overlap between gibbons (*Hylobates albibarbis*) and other frugivorous vertebrates in Gunung Palung National Park, West Kalimantan, Indonesia. In S. Lappan, D. Whittaker, and T. Geissmann, eds. *The Gibbons: New Perspectives on Small Ape Socioecology and Population Biology*, pp. 161–88. Springer, New York.

Marshall, A. J., and Leighton, M. (2006). How does food availability limit the population density of white-bearded gibbons? In G. Hohmann, M. M. Robbins, and C. Boesch, eds. *Feeding Ecology of the Apes and other Primates*, pp. 311–33. Cambridge University Press, Cambridge.

Marshall, A. J., and Wrangham, R. W. (2007). The ecological significance of fallback foods. *International Journal of Primatology*, **28**, 1219–35.

Marshall, T. C., Slate, J., Kruuk, L. E. B., and Pemberton, J. M. (1998). Statistical confidence for likelihood-based paternity inference in natural populations. *Molecular Ecology*, **7**, 639–55.

Martin, P., and Bateson, P. (2007). *Measuring Behaviour*, 2nd edition. Cambridge University Press, Cambridge.

Martin, R. (1976). Breeding great apes in captivity. *New Scientist*, **72**, 100–2.

Matheson, M., Sheeran, L. K., Li, J.-H., and Wagner, R. S. (2006). Tourist impact on Tibetan macaques. *Anthrozoös*, **19**, 158–86.

Mathewson, P. D., Spehar, S. N., Meijaard, E., *et al.* (2008). Evaluating orangutan census techniques using nest decay rates: Implications for population estimates. *Ecological Applications*, **18**, 208–21.

Mathy, J. W., and Isbell, L. A. (2001). The relative importance of size of food and interfood distance in eliciting aggression in captive rhesus macaques (*Macaca mulatta*). *Folia Primatologica*, **72**, 268–77.

Matsuzawa, T. (1985). Use of numbers by a chimpanzee. *Nature*, **315**, 57–9.

Mau, M., Martinho de Almeida, A., Coelho, A. V., and Sudekum, K. (2011). First identification of tannin-binding proteins in saliva of *Papio hamadryas* using MS/MS mass spectroscopy. *American Journal of Primatology*, **73**, 1–7.

Mawdsley, J. R., O'Malley, R., and Ojima, D. S. (2009). A review of climate-change adaptation strategies for wildlife management and biodiversity conservation. *Conservation Biology*, **23**, 1080–9.

Maxwell, J. C. (1860). Illustrations of the dynamical theory of gases. Part I. On the motions and collisions of perfectly elastic spheres. *Philosophical Magazine*, **19**, 19–32.

Mayden, R. L. (1997). A hierarchy of species concepts: The denouement in the saga of the species problem. In M. F. Claridge, H. A. Dawah, and M. R. Wilson, eds. *Species: The Units of Biodiversity*, pp. 381–4. Chapman and Hall, London.

Mayes, R. W. (2006). The possible application of novel marker methods for estimating dietary intake and nutritive value in primates. In G. Hohmann, M. M. Robbins, and C. Boesch, eds. *Feeding Ecology in Apes and Other Primates*, pp. 421–44. Cambridge University Press, Cambridge.

Maynard Smith, J., and Parker, G. A. (1976). The logic of asymmetric contests. *Animal Behaviour*, **24**, 159–75.

Maynard Smith, J., and Price, G. R. (1973). The logic of animal conflict. *Nature*, **246**, 15–18.

Mayr, E. (1963). *Animal Species and Evolution*. Oxford University Press, New York.

Mbete, R. A., Banga-Mboko, H., Racey, P. *et al.* (2011). Household bushmeat consumption in Brazzaville, the Republic of the Congo. *Tropical Conservation Science*, **4**, 187–202.

McCann, C. (2012). Great Apes in Zoos. In L. Penn, M. Gussett, and G. Dick, eds. *77 Years: The History and Evolution of the World Association of Zoos and Aquariums 1935–2012*. World Association of Zoos and Aquariums (WAZA), Gland, Switzerland.

McComb, K. (1992). Playback as a tool for studying contests between social groups. In P. K. McGregor, ed. *Playback and Studies of Animal Communication* Series A: Life Sciences Vol. 228, pp. 111–19. Plenum Press, New York.

McConkey, K. R., Aldy, F., Ario, A., and Chivers, D. J. (2002). Selection of fruit by gibbons (*Hylobates muelleri x agilis*) in the rain forests of central Borneo. *International Journal of Primatology*, **23**, 123–45.

McGregor, P. K. (2000). Playback experiments: design and analysis. *Acta Ethologica*, **3**, 3–8.

McGuire, J. T., Dierenfeld, E. S., Poppengga, R. H., and Brazelton, W. E. (1989). Plasma alpha-tocopherol, retinol, cholesterol, and mineral concentrations in captive gorillas. *Journal of Medical Primatology*, **18**, 155–61.

McNab, B. K. (1992). Energy expenditure: a short history. In T. E. Tomasi, and T. H. Horton, eds. *Mammalian Energetics: Interdisciplinary Views of Metabolisms and Reproduction*, pp. 1–15. Cornell University Press, Ithaca NY.

McRae, B. H. (2006). Isolation by resistance. *Evolution*, **60**, 1551–61.

McRae, B. H., and Shah, V. B. (2009). *CIRCUITSCAPE User Guide*. The University of California Santa Barbara, Santa Barbara, CA. Available at <http://www.circuitscape.org> (accessed October 2012).

Melfi, V. A., and Feistner, A. T. C. (2002). A comparison of the activity budgets of wild and captive Sulawesi crested black macaques (*Macaca nigra*). *Animal Welfare*, **11**, 213–22.

Mendes, S. L., de Melo, F. R., Boubli, J. P., *et al.* (2005). Directives for the conservation of the northern muriqui, *Brachyteles hypoxanthus* (Primates, Atelidae). *Neotropical Primates*, **13**, 7–18.

Mendoza, S. P., and Mason, W. A. (1997). Attachment relationships in New World primates. *Annals of the New York Academy of Sciences*, **807**, 203–9.

Menzel, C. (2005). Progress in the study of chimpanzee recall and episodic memory. In H. S. Terrace, and J. Metcalfe, eds. *The Missing Link in Cognition: Origins of Self-Reflective Consciousness*, pp. 188–224. Oxford University Press, New York.

Menzel, C. R., and Beck, B. B. (2000). Homing and detour behavior in golden lion tamarin social groups. In S. Boinski, and P. A. Garber, eds. *On the Move: How and Why Animals Travel in Groups*, pp. 299–326. University of Chicago Press, Chicago.

Menzel, E. W., Jr. (1973). Chimpanzee spatial memory organization. *Science*, **182**, 943–5.

Merker, S. (2003). Vom Aussterben bedroht oder anpassungsfähig? Der Koboldmaki *Tarsius dianae* in den Regenwäldern Sulawesis (Ph.D. Dissertation). Georg-August-Universitat Goettingen, Goettingen.

Merker, S., Yustian, I., and Mühlenberg, M. (2004). Losing ground but still doing well— *Tarsius dianae* in human-altered rainforests of Central Sulawesi, Indonesia. In G. Gerold, M. Fremerey, and E. Guhardja, eds. *Land Use, Nature Conservation and the Stability of Rainforest Margins in Southeast Asia*, pp. 299–311. Springer, Heidelberg.

Merker, S., Yustian, I., and Mühlenberg, M. (2005). Responding to forest degradation: altered habitat use by Dian's tarsier *Tarsius dianae* in Sulawesi, Indonesia. *Oryx*, 39, 189–95.

Merrill, R. M. (2009). *Introduction to Epidemiology*, 5th edition. Jones & Bartlett, Sudbury, MA.

Mertl-Millhollen, A. S., Moret, E. S., Felantsoa, D., *et al.* (2003). Ring-tailed lemur home ranges correlate with food abundance and nutritional content at a time of environmental stress. *International Journal of Primatology*, 24, 969–85.

Meyler, S. V., Salmona, J., Ibouroi, M. T., *et al.* (2012). Density estimates of two endangered nocturnal lemur species from northern Madagascar: new results and a comparison of commonly used methods. *American Journal of Primatology*, 74, 414–22.

Milius, S. (2005). Bushmeat on the Menu: untangling the influences of hunger, wealth, and international commerce. *Science News*, 167, 138.

Miller, C. R., Joyce, P., and Waits, L. P. (2005). A new method for estimating the size of small populations from genetic mark-recapture data. *Molecular Ecology*, 14, 1991–2005.

Milner-Gulland, E. J., Bennett, E., and the SCB bushmeat working group (2003). Wild meat: The big picture. *Trends in Ecology & Evolution*, 18, 351–7.

Milton, K. (1979). Factors influencing leaf choice by howler monkeys: a test of some hypotheses of food selection by generalist herbivores. *American Naturalist*, 114, 362–78.

Milton, K. (1980). *The Foraging Strategies of Howler Monkeys: A Study in Primate Economics*. Columbia University Press, New York.

Milton, K., and Dintzis, F. (1981). Nitrogen-to-protein conversion factors for tropical plant samples. *Biotropica*, 12, 177–81.

Milton, K., and McBee, R. H. (1983). Rates of fermentative digestion in the howler monkey, *Alouatta palliata* (Primates, Ceboidea). *Comparative Biochemistry and Physiology A-Molecular & Integrative Physiology*, 74, 29–31.

MINAE. (1997). *Lista de Especies de Fauna Silvestre con Poblaciones Reducidas y en Peligro de Extincion para Costa Rica. Decreto No. 26453-MINAE. Publicado en La Gaceta el 3 de diciembre de 1997*. Ministerio Nacional de Ambiente y Energia (MINAE), San Jose, Costa Rica.

Mineka, S., and Cook, M. (1988). Social learning and the acquisition of snake fear in monkeys. In T. Zentall, and B. Galef, eds. *Social learning: Psychological and Biological Perspectives*, pp. 51–73. Erlbaum, Hillsdale, NJ.

Minta, S. (1992). Tests of spatial and temporal interaction among animals. *Ecological Applications*, 2, 178–88.

Mitani, J. C., Gros-Louis, J., and Richards, A. F. (1996). Sexual Dimorphism, the Operational Sex Ratio, and the Intensity of Male Competition in Polygynous Primates. *The American Naturalist*, 147, 966–80.

Mitani, J. C., and Rodman, P. S. (1979). Territoriality: the relation of ranging pattern and home range size to defendability, with an analysis of territoriality among primate species. *Behavioral Ecology and Sociobiology*, 5, 241–51.

Mitani, J. C., Struhsaker, T. T., and Lwanga, J. S. (2000). Primate community dynamics in old growth forest over 23.5 years at Ngogo, Kibale National Park, Uganda: implications for conservation and census methods. *International Journal of Primatology*, 21, 269–86.

Mitani, J. C., Watts, D., and Amsler, S. J. (2010). Lethal intergroup aggression leads to territorial expansion in wild chimpanzees. *Current Biology*, 20, R507–8.

Mittermeier, R. A. (1987). Effects of hunting on rain forest primates. In C. W. Marsh, and R. A. Mittermeier, eds. *Monographs in Primatology, Vol. 9. Primate Conservation in the Tropical Rain Forest*, pp. 109–46. Alan R. Liss, Inc., New York.

Mittermeier, R. A. (1997). Foreword. In J. Wallis, ed. *Special Topics in Primatology Volume 1, Primate Conservation: The Role of Zoological Parks, pp.* xi-xiii. American Society of Primatologists, USA.

Mittermeier, R. A., and Konstant, W. R. (1996). Primate Conservation: A retrospective and look into the 21st Century. *Primate Conservation*, 17, 7–17.

Mittermeier, R. A., Konstant, W. R., and Mast, R. B. (1994). Use of Neotropical and Malagasy primate species in biomedical research. *American Journal of Primatology*, 34, 73–80.

Mittermeier, R. A., Konstant, W. R., Nicoll, M. E., and Langrand, O. (1992). *Lemurs of Madagascar: An action plan for their conservation, 1993–1999*. IUCN, Gland, Switzerland.

Mittermeier, R. A., Louis, E. E. Jr., Richardson, M., *et al.* (2010). *Conservation International Tropical Field Guide Series Lemurs of Madagascar*, 3rd edition. Pan Americana Formas e Impresos, S.A., Bogota, Columbia.

Mittermeier, R. A., Wallis, J., Rylands, A. B., *et al.* (2009). *Primates in Peril: The World's 25 Most Endangered Primates 2008–2010*. IUCN/SSC Primate Specialist Group, International Primatological Society, and Conservation International, Arlington, VA.

Moberg, G. P., and Mench, J. A. (2000). *The Biology of Animal Stress: Basic Principles and Implications for Animal Welfare*. CABI Publishing, New York.

Moorcroft, P. R., and Lewis, M. A. (2006). *Mechanistic Home Range Analysis*. Princeton University Press, Princeton, NJ.

Moore, R. S., Nekaris, K. A. I., and Eschmann, C. (2010). Habitat use by western purple-faced langurs *Trachypithecus vetulus nestor* (Colobinae) in a fragmented suburban landscape. *Endangered Species Research*, 12, 227–34.

Morales-Jimenez, A. L., and Link, A. (2006). The brown spider monkey (*Ateles hybridus*) conservation program 2006–2010. *International Journal of Primatology*. 27, 546.

Morales-Jimenez, A. L., Nekaris, K. A. I., Lee, J., and Thompson, S. (2005). Modeling distributions for Colombian spider monkeys (*Ateles* ssp.) to find priorities for conservation. *American Journal of Primatology*, 66, 131.

Morellato, L. P. C., Camargo, M. G. G., D'Eça Neves, F. F., *et al.* (2010). The influence of sampling method, sample size, and frequency of observations on plant phenological patterns and interpretatoin in tropical forest trees. In I. L. Hudson, and M. R. Keatley, eds. *Phenological Research*, pp. 99–121. Springer, Dordrecht.

Morin, P. A., Chambers, K. E., Boesch, C., and Vigilant, L. (2001). Quantitative polymerase chain reaction analysis of DNA from noninvasive samples for accurate microsatellite genotyping of wild chimpanzees (*Pan troglodytes verus*). *Molecular Ecology,* **10**, 1835–44.

Moritz, C. (1994). Defining "evolutionarily significant units" for conservation. *Trends in Ecology & Evolution,* **9**, 373–5.

Morris, W. F., and Doak, D. F. (2002). *Quantitative Conservation Biology. Theory and Practice of Population Viability Analysis.* Sinauer, Sunderland, MA.

Morrison, D. A. (2009). Why would phylogeneticists ignore computerized sequence alignment? *Systematic Biology,* **58**, 150–8.

Morton, W., Agy, M., Capuano, S., and Grant, R. (2008). Specific pathogen-free macaques: Definition, history, and current production. *Institute for Lab Animal Research Journal,* **49**, 137–44.

Moscovice, L. R., Issa, M. H., Petrelkova, K. J., *et al.* (2007). Fruit availability, chimpanzee diet, and grouping patterns on Rubondo Island, Tanzania. *American Journal of Primatology,* **59**, 487–502.

Moses, K. L., and Semple, S. (2011). Primary seed dispersal by the black-and-white ruffed lemur (*Varecia variegata*) in the Manombo forest, south-east Madagascar. *Journal of Tropical Ecology,* **27**, 529–38.

Mowat, G., and Strobeck, C. (2000). Estimating population size of Grizzly Bears using hair capture, DNA profiling and mark-recapture analysis. *Journal of Wildlife Management,* **64**, 183–93.

Muehlenbein, M. P. (2005). Parasitological analyses of the male chimpanzees (*Pan troglodytes schweinfurthii*) at Ngogo, Kibale National Park, Uganda. *American Journal of Primatology,* **65**, 167–79.

Muehlenbein, M. P. (2006). Intestinal parasite infections and fecal steroid levels in wild chimpanzees. *American Journal of Physical Anthropology,* **130**, 546–50.

Muehlenbein, M. P. (2009). The application of endocrine measures in primate parasite ecology. In M. A. Huffman, and C. A. Chapman, eds. *Primate Parasite Ecology: The Dynamics of Host-Parasite Relationships,* pp. 63–81. Cambridge University Press, Cambridge.

Muehlenbein, M. P. (2010). Do the benefits of primate tourism outweigh the costs of potential anthropozoonoses and stressed animals? 23rd Congress of the International Primatological Society, Kyoto, Japan.

Muehlenbein, M. P., and Ancrenaz, M. (2009). Minimizing pathogen transmission at primate ecotourism destinations: The need for input from travel medicine. *Journal of Travel Medicine,* **16**, 229–32.

Muehlenbein, M. P., Martinez, L. A., Lemke, A. A., *et al.* (2010). Unhealthy travelers present challenges to sustainable ecotourism. *Travel Medicine and Infectious Disease,* **8**, 169–75.

Muehlenbein, M. P., Martinez, L. A., Lemke, A. A., *et al.* (2008). Perceived vaccination status in ecotourists and risks of anthropozoonoses. *EcoHealth,* **5**, 371–8.

Muehlenbein, M. P., Schwartz, M., and Richard, A. (2003). Parasitological analyses of the sifaka (*Propithecus verreauxi verreauxi*) at Beza Mahafaly, Madagascar. *Journal of Zoo and Wildlife Medicine,* **34**, 274–7.

Muehlenbein, M. P., and Watts, D. P. (2010). The costs of dominance: testosterone, cortisol and intestinal parasites in wild male chimpanzees. *BioPsychoSocial Medicine,* **4**, 21.

Mueller-Dombois, D. and Ellenberg, H. (1974). *Aims and Methods of Vegetation Ecology*. Wiley, London.

Muir, W. W. and Mason, D. E. (1993). Effects of diazepam, acepromazine, detomidine, and xylazine on thiamylal anesthesia in horses. *Journal of the American Veterinary Medical Association*, **203**, 1031–8.

Müller, E. F., Kamau, J. M., Maloiy, G. M. (1983). A comparative study of basal metabolism and thermoregulation in a folivorous (*Colobus guereza*) and an omnivorous (*Cercopithecus mitis*) primate species. *Comparative Biochemistry and Physiology*, **74**, 319–22.

Muller, A., and Thalmann, U. (2000). Origin and evolution of primate social organisation: a reconstruction *Biological Reviews*, **75**, 405–35.

Muller, M. N., and Wrangham, R. W. (2004). Dominance, aggression and testosterone in wild chimpanzees: a test of the "challenge hypothesis." *Animal Behaviour*, **67**, 113–23.

Mullin, M. H. (1999). Mirrors and windows: Sociocultural studies of human-animal relationships. *Annual Review of Anthropology*, **28**, 201–24.

Mundry, R., and Nunn, C. L. (2009). Stepwise model fitting and statistical inference: turning noise into signal pollution. *The American Naturalist*, **173**, 119–23.

Munro, C. J., Stabenfeldt, G. H., Cragun, J. R., *et al.* (1991). Relationship of serum estradiol and progesterone concentrations to the excretion profiles of their major urinary metabolites as measured by enzyme-immunoassay and radioimmunoassay. *Clinical Chemistry*, **37**, 838–44.

Muoria, P. K., Karere, G. M., Moinde, N. N., and Suleman, M. A. (2003). Primate census and habitat evaluation in the Tana delta region, Kenya. *African Journal of Ecology*, **41**, 157–63.

Murnane, R. D., Zdziarski, J. M., Walsh, T. F., *et al.* (1996). Melengestrol Acetate-Induced Exuberant Endometrial Decidualization in Goeldi's Marmosets (*Callimico goeldii*) and Squirrel Monkeys (*Saimiri sciureus*). *Journal of Zoo and Wildlife Medicine*, **27**, 315–24.

Murphy, H. M., Dennis, P., Devlin, W., *et al.* (2011). Echocardiographic parameters of captive western lowland gorillas (*Gorilla gorilla gorilla*). *Journal of Zoo and Wildlife Medicine*, **42**, 572–9.

Myers, D. M., and Wright, P. C. (1993). Resource tracking: food availability and *Propithecus* seasonal reproduction. In P. M. Kappeler, and J. U. Ganzhorn, eds. *Lemur Social Systems and their Ecological Basis*, pp. 179–92. Plenum Press, New York.

Myneni, R. B., Hall, F. G., Sellers, P. J., and Marshak, A. L. (1995). The interpretation of spectral vegetation indexes. *IEEE Transactions on Geoscience and Remote Sensing*, **33**, 481–6.

Nadler, T., Vu Ngoc Thanh, and Streicher, U. (2007). Conservation status of Vietnamese primates. *Vietnamese Journal of Primatology*, **1**, 7–26.

Nagy, K. A. (1980). CO_2 production in animals: analysis of potential errors in the doubly labelled water method. *American Journal of Physiology*, **238**, R466–73.

Nagy, K. A. (1983). *Doubly-Labelled Water: A Guide to its Use*. UCLA publications, Los Angeles, CA.

Nash, L. (2007). Moonlight and behavior in nocturnal and cathemeral primates, especially *Lepilemur leucopus*: illuminating possible anti-predator efforts primate anti-predator strategies. In S. L. Gursky, and K. A. I. Nekaris, eds. *Primate Anti-Predator Strategies*, pp. 173–205. Springer, New York.

Nathan, R., Getz, W. M., Revilla, E., *et al.* (2008). A movement ecology paradigm for unifying organismal movement research. *Proceedings of the National Academy of Sciences USA*, **105**, 19052–9.

National Research Council. (1981). *Techniques for the Study of Primate Population Ecology*. National Academy Press, Washington, DC.

National Research Council. (2003). *Nutrient Requirements of Nonhuman Primates*, 2nd edition. National Academy Press, Washington, DC.

Naughton-Treves, L., Treves, A., Chapman, C., and Wrangham, R. (1998). Temporal patterns of crop-raiding by primates: Linking food availability in croplands and adjacent forest. *Journal of Applied Ecology*, **35**, 596–606.

Nchanji, A. C., and Plumptre, A. J. (2001). Elephant dung decay rates in Cameroon: problems for census methods. *African Journal of Ecology*, **39**, 24–32.

Neilson, E. (2010). Estimating Pileated Gibbon (Hylobates pileatus) Occupancy in the Cardamom Mountains, Cambodia (Masters thesis), Oxford Brookes University, Oxford.

Nekaris, K. A. I. (2001). Activity budget and positional behavior of the Mysore Slender Loris (*Loris tardigradus lydekkerianus*): implications for slow climbing locomotion. *Folia Primatologica*, **72**, 228–41.

Nekaris, K. A. I. (2003). Observations on mating, birthing and parental care in three taxa of slender loris in India and Sri Lanka (*Loris tardigradus* and *Loris lydekkerianus*). *Folia Primatologica*, **74**, 312–36.

Nekaris, K. A. I. (2006). Horton Plains slender loris, Ceylon mountain slender loris *Loris tardigradus nycticeboides* (Hill, 1942). In R. A. Mittermeier, C. Valladares-Padua, A. B. Rylands, *et al.*, eds. *Primates in Peril: The World's 25 Most Endangered Primates, 2004–2006*, pp.10–11.

Nekaris, K. A. I., Blackham, G., and Nijman, V. (2008). Conservation implications of low encounter rates of five nocturnal primate species (*Nycticebus* sp.) in Southeast Asia. *Biodiversity and Conservation*, **17**, 733–47.

Nekaris, K. A. I., and Nijman, V. (2007). CITES proposal highlights rarity of Asian nocturnal primates (Lorisidae: *Nycticebus*). *Folia Primatologica*, **78**, 211–14.

Nekaris, K. A. I., Shepherd, C. R., Starr, C. R., and Nijman, V. (2010). Exploring cultural drivers for wildlife trade via an ethnoprimatological approach: a case study of slender and slow lorises (*Loris* and *Nycticebus*) in South and Southeast Asia. *American Journal of Primatology*, **72**, 877–86.

Nelson, R. J. (2011). *An Introduction to Behavioral Endocrinology*, 4th edition. Sinauer Associates, Inc., Sunderland, MA.

Neri-Arboleda, I., Stott, P., and Arboleda, N. P. (2002). Home ranges, spatial movements, and habitat associations of the Phillippine tarsier (*Tarsius syrichta*) in Corella, Bohol. *Journal of Zoology*, **257**, 387–402.

Newstrom, L. E., Frankie, G. W., and Baker, H. G. (1994). A new classification for plant phenology based on flowering patterns in lowland tropical rain forest trees at La Selva, Costa Rica. *Biotropica*, **26**, 141–59.

Newton-Fisher, N. E., and Lee, P. C. (2011). Grooming reciprocity in wild male chimpanzees. *Animal Behaviour*, **81**, 439–46.

Newton-Fischer, N. E., Reynolds, V., and Plumptre, A. J. (2000). Food supply and chimpanzee (*Pan troglodytes schweinfurthii*) party size in the Budongo Forest Reserve, Uganda. *International Journal of Primatology*, **21**, 613–28.

Neyman, P. (1979). Ecology and Social Organization of the Cotton-top Tamarin (*Saguinus oedipus*) (Ph.D. Dissertation), University of California, Berkeley.

Nguyen, N., Gesquiere, L., Alberts, S. C., and Altmann, J. (2012). Sex differences in the mother-neonate relationship in wild baboons: social, experiential and hormonal correlates. *Animal Behaviour*, **83**, 891–903.

Nguyen, N., Gesquiere, L. R., Wango, E. O., *et al.* (2008). Late pregnancy glucocorticoid levels predict responsiveness in wild baboon mothers (*Papio cynocephalus*). *Animal Behaviour*, **75**, 1747–56.

Nguyen Van Song. (2008). Wildlife trading in Vietnam: situation, causes, and solutions. *Journal of Environment and Development*, **17**, 145–165.

Nicholls, J. A., and Goldizen, A. W. (2006). Habitat type and density influence vocal signal design in satin bowerbirds. *Journal of Animal Ecology*, **75**, 549–58.

Nicholson, A. J. (1954). An outline of the dynamics of animal populations. *Australian Journal of Zoology*, **2**, 9–65.

Nielsen, M. R., and Treue, T. (2012). Hunting for the benefits of joint forest management in the eastern Afromontane Biodiversity Hotspot: effects on bushmeat hunters and wildlife in the Udzungwa Mountains. *World Development*, **40**, 1224–39.

Nijman, V. (2005). *Hanging in the Balance: An Assessment of Trade in Orang-utans and Gibbons on Kalimantan, Indonesia*. TRAFFIC Southeast Asia, Cambridge, UK.

Nijman, V. (2009). *An Assessment of Trade in Gibbons and Orangutans in Sumatra, Indonesia*. TRAFFIC Southeast Asia, Cambridge, UK.

Nijman, V., and Nekaris, K. A. I. (2010). Testing a model for predicting primate crop-raiding using crop- and farm-specific risk values. *Applied Animal Behaviour Science*, **127**, 125–9.

Nijman, V., Nekaris, K. A. I., Donati, G., *et al.* (2011). Primate conservation: measuring and mitigating trade in primates. *Endangered Species Research*, **13**, 159–61.

Niklas, K. J. (1993). The allometry of plant reproductive biomass and stem diameter. *American Journal of Botany*, **80**, 461–7.

Nishida, T., Corp, N., Hamai, M., *et al.* (2003). Demography, female life history, and reproductive profiles among the chimpanzees of Mahale. *American Journal of Primatology*, **59**, 99–121.

Noë, R., and Hammerstein, P. (1995). Biological markets. *Trends in Ecology & Evolution*, **10**, 336–40.

Noser, R., and Byrne, R. W. (2010). How do wild baboons (*Papio ursinus*) plan their routes? Travel among multiple high-quality food sources with inter-group competition. *Animal Cognition*, **13**, 145–55.

Nowack, J., Mzilikazi, N., and Dausmann, K. H. (2010). Torpor on Demand: Heterothermy in the Non-Lemur Primate *Galago moholi*. *PLoS ONE*, **5**, 1–6.

Nsubuga, A. M., Robbins, M. M., Roeder, A. D., *et al.* (2004). Factors affecting the amount of genomic DNA extracted from ape faeces and the identification of an improved sample storage method. *Molecular Ecology*, **13**, 2089–94.

Nunn, C. L. (2000). Collective action, free-riders, and male extra-group conflicts. In P. Kappeler, ed. *Primate Males*, pp. 192–204. Cambridge University Press, Cambridge.

Nunn, C. (2009). Using agent-based models to investigate primate disease ecology. In M. A. Huffman, and C. A. Chapman, eds. *Primate Parasite Ecology: The Dynamics and Study of Host-Parasite Relationships*, pp. 83–110. Cambridge University Press, Cambridge.

Nunn, C., and Altizer, S. (2004). Sexual selection, behaviour and sexually transmitted diseases. In P. M. Kappeler, and C. P. van Schaik, eds. *Sexual Selection in Primates: New and Comparative Perspectives.* Cambridge University Press, Cambridge.

Nunn, C., and Altizer, S. (2006). *Infectious Diseases in Primates: Behavior, Ecology and Evolution.* Oxford University Press, New York.

O'Brien, J. (2006). Animal conservation genetics—an overview with relevance to captive breeding programmes. *EAZA News*, 57, 26.

O'Connell, T., and Hedges, R. (1999). Investigations into the effect of diet on modern human hair isotopic values. *American Journal of Physical Anthropology*, 108, 409–25.

O'Leary, M. (1981). Carbon isotope fractionation in plants. *Phytochemistry*, 20, 553–67.

Oates, J. F. (1977). The guereza and man: how man has affected the distribution and abundance of *Colobus guereza* and other black colobus monkeys. In G. H. Bourne, ed. *Primate Conservation.* Academic Press, London and New York.

Oates, J. F. (1996). *African Primates. Status Survey and Conservation Action Plan.* IUCN, Gland, Switzerland.

Oates, J. F., Davies, A. G., and Delson, E. (1994). The diversity of living colobines. In A. G. Davies, and J. F. Oates, eds. *Colobine Monkeys: Their Ecology, Behaviour and Evolution*, 45–73. Cambridge University Press.

Oelze, V., Fuller, B., Richards, M., *et al.* (2011). Exploring the contribution and significance of animal protein in the diet of bonobos by stable isotope ratio analysis of hair. *Proceedings of the National Academy of Sciences USA*, 108, 9792–7.

Odling-Smee, F. J., Laland, K. N., and Feldman, M. W. (2003). *Niche Construction: The Neglected Process in Evolution.* Princeton University Press, Princeton, NJ.

Ohsawa, H. (2003). Long-term study of the social dynamics of patas monkeys (*Erythrocebus patas*): group male supplanting and changes to the multi-male situation. *Primates*, 44, 99–107.

Oliviera, M. M., Marini-Filho, O. J., and Campos, V. O. (2005). The international committee for the conservation and management of Atlantic Forest atelids. *Neotropical Primates*, 13, 101–4.

Onibala, J. S. I. T., and Laatung, S. (2007). Bushmeat hunting in North Sulawesi and related conservation strategies (a case study at the Tangkoko Nature Reserve). *Journal of Agriculture and Rural Development*, S90, 110–16.

Onyango, P. O., Gesquiere, L. R., Wango, E. O., *et al.* (2008). Persistence of maternal effects in baboons: Mother's dominance rank at son's conception predicts stress hormone levels in subadult males. *Hormones and Behavior*, 54, 319–24.

Ortmann, S., Bradley, B., Stolter, C., and Ganzhorn, J. U. (2006). Estimating the quality and composition of wild animal diets: a critical survey of methods. In G. Hohmann, M. M. Robbins, and C. Boesch, eds. *Feeding Ecology in Apes and Other Primates*, pp. 396–420. Cambridge University Press, Cambridge.

Ostfeld, R. S. (1990). The ecology of territoriality in small mammals. *Trends in Ecology & Evolution*, 5, 411–15.

Ostner, J., and Kappeler, P. M. (2004). Male life history and the unusual adult sex ratios of redfronted lemur, *Eulemur fulvus rufus*, groups. *Animal Behaviour*, **67**, 249–59.

Ottoni, E. B., and Izar, P. (2008). Capuchin monkey tool use: Overview and implications. *Evolutionary Anthropology*, **17**, 171–8.

Overdorff, D. J. (1992). Differential Paterns in Flower Feeding by *Eulemur fulvus rufus* and *Eulemur rubriventer* in Madagascar. *American Journal of Primatology*, **28**, 191–203.

Owusu-Ansah, N. (2010). Evaluation of Wildlife Hunting Restrictions on Bushmeat Trade in Five Major Markets around Digya National Park (Masters thesis), University of Cape Coast, Ghana.

Paetkau, D. (2003). An empirical exploration of data quality in DNA-based population inventories. *Molecular Ecology*, **12**, 1375–87.

Palme, R. (2005). Measuring fecal steroids—Guidelines for practical application. Bird hormones and bird migrations: analyzing hormones in droppings and egg yolks and assessing adaptations in long-distance migration. *Annals of the New York Academy of Sciences*, **1046**, 75–80.

Parker, L., Nijman, V., and Nekaris, K. A. I. (2008). When there is no forest left: fragmentation, local extinction, and small population sizes in the Sri Lankan western purple-faced langur. *Endangered Species Research*, **5**, 29–36.

Parmesan, C. (2006). Ecological and evolutionary responses to recent climate change. *Annual Review of Ecology, Evolution, and Systematics*, **37**, 637–69.

Parmesan, C., Ryrholm, N., Stefanescu, C., *et al.* (1999). Poleward shifts in geographical ranges of butterfly species associated with regional warming. *Nature*, **399**, 579–83.

Parmesan, C., and Yohe, G. (2003). A globally coherent fingerprint of climate change impacts across natural systems. *Nature*, **421**, 37–42.

Paterson, J. D., and Wallis, J., eds. (2005). *Commensalism and conflict: The Human-Primate Interface* (Vol. 4). American Society of Primatologists, Norman, OK.

Pearce, J. L., Cherry, K., Drielsma, M., Ferrier, S. and Whish, G. (2001). Incorporating expert opinion and fine-scale vegetation mapping into statistical models of faunal distribution. *Journal of Applied Ecology*, **38**, 412–24.

Pearson, R. G., Raxworthy, C. J., Nakmura, M., and Peterson, A. T. (2007). Predicting species distributions from small numbers of occurrence records: a test case using cryptic geckos in Madagascar. *Journal of Biogeography*, **34**, 102–17.

Peck, M. R., Tirira D., Thorne J., *et al.* (2011). Focusing conservation efforts for the critically endangered brown-headed spider monkey (*Ateles fusciceps*) using remote sensing, modeling and playback survey methods. *International Journal of Primatology*, **32**, 134–48.

Pepper, J. W., Mitani, J. C., and Watts, D. P. (1999). General gregariousness and specific social preferences among wild chimpanzees. *International Journal of Primatology*, **20**, 613–32.

Pepperberg, I. M. (2002). *The Alex Studies: Cognitive and Communicative Abilities of Grey Parrots*. Harvard University Press, Cambridge, MA.

Peres, C. A. (1989). Costs and benefits of territorial defense in wild golden lion tamarins, *Leontopithecus rosalia*. *Behavioral Ecology and Sociobiology*, **25**, 227–33.

Peres, C. A. (2000). Effects of subsistence hunting on vertebrate community structure in Amazonian forests. *Conservation Biology*, **14**, 240–53.

Perry, J., Bridgwater, D. D., and Horsemen, D. L. (1972). Captive Propagation: A Progress Report. *Zoologica*, 57, 109–17.

Peterson, A. T. (2001). Predicting species' geographic distributions based on ecological niche modeling. *Condor*, 103, 599–605.

Peterson, A. T., Soberón, J., Pearson, R. G., *et al.* (2011). *Ecological Niches and Geographic Distributions*. Princeton University Press, Princeton, NJ.

Petrů, M., Špinka, M., Charvátová V., and Lhota, S. (2009). Revisiting Play Elements and Self-Handicapping in Play. *Journal of Comparative Psychology*, 123, 250–63.

Petter, J. J. (1977). The aye-aye. In Prince Rainier, and G. Bourne, eds. *Primate Conservation*, pp. 37–57. Academic Press, New York.

Petter, J. J., and Peyrieras, A. (1970). Nouvelle contribution à l'étude d'un lémurien malgache, le aye-aye (*Daubentonia madagascariensis* E. Geoffroy). *Mammalia*, 34, 167–93.

Pettorelli, N., Ryan, S., Mueller, T., *et al.* (2011). The normalized difference vegetation index (NDVI): unforeseen successes in animal ecology. *Climate Research*, 46, 15–27.

Phoenix, C. H., Goy, R. W., Gerall, A. A., and Young, W. C. (1959). Organizing action of prenatally administered testosterone propionate on the tissues mediating mating behavior in the female guinea pig. *Endocrinology*, 65, 369–82.

Phillips, K. A., Elvey, C. R., and Abercrombie, C. L. (1998). Applying GPS to the study of primate ecology: a useful tool? *American Journal of Primatology*, 46, 167–72.

Phillips, S. J., Anderson, R. P. and Schapire, R. E. (2006). Maximum entropy modeling of species geographic distributions. *Ecological Modeling*, 190, 231–59.

Phillips, S. J., Dudík, M., Elith, J., *et al.* (2009). Sample selection bias and presence-only distribution models: implications for background and pseudoabsence data. *Ecological Applications*, 19, 181–97.

Phillips-Conroy, J. E., Jolly, C. J., and Brett, F. L. (1991). Characteristics of hamadryas-like male baboons living in anubis baboon troops in the Awash Hybrid Zone, Ethiopia. *American Journal of Physical Anthropology*, 86, 353–68.

Piggott, M. P., Bellemain, E., Taberlet, P., and Taylor, A. C. (2004). A multiplex pre-amplification method that significantly improves microsatellite amplification and error rates for faecal DNA in limiting conditions. *Conservation Genetics*, 5, 417–20.

Pinder, N. J., and Barkham, J. P. (1978). An assessment of the contribution of captive breeding to the conservation of rare mammals. *Biological Conservation*, 13, 187–245.

Piry, S., Luikart, G., and Cornuet, J. M. (1999). Bottleneck: a computer program for detecting recent reductions in the effective size using allele frequency data. *Journal of Heredity*, 90, 502–3.

Platt, M. L., and Ghazanfar, A. A. (2010). *Primate Neuroethology*. Oxford University Press, New York.

Pledger, S., and Geange, S. W. (2009). *Niche Overlap: A unified definition and analysis for data of different types. School of Mathematics, Statistics and Operations Research, Research Report 2009–05*. Victoria University of Wellington, Wellington, New Zealand.

Plumptre, A. J. (2000). Monitoring mammal populations with line transect techniques in African forests. *Journal of Applied Ecology*, 37, 356–68.

Plumptre, A. J., and Reynolds, V. (1996). Censusing chimpanzees in the Budongo forest. *International Journal of Primatology*, 17, 85–99.

Plumptre, A. J., and Cox, D. (2005). Counting primates for conservation: primate surveys in Uganda. *Primates, 47*, 65–73.

Pollick, A. S., Gouzoules, H., and de Waal, F. B. M. (2005). Audience effects on food calls in captive brown capuchin monkeys (*Cebus apella*). *Animal Behavior, 70*, 1273–81.

Porter, L. M., Garber, P. A., and Nacimento, E. (2009). Exudates as a fallback food for *Callimico goeldii*. *American Journal of Primatology, 71*, 120–9.

Posada, D. (2008). jModelTest: Phylogenetic model averaging. *Molecular Biology and Evolution, 25*, 1253–6.

Poss, S. R., and Rochat, P. (2003). Referential Understanding of Videos in Chimpanzees (*Pan troglodytes*), Orangutans (*Pongo pygmaeus*), and Children (*Homo sapiens*). *Journal of Comparative Psychology, 117*, 420–8.

Possingham, H. P., and Davies, I. (1995). ALEX: A model for the viability analysis of spatially structured populations. *Biological Conservation, 73*, 143–50.

Poti, P. (2000). Aspects of spatial cognition in capuchins (*Cebus apella*): frames of reference and scale of space. *Animal Cognition, 3*, 69–77.

Poulin, R. (2006). *Evolutionary Ecology of Parasites*, 2nd edition. Princeton University Press, Princeton.

Poulsen, J. R., Clark, C. J., and Smith, T. B. (2001). Seasonal variation in the feeding ecology of the Grey-Cheeked Mangabey (*Lophocebus albigena*) in Cameroon. *American Journal of Primatology, 54*, 91–105.

Poulsen, J. R., Clark, C. J., Mavah, G., and Elkan, P. W. (2009). Bushmeat supply and consumption in a tropical logging concession in northern Congo. *Conservation Biology, 23*, 1597–608.

Povey, K. D., and Rios, J. (2002). Using interpretive animals to deliver affective messages in zoos. *Journal of Interpretation Research, 7*, 19–28.

Powell, R. A. (2000). Animal home range and territories and home-range estimators. In L. Boitani, and T. K. Fuller, eds. *Research techniques in animal ecology: controversies and consequences*, pp. 65–110. Columbia University Press, New York.

Pozo-Montiy, G., Serio-Silva, J. C., Bonilla-Sánchez, Y. M., *et al.* (2008). Current status of the habitat and population of the black howler monkey (*Alouatta pigra*) in Balancán, Tabasco, Mexico. *American Journal of Primatology, 70*, 1169–76.

Priston, N. E. C. (2005). Crop-Raiding by *Macaca ochreata brunnescens* in Sulawesi: Reality, Perceptions and Outcomes for Conservation. (Ph.D. Dissertation), University of Cambridge, Cambridge.

Priston, N. E. C. (2009). Exclosure plots as a mechanism for quantifying damage to crops by primates. *International Journal of Pest Management, 55*, 243–9.

Priston, N. E. C., and Underdown, S. J. (2009). A simple method for calculating the likelihood of crop damage by primates: An epidemiological approach. *International Journal of Pest Management, 55*, 51–6.

Pritchard, J. K., Stephens, M., and Donnelly, P. (2000). Inference of population structure using multilocus genotype data. *Genetics, 155*, 945–59.

Pruetz, J. D. (1999). Socioecology of Adult Female Vervet (*Cercopithecus aethiops*) and Patas Monkeys (*Erythrocebus patas*) in Kenya: Food Availablity, Feeding Competition, and Dominance Relationships (Ph.D. Dissertation), University of Illinois, Urbana, IL.

Prugnolle, F., Durand, P., Neel, C., *et al.* (2010). African great apes are natural hosts of multiple related malaria species, including *Plasmodium falciparum*. *Proceedings of the National Academy of Sciences USA,* **107**, 1458–63.

Pusey, A. E., Oehlert, G. W., Williams, J. M., and Goodall, J. (2005). Influence of ecological and social factors on body mass of wild chimpanzees. *International Journal of Primatology,* **26**, 3–31.

Pyke, G. H., Pulliam, H. R., and Charnov, E. L. (1977). Optimal foraging: a selective review of theory and tests. *Quarterly Review of Biology,* **52**, 137–54.

Queller, D. C., and Goodnight, K. F. (1989). Estimating relatedness using genetic markers. *Evolution,* **43**, 258–75.

Quemere, E., Crouau-Roy, B., Rabarivola, C., Louis, E. E., and Chikhi, L. (2010). Landscape genetics of an endangered lemur (*Propithecus tattersalli*) within its entire fragmented range. *Molecular Ecology,* **19**, 1606–21.

R Core Development Team. (2011). *R: A language and environment for Statistical Computing.* R Foundation for Statistical Computing, Vienna, Austria.

Radespiel, U., Cepok, S., Zietemann, V., and Zimmermann, E. (1998). Sex-specific usage patterns of sleeping sites in grey mouse lemurs (*Microcebus murinus*) in northwestern Madagascar. *American Journal of Primatology,* **46**, 77–84.

Railsback S. F., and Grimm, V. (2011). *Agent-Based and Individual-Based Modeling: A Practical Introduction.* Princeton University Press, Princeton, NJ.

Railsback, S. F., Lytinen, S. L., and Jackson, S. K. (2006). Agent-based simulation platforms: review and development recommendations. *Simulation-Transactions of the Society for Modeling and Simulation International,* **82**, 609–23.

Ramos-Fernandez, G., Boyer, D., and Gomez, V. P. (2006). A complex social structure with fission-fusion properties can emerge from a simple foraging model. *Behavioral Ecology and Sociobiology,* **60**, 536–49.

Randrianandrianina, F. H., Racey, P. A., and Jenkins, R. K. B. (2010). Hunting and consumption of mammals and birds by people in urban areas of western Madagascar. *Oryx,* **44**, 411–15.

Rao, M., Myint, T., Zaw, T., and Htun, S. (2005). Hunting patterns in tropical forests adjoining the Hkakaborazi National Park, north Myanmar. *Oryx,* **39**, 292–300.

Rao, M., Zaw, T., Htun, S., and Myint, T. (2011). Hunting for a living: wildlife trade, rural livelihoods and declining wildlife in the Hkakaborazi National Park, North Myanmar. *Environmental Management,* **48**, 158–67.

Rasmussen, D. R. (1980). Clumping and consistency in primate patterns of range use—definitions, sampling, assessment and applications. *Folia Primatologica,* **34**, 111–39.

Ravaloharimanitra, M., Ratolojanahary, T., Rafalimandimby, J., *et al.* (2011). Gathering local knowledge in Madagascar results in a major increase in the known range and number of sites for critically endangered greater bamboo lemurs (*Prolemur simus*). *International Journal of Primatology,* **32**, 776–92.

Raynal, D. J., Gibbs, J. P., Ringler, N. H., and Leopold, D. J. (1998). Ecological surveys: the basis for natural area management. In J. P. Gibbs, M. L. Hunter Jr., and E. J. Sterling, eds. *Problem-Solving In Conservation Biology And Wildlife Management: Exercises For Class, Field, And Laboratory,* pp. 141–60. Blackwell Science, Massachusetts.

Refisch, J., and Koné, I. (2005). Market hunting in the Taï Region, Côte d'Ivoire and implications for monkey populations. *International Journal of Primatology*, **26**, 621–9.

Reid, J., Morra, W., Posa Bohome, C., and Fernandez Sobrado, D. (2005). *The economics of the primate trade in Bioko, Equatorial Guinea.* Report to Conservation International, Washington, D.C.

Remis, M. J. (1997). Western lowland gorillas (*Gorilla gorilla gorilla*) as seasonal frugivores: Use of variable resources. *American Journal of Primatology*, **43**, 87–109.

Ren, B., Li, D., Garber, P. A., and Li, M. (2012). Fission–Fusion Behavior in Yunnan Snub-Nosed Monkeys (*Rhinopithecus bieti*) in Yunnan, China. *International Journal of Primatology*, **33**, 1096–1109.

Rendall, D., Rodman, P. S., and Emond, R. E. (1996). Vocal recognition of individuals and kin in free-ranging rhesus monkeys. *Animal Behaviour,* **51**, 1007–15.

Reynolds, V., Plumptre, A. J., Greenham, J., and Harborne, J. (1998). Condensed tannins and sugars in the diet of chimpanzees (*Pan troglodytes schweinfurthii*) in the Budongo Forest, Uganda. *Oecologia,* **115**, 331–6.

Richard, A. F., Dewar, R. E., Schwartz, M., and Ratsirarson J. (2002). Life in the slow lane? Demography and life histories of male and female sifaka (*Propithecus verreauxi verreauxi*). *Journal of Zoology*, **256**, 421–36.

Richards, P. W. (1996). *The Tropical Rain Forest*, 2nd edition. Cambridge University Press, Cambridge.

Richards, S. A., Whittingham, M. J., and Stephens, P. A. (2011). Model selection and model averaging in behavioural ecology: the utility of the IT-AIC framework. *Behavioral Ecology and Sociobiology,* **65**, 77–89.

Riley, E. P. (2006). Ethnoprimatology: toward reconciliation between biological and cultural anthropology. *Ecological and Environmental Anthropology*, **2**, 75–86.

Riley, E. P. (2007). The human-macaque interface: Conservation implications of current and future overlap and conflict in Lore Lindu National park, Sulawesi, Indonesia. *American Anthropologist,* **109**, 473–84.

Riley, E. P. (2010). The importance of human-macaque folklore for conservation in Lore Lindu National Park, Sulawesi, Indonesia. *Oryx*, **44**, 235–40.

Riley, E. P., and Fuentes, A. (2011). Conserving social-ecological systems in Indonesia: Human-nonhuman primate interconnections in Bali and Sulawesi. *American Journal of Primatology*, **73**, 62–74.

Riley, E. P., Fuentes, A., and Wolfe, L. (2011). Ethnoprimatology: Contextualizing human and nonhuman primate interactions. In C. Campbell, A. Fuentes, K. MacKinnon, *et al.*, eds. *Primates in Perspective,* 2nd edition, pp. 676–86. Oxford University Press, New York.

Robbins, M. M. (1996). Male-male interactions in heterosexual and all-male wild mountain gorilla groups. *Ethology,* **102**, 942–65.

Robbins, M. M., Gray, M., Kagoda, E., and Robbins, A. M. (2009). Population dynamics of the Bwindi mountain gorillas. *Biological Conservation*, **142**, 2886–95.

Roberts, E. K., Lu, A., Bergman, T. J., and Beehner, J. C. (2012). A Bruce effect in wild geladas. *Science,* **335**, 1222–5.

Robinson, J. G. (1988). Group size in wedge-capped capuchin monkeys *Cebus olivaceus* and the reproductive success of males and females. *Behavioral Ecology and Sociobiology*, **23**, 187–97.

Rocha, V. J., Aguiar, L. M., Ludwig, G., *et al.* (2007). Techniques and trap models for capturing wild tufted capuchins. *International Journal of Primatology, 28*, 231–43.

Rodgers, A. R. (2001). Recent telemetry technology. In J. J. Millspaugh, and J. M. Marzluff, eds. *Radio Tracking and Animal Populations*, pp. 82–121. Academic Press, San Diego, CA.

Rodrigues, A., Pilgrim, J., Lamoreaux, J., *et al.* (2006). The value of the IUCN Red List for conservation. *Trends in Ecology & Evolution, 21*, 71–6.

Rodriguez, J. P., Brotons, L., Bustamante, J., and Seoane, J. (2007). The application of predictive modeling of species distribution to biodiversity conservation. *Diversity and Distributions, 13*, 243–51.

Rogers, C. (2011). *The Story of Mountain Gorilla Tourism: Finding a Place for Endangered Species in a Human Dominated Landscape. Case study*. Network for Conservation Educators and Practitioners, Center for Biodiversity and Conservation, American Museum of Natural History, New York. Available from <http://ncep.amnh.org/> (accessed October 2012).

Rogers, M. E., Abernethy, K., Bermejo, M., *et al.* (2004). Western gorilla diet: A synthesis from six sites. *American Journal of Primatology, 64*, 173–92.

Rogers, M. E., Abernethy, K. A., Fontaine, B., *et al.* (1996). Ten days in the life of a mandrill horde in the Lope Reserve, Gabon. *American Journal of Primatology, 40*, 297–313.

Romero, L. M. (2004). Physiological stress in ecology: lessons from biomedical research. *Trends In Ecology & Evolution, 19*, 249–55.

Ronquist, F., and Huelsenbeck, J. P. (2003). MRBAYES 3: Bayesian phylogenetic inference under mixed models. *Bioinformatics, 19*, 1572–4.

Root, T. L., Price, J. T., Hall, K. R., *et al.* (2003). Fingerprints of global warming on wild animals and plants. *Nature, 421*, 57–60.

Rose, L. M. (2000). Behavioral sampling in the field: continuous focal versus focal interval sampling. *Behaviour, 137*, 153–80.

Rose, A. L. (2011). Bonding, biophilia, biosynergy, and the future of primates in the wild. *American Journal of Primatology, 73*, 245–52.

Rosenberger, A. L., and Kinzey, W. G. (1976). Functional patterns of molar occlusion in platyrrhine primates: comparative study. *American Journal of Physical Anthropology, 45*, 281–98.

Rosenzweig, C., Karoly, D., Vicarelli, M., *et al.* (2008). Attributing physical and biological impacts to anthropogenic climate change. *Nature, 453*, 353–7.

Rothman, J. M., Chapman, C. A., Twinomugisha, D., *et al.* (2008). Measuring physical traits of primates remotely: the use of parallel lasers. *American Journal of Primatology, 70*, 1191–5.

Rothman, J. M., Chapman, C. A., Hansen, J. L., *et al.* (2009b). Rapid assessment of the nutritional value of foods eaten by mountain gorillas: applying near-infrared reflectance spectroscopy to primatology. *International Journal of Primatology, 30*, 729–42.

Rothman, J. M., Chapman, C. A., and van Soest, P. J. (2012). Methods in primate nutritional ecology: a user's guide. *International Journal of Primatology, 33*, 542–66.

Rothman, J. M., Dierenfeld, E. S., Hintz, H. F., and Pell, A. N. (2008). Nutritional quality of gorilla diets: consequences of age, sex and season. *Oecologia, 155*, 111–22.

Rothman, J. M., Dusinberre, K., and Pell, A. N. (2009a). Condensed tannins in the diets of primates: A matter of methods. *American Journal of Primatology, 71*, 70–6.

Rothman, J. M., Raubenheimer, D., and Chapman, C. A. (2011). Nutritional geometry: gorillas prioritize non-protein energy while consuming surplus protein. *Biology Letters*, 7, 847–9.

Rothman, J. M., Van Soest, P. J., and Pell, A. N. (2006). Decaying wood is a sodium source for mountain gorillas. *Biology Letters* 2, 321–4.

Rothman, K. J., ed. (1988). *Causal Inference*. Epidemiology Resources Inc, Newton Lower Falls, MA.

Rothman, K. J., Greenland, S., and Lash, T. L. (2009). *Modern Epidemiology*, 3rd edition. Lippincott Williams & Wilkins, Philadelphia, PA.

Rowe, N., and Myers, M., eds. (2011). *All the World's Primates*. Primate Conservation, Inc., Charlestown, RI. Available at: <http://www.alltheworldsprimates.org> (accessed October 2012).

Royle, J. A., and Nichols, J. D. (2003). Estimating abundance from repeated presence-absence data or point counts. *Ecology*, **84**, 777–90.

Rubenstein, D. I., and Hack, M. (2004). Natural and sexual selection and the evolution of mulit-level societies: insights from zebras with comparisons to primates. In P. M. Kappeler, and C. van Schaik, eds. *Sexual Selection in Primates: New and Comparative Perspectives.* Cambridge University Press, Cambridge.

Ruperti, F. S. (2007). Population Density and Habitat Preferences of the Sahamalaza Sportive Lemur (*Lepilemur sahamalazensis*) at the Ankarafa Research Site, NW Madagascar (Masters thesis), Oxford Brookes University, Oxford.

Rwego, I. B., Isabirye-Basuta, G., Gillespie, T. R., and Goldberg, T. L. (2008). Gastrointest-inal bacterial transmission among humans, mountain gorillas, and livestock in Bwindi Impenetrable National Park, Uganda, *Conservation Biology*, **22**, 1600–7.

Ryder, O. A. (1986). Species conservation and systematics—the dilemma of subspecies. *Trends in Ecology & Evolution*, **1**, 9–10.

Rylands, A. B., Strier, K. B., Mittermeier, R. A., *et al.* (1998). *Population and Habitat Viability Assessment Workshop of the Muriqui (Brachyteles arachnoides)*. IUCN/SSC Conservation Breeding Specialist Group, Apple Valley, MN.

Safner, T., Miller, M. P., McRae, B. H., *et al.* (2011). Comparison of bayesian clustering and edge detection methods for inferring boundaries in landscape genetics. *International Journal of Molecular Sciences*, **12**, 865–89.

Salafsky, N. (1998) Dought in the rainforest, Part II: An update based on the 1994 ENSO event. *Climatic Change*, **39**, 601–3.

Salzer, D. (2007). *The Nature Conservancy's Threat Ranking System*. The Nature Conservancy, Washington, DC.

Sanchez-Azofeifa, A., Rivard, B., Wright, J., *et al.* (2011). Estimation of the distribution of *Tabebuia guayacan* (Bignoniaceae) using high-resolution remote sensing imagery. *Sensors*, **11**, 3831–51.

Santiago, M. L., Bibollet-Ruche, F., Bailes, E., *et al.* (2003). Amplification of a complete simian immunodeficiency virus genome from fecal RNA of a wild chimpanzee. *Journal of Virology*, 77, 2233–42.

Santiago, M. L., Range, F., Keele, B. F., *et al.* (2005). Simian immunodeficiency virus infection in free-ranging sooty mangabeys (*Cercocebus atye atys*) from the Tai Forest,

Cote d'Ivoire: implications for the origin of epidemic human immunodeficiency virus type 2. *Journal of Virology,* 79, 12515–27.

Santos, L. R., Nissen, A. G., and Ferrugia, J. A. (2006). Rhesus monkeys, *Macaca mulatta,* know what others can and cannot hear. *Animal Behaviour,* 71, 1175–81.

Sapolsky, R. M. (1993). Endocrinology Alfresco: Psychoendocrine studies of wild baboons. *Recent Progress in Hormone Research,* 48, 437–68.

Sapolsky, R. M. (2005). The influence of social hierarchy on primate health. *Science,* 308, 648–52.

Sapolsky, R. M., Romero, L. M., and Munck, A. U. (2000). How do glucocorticoids influence stress responses? Integrating permissive, suppressive, stimulatory, and preparative actions. *Endocrine Reviews,* 21, 55–89.

Sato, N., Sakata, H., Tanaka, Y., and Taira, M. (2004). Navigation in virtual environment by the macaque monkey. *Behavioural Brain Research,* 153, 287–91.

Sauther, M. L., Sussman, R. W., and Cuozzo F. (2002). Dental and general health in a population of wild ring-tailed lemurs: a life history approach. *American Journal of Physical Anthropology,* 117, 122–32.

Savage, A., Giraldo, H., Blumer, E. S., *et al.* (1993). Field techniques for monitoring cotton-top tamarins (*Saguinus oedipus oedipus*) in Columbia. *American Journal of Primatology,* 31, 189–96.

Savage, A., Thomas, L., Leighty, K. A., *et al.* (2010). Novel survey method finds dramatic decline of wild cotton-top tamarin population. *Nature Communications,* 1, 1–7.

Sayer, J. (2009). Can conservation and development really be integrated? *Madagascar Conservation and Development,* 4, 9–12.

Schaller, G. B. (1963). *The Mountain Gorilla: Ecology and Behavior.* University of Chicago Press, Chicago.

Schlichting, C. D., and Piggliucci, M. (1998). *Phenotypic Evolution: A Reaction Norm Perspective.* Sinauer Associates, Sunderland, MA.

Schmid, J. (2000). Daily torpor in the gray mouse lemur (*Microcebus murinus*) in Madagascar: energetical consequences and biological significance. *Oecologia,* 123, 175–83.

Schmid, J., and Ganzhorn, J. U. (2009). Optional strategies for reduced metabolism in gray mouse lemurs. *Naturwissenschaften,* 96, 737–41.

Schmid, J., Ruf, T., and Heldmaier, G. (2000). Metabolism and temperature regulation during daily torpor in the smallest primate, the pygmy mouse lemur (*Microcebus myoxinus*) in Madagascar. *Journal of Comparative Physiology B,* 170, 59–68.

Schmid, J., and Speakman, J. R. (2009). Torpor and energetic consequences in free-ranging gray mouse lemurs (*Microcebus murinus*): a comparison of dry and wet forests. *Naturwissenschaften,* 96, 609–20.

Schmidt, C. R. (1986). A review of zoo breeding programmes for primates. *International Zoo Yearbook,* 24/25, 107–23.

Schmidt-Nielsen, K. (1997). *Animal Physiology: Adaptation and Environment.* Cambridge University Press, Cambridge.

Schnitzer, S. A., Rutishauser, S., and Aguiar, S. (2008). Supplemental protocol for liana censuses. *Forest Ecology and Management,* 255, 1044–9.

Schoeninger, M., and DeNiro, M. J. (1984). Nitrogen and carbon isotopic composition of bone collagen from marine and terrestrial animals. *Geochimica et Cosmochimica Acta*, **48**, 625–39.

Schoeninger, M., Iwaniec, U., and Nash, L. (1998). Ecological attributes recorded in stable isotope ratios of arboreal prosimian hair. *Oecologia*, **113**, 222–30.

Schoeninger, M., Moore, J., and Sept, J. (1999). Subsistence strategies of two "savanna" chimpanzee populations: the stable isotope evidence. *American Journal of Primatology*, **49**, 297–314.

Scholz, F., and Kappeler, P. M. (2004). Effects of seasonal water scarcity on the ranging behavior of *Eulemur fulvus rufus*. *International Journal of Primatology*, **25**, 599–613.

Schupp, E. W. (1993). Quantity, quality, and the effectiveness of seed dispersal by animals. *Plant Ecology*, **107/108**, 15–29.

Schurr, M. R., Fuentes, A., Luecke, E. C. J., and Shaw, E. (2012). Intergroup variation in stable isotope ratios reflects anthropogenic impact on the Barbary macaques (*Macaca sylvanus*) of Gibraltar. *Primates*, **53**, 31–40.

Sealy, J., van der Merwe, N., Thorp, J., and Lanham, J. (1987). Nitrogen isotopic ecology in southern Africa: implications for environmental and dietary tracking. *Geochimica et Cosmochimica Acta*, **51**, 2707–17.

Seaman, D. E., Millspaugh, J. J., Kernohan, B. J., *et al.* (1999). Effects of sample size on kernel home range estimates. *Journal of Wildlife Management*, **63**, 739–47.

Segelbacher, G., Cushman, S. A., Epperson, B. K., *et al.* (2010). Applications of landscape genetics in conservation biology: concepts and challenges. *Conservation Genetics*, **11**, 375–85.

Seidel, K. S. (1992). Statistical Properties and Applications of a New Measure of Joint Space use for Wildlife. (Masters thesis), University of Washington, Seattle, WA.

Serio-Silva, J. C., Rico-Gray, V. and Ramos-Fernandez, G. (2006). Mapping primate populations in the Yucatan Peninsula, Mexico: a first assessment. In A. Estrada, P. A. Garber, M. S. M. Pavelka, and L. Luecke, eds. *New perspectives in the Study of Mesoamerican Primates*, pp. 489–512. Springer, New York.

Sha, J. C. M., Gumert, M. D., Lee, B. P., *et al.* (2009). Macaque-human interactions and the societal perceptions of macaques in Singapore. *American Journal of Primatology*, **71**, 825–39.

Shekelle, M., and Salim, A. (2009). An acute conservation threat to two tarsier species in the Sangihe Island chain, North Sulawesi, Indonesia. *Oryx*, **43**, 419–26.

Shelmidine, N., Borres, C., and McCann, C. (2009). Patterns of reproduction in Malayan silvered leaf monkeys at the Bronx Zoo. *American Journal of Primatology*, **71**, 852–9.

Shendure, J., and Ji, H. (2008). Next-generation DNA sequencing. *Nature Biotechnology*, **26**, 1135–45.

Shepherd, C. R. (2010). Illegal primate trade in Indonesia exemplified by surveys carried out over a decade in North Sumatra. *Endangered Species Research*, **11**, 201–5.

Shepherd, C. R., Sukumaran, J., and Wich, S. A. (2004). *Open Season: An analysis of the pet trade in Medan, Sumatra 1997–2001*. TRAFFIC Southeast Asia, Petaling Jaya, Malaysia.

Sheres, A. (2010). Ecological and Cultural Interconnections Between the Guizhou Snub-nosed Monkey (*Rhinopithecus brelichi*) and Local Communities at Fanjingshan National Nature Reserve, China (Masters thesis), San Diego State University, San Diego.

Sicotte, P. (1993). Inter-group encounters and female transfer in mountain gorillas: influence of group composition on male behavior. *American Journal of Primatology,* **30**, 21–36.

Sievers, F., Wilm, A., Dineen, D., *et al.* (2011). Fast, scalable generation of high quality protein multiple sequence alignments using Clustal Omega. *Molecular Systems Biology,* **7**, 539.

Sih, A., Hanser, S. F., and McHugh, K. A. (2009). Social network theory: new insights and issues for behavioral ecologists. *Behavioral Ecology and Sociobiology,* **63**, 975–88.

Sillentullberg, B., and Moller, A. P. (1993). The relationship between concealed ovulation and mating systems in anthropoid primates—a phylogenetic analysis. *American Naturalist,* **141**, 1–25.

Singh, M., Erinjery, J. J., Kavana, T. S., *et al.* (2011a). Drastic population decline and conservation prospects of roadside dark-bellied bonnet macaques (*Macaca radiata radiata*) of southern India. *Primates,* **52**, 149–54.

Singh, M., Roy, K., and Singh, M. (2011b). Resource partitioning in sympatric langurs and macaques in tropical rainforests of the central Western Ghats, south India. *American Journal of Primatology,* **73**, 335–46.

Sinha, A., Mukhopadhyay, K., Datta-Roy, A., and Ram, S. (2005). Ecology proposes, behaviour disposes: ecological variability in social organization and male behavioural strategies among wild bonnet macaques. *Current Science,* **89**, 1166–79.

Smith, B. H., and Boesch, C. (2011). Mortality and the magnitude of the "wild effect" in chimpanzee tooth emergence. *Journal of Human Evolution,* **60**, 34–46.

Smith, T. M. (2006). Experimental determination of the periodicity of incremental features in enamel. *Journal of Anatomy,* **208**, 99–113.

Smouse, P. E., Long, J. C., and Sokal, R. R. (1986). Multiple regression and correlation extensions of the Mantel test of matrix correspondence. *Systematic Zoology,* **35**, 627–32.

Snowdon, C. T. (1993). The rest of the story: grooming, group size and vocal exchanges in neotropical primates. *Behavioural and Brain Science,* **16**, 718.

Soewu, D. A. (2008). Wild animals in ethnozoological practices among the Yorubas of southwestern Nigeria and the implications for biodiversity conservation. *African Journal of Agricultural Research,* **3**, 421–7.

Soorae, P. S., ed. (2008). *Global Re-Introduction Perspectives: Re-introduction Case-Studies from Around the Globe.* IUCN/SSC Re-introduction Specialist Group, Abu Dhabi, UAE.

Soorae, P. S., ed. (2010). *IUCN Global Re-Introduction Perspectives: Additional Case-Studies from Around the Globe.* IUCN/SSC Re-introduction Specialist Group, Abu Dhabi, UAE.

Southwick, C. H., and Siddiqi, M. F. (1994). Population status of nonhuman primates in Asia with emphasis on rhesus macaques in India. *American Journal of Primatology,* **34**, 51–9.

Southwick, C. H., and Siddiqi, M. F. (2001). Status conservation and management of primates in India. *ENVIS bulletin: wildlife and protected areas,* **1**, 81–91.

Speakman, J. R. (1993). How should we calculate CO_2 production in doubly labeled water studies of animals? *Functional Ecology,* **7**, 746–50.

Speakman, J. R. (1997). *Doubly Labelled Water. Theory and Practise.* Chapman and Hall, London.

Speakman, J. R., and Krol, E. (2005). Validation of the doubly-labelled water method in a small mammal. *Physiological and Biochemical Zoology,* **78**, 650–67.

Speakman, J. R., and Racey, P. A. (1988). Consequences of non steady-state CO_2 production for accuracy of the doubly labeled water technique—the importance of recapture interval. *Comparative Biochemistry and Physiology A,* **90**, 337–40.

Sponheimer, M., Codron, D., Passey, B. H., *et al.* (2009). Using carbon isotopes to track dietary change in modern, historiacal, and ancient primates. *American Journal of Physical Anthropology,* **140**, 661–70.

Sponsel, L. E. (1997). The human niche in Amazonia: Explorations in ethnoprimatology. In W. G. Kinzey, ed. *New World Primates: Ecology, Evolution, and Behavior,* pp. 143–65. Aldine Gruyter, New York.

Sponsel, L. E., Ruttanadakul, N., and Natadecha-Sponsel, P. (2002). Monkey Business? The Conservation Implications of Macaque Ethnoprimatology in Southern Thailand. In A. Fuentes, and L. D. Wolfe, eds. *Primates face to face: The Conservation Implications of Human-Nonhuman Primate Interconnections,* pp. 288–309. Cambridge University Press, Cambridge.

Sprague, D. S. (2002). Monkeys in the Backyard: Encroaching Wildlife and Rural Communities in Japan. In A. Fuentes, and L. D. Wolfe, eds. *Primates Face to Face: The Conservation Implications of Human-Nonhuman Primate Interconnections,* pp. 254–72. Cambridge University Press, Cambridge.

Stamatakis, A., Ludwig, T., and Meier, H. (2005). RAxML-III: a fast program for maximum likelihood-based inference of large phylogenetic trees. *Bioinformatics,* **21**, 456–63.

Stammbach, E. (1986). Desert, forest and montane baboons: multilevel-societies. In N. B. Smuts, D. L. Cheney, R. M. Seyfarth, *et al.* eds. *Primate Societies.* The University of Chicago Press, Chicago.

Starr, C. R., Nekaris, K. A. I., and Leung, L. (2012). Why are pygmy slow lorises afraid of the light? Moonlight and temperature affect pygmy slow loris activity in a mixed deciduous forest in Cambodia. *PLoS ONE,* 7, e36396.

Stear, M. J., Bishop, S. C., Doligalska, M., *et al.* (1995). Regulation of egg production, worm burden, worm length and worm fecundity by host responses in sheep infected with *Ostertagia circumcincta. Parasite Immunology,* **17**, 643–52.

Steenbeek, R. (2000). Infanticide by males and female choice in wild Thomas's langurs. In C. P. van Schaik, and C. H. Janson, eds. *Infanticide by males and its implications,* pp. 153–77. Cambridge University Press, Cambridge.

Stepien, R. L., Bonagura, J. D., Bednarski, R. M., *et al.* (1995). Cardiorespiratory effects of acepromazine maleate and buprenorphine hydrochloride in clinically normal dogs. *American Journal of Veterinary Research,* **56**, 78–84.

Sterck, E. H. M., Watts, D. P., and van Schaik, C. P. (1997). The evolution of female social relationships in nonhuman primates. *Behavioral Ecology and Sociobiology,* **41**, 291–309.

Sterling, E. J. (1993a). Behavioral Ecology of the Aye-Aye (*Daubentonia madagascariensis*) on Nosy Mangabe, Madagascar (Ph.D. Dissertation), Yale University, New Haven, CT.

Sterling, E. J. (1993b). Patterns of range use and social organization in aye-ayes (*Daubentonia madagascariensis*) on Nosy Mangabe. In P. M. Kappeler, J. U. Ganzhorn, eds. *Lemur Social Systems and their Ecological Basis,* pp. 1–10. Plenum Press, New York.

Sterling, E. J. (1994a). Taxonomy and distribution of *Daubentonia*: A historical perspective. *Folia Primatologica,* **62**, 8–13.

Sterling, E. J. (1994b). Aye-ayes: Specialists on structurally defended resources. *Folia Prima-tologica*, **62**, 142–54.

Sterling, E. J. (2003). *Daubentonia madagascariensis*, Aye-aye. In S. Goodman and J. Benstead, eds. *The Natural History of Madagascar*, pp.1348–51. University of Chicago Press, Chicago.

Sterling, E. J., Dierenfeld, E. S., Ashbourne, C. J., and Feistner, A. T. C. (1994). Dietary intake, food composition and nutrient intake in wild and captive populations of *Daubentonia madagascariensis*. *Folia Primatologica*, **62**, 115–24.

Sterling, E., and Feistner, A. T. (2000). Aye-aye. In R. Reading, and B. Miller, eds. *Endangered Animals: a Reference Guide to Conflicting Issues*, pp. 45–8. Greenwood Press, Westport, CT.

Sterling, E. J., Gomez, A., and Porzecanski, A. L. (2010). A systemic view of biodiversity and its conservation: Processes, interrelationships, and human culture. *Bioessays*, **32**, 1090–8.

Sterling, E. J., and McCreless, E. E. (2006). Adaptations in the aye-aye: A review. In L. Gould, and M. L. Sauther, eds. *Lemurs: Ecology and Adaptation*, pp. 158–84. Springer, New York.

Sterling, E. J., Nguyen, N., and Fashing, P. J. (2000). Spatial patterning in nocturnal prosimians: a review of methods and relevance to studies of sociality. *American Journal of Primatology*, **51**, 3–19.

Stevens, J., Rosati, A., Ross, K., and Hauser, M. (2005). Will travel for food: spatial discounting in two new world monkeys. *Current Biology*, **15**, 1855–60.

Stevenson, M. F. (1983). Effectiveness of primate captive breeding. In D. Harper, ed. *Symposium on the Conservation of Primates and Their Habitats, Vaughan Paper* No 31, Volume II, pp. 202–32. University of Leicester, Leicester, UK.

Stewart, P. D., Anderson, C., and Macdonald, D. W. (1997). A mechanism for passive range exclusion: evidence from the European badger (*Meles meles*). *Journal of Theoretical Biology*, **184**, 279–89.

Stiling, P. D. (1999). *Ecology. Theories and Applications*, 3rd edition. Prentice Hall, Upper Saddle River.

Stockwell, D. R. B., and Noble, I. R. (1992). Induction of sets of rules from animal distribution data—a robust and informative method of data-analysis. *Mathematics and Computers in Simulation*, **33**, 385–90.

Stockwell, D., and Peters, D. (1999). The GARP modeling system: problems and solutions to automated spatial prediction. *International Journal of Geographical Information Science*, **13**, 143–58.

Stockwell, D. R. B., and Peterson, A. T. (2002). Effects of sample size on accuracy of species distribution models. *Ecological Modeling*, **148**, 1–13.

Stoinski, T., Lukas, K. E., Kuhar, C. W., and Maple, T. L. (2004). Factors influencing the formation and maintenance of all-male gorilla groups in captivity. *Zoo Biology*, **23**, 189–203.

Stoner, K. (1996). Prevalence and intensity of intestinal parasites in mantled howling monkeys (*Alouatta palliata*) in northeastern Costa Rica: Implications for conservation biology. *Conservation Biology*, **10**, 539–46.

Strier, K. B. (1992). Atelinae adaptations: behavioral strategies and ecological constraints. *American Journal of Physical Anthropology*, **88**, 515–24.

Strier, K. B. (1993). Viability analyses of an isolated population of muriqui monkeys (*Brachyteles arachnoides*): Implications for primate conservation and demography. *Primate Conservation*, **14–15**, 43–52.

Strier, K. B. (2000). Population viabilities and conservation implications for muriquis (*Brachyteles arachnoides*) in Brazil's Atlantic Forest. *Biotropica*, **32**, 903–13.

Strier, K. B. (2003). *Primate Behavioral Ecology*. Allyn and Bacon, Boston, MA.

Strier, K. B., Altmann, J., Brockman, D. K., *et al.* (2010). The Primate Life History Database: a unique shared ecological data resource. *Methods in Ecology and Evolution*, **1**, 199–211.

Strier K. B., and Ives, A. R. (2012). Unexpected demography in the recovery in the recovery of an endangereal primate population. *PLoS ONE*, 7, e 44407.

Strier, K. B., and Ziegler, T. E. (2005). Advances in field-based studies of primate behavioral endocrinology. *American Journal of Primatology*, **67**, 1–4.

Strindberg, S., Buckland, S. T., and Thomas, L. (2004). Design of distance sampling surveys and Geographic Information Systems. In S. T. Buckland, D. R. Anderson, K. P. Burnham, *et al.* eds. *Advanced Distance Sampling*, pp. 190–228. Oxford University Press, Oxford.

Struhsaker, T. T. (1975). *The Red Colobus Monkey*. University of Chicago Press, Chicago.

Struhsaker, T. T. (1981). Census methods for estimating densities. In National Research Council, *Techniques for the Study of Primate Population Ecology*, pp. 36–80. National Academy Press, Washington.

Strum, S. C. (2005). Measuring Success in Primate Translocations: A baboon case study. *American Journal of Primatology*, **65**, 117–40.

Strum, S. C., and Southwick, C. (1986). Translocation of Primates. In K. Benirschke, ed. *Primates: The Road to Self-Sustaining Populations*, pp. 951–7. Springer-Verlag, New York.

Suarez, E., Morales, M., Cueva, R., *et al.* (2009). Oil industry, wild meat trade and roads: indirect effects of oil extraction activities in a protected area in north-eastern Ecuador. *Animal Conservation*, **12**, 364–73.

Sueur, C., Petit, O., and Deneubourg, J. (2010). Short-term group fission processes in macaques: a social networking approach. *Journal of Experimental Biology*, **213**, 1338–46.

Sueur, C., Jacobs, A., Amblard, F., *et al.* (2011a). How can social network analysis improve the study of primate behavior? *American Journal of Primatology*, **73**, 703–19.

Sueur, C., Salze, P., Weber, C., and Petit, O. (2011b). Land use in semi-free ranging Tonkean macaques *Macaca tonkeana* depends on environmental conditions: A geographical information system approach. *Current Zoology*, **57**, 8–17.

Sugiura, H., Saito, C., Sato, S., *et al.* (2000). Variation in intergroup encounters in two populations of Japanese macaques. *International Journal of Primatology*, **21**, 519–35.

Sunderland-Groves, J., Ekinde, A., Mboh, H. (2009). Cross River gorilla (*Gorilla gorilla diehli*) nesting behaviour at Kagwene Mountain, Cameroon: implications for assessing group size and density. *International Journal of Primatology*, **30**, 253–66.

Sussman, R. W., Garber, P. A., and Cheverud, J. M. (2005). Importance of cooperation and affiliation in the evolution of primate sociality. *American Journal of Physical Anthropology*, **128**, 84–97.

Sussman, R. W., Green, G. M., Porton, I., *et al.* (2003). A survey of the habitat of *Lemur catta* in southwestern and southern Madagascar. *Primate Conservation*, **19**, 32–57.

Sutherland, W. J., Adams, W. M., Aronson, R. B. Aveling, R., *et al.* (2009). One hundred questions of importance to the conservation of global biological diversity. *Conservation Biology*, **23**, 557–67.

Svensson, M. S., Samudio, R., Bearder, S. K., and Nekaris, K. A. I. (2010). Density estimates of Panamanian owl monkeys (*Aotus zonalis*) in Three Habitat Types. *American Journal of Primatology*, **72**, 187–92.

Swedell, L. (2002). Ranging behavior, group size and behavioral flexibility in Ethiopian Hamadryas baboons (*Papio hamadryas hamadryas*). *Folia primatologica*, **73**, 95–103.

Swedell, L., Saunders, J., Schreier, A., *et al.* (2011). Female dispersal in hamadryas baboons: transfer among social units in a multilevel society. *American Journal of Physical Anthropology*, **145**, 360–70.

Swihart, R. K., and Slade, N. A. (1985). Testing for independence of observation in animal movements. *Ecology*, **69**, 393–9.

Swindler, D. R. (2002). *Primate dentition. An introduction to the Teeth of Non-Human Primates.* Cambridge Press, Cambridge.

Swofford, D. L. (2002). *PAUP*. Phylogenetic Analysis Using Parsimony (*and Other Methods).* Version 4. Sinauer, Sunderland, MA.

Takahashi, H. (2001). Influence of fluctuation in the operational sex ratio to mating of troop and non-troop male Japanese Macaques for four years on Kinkazan Island, Japan. *Primates*, **42**, 183–91.

Terborgh, J. (1983). *Five New World Primates.* Princeton Univ. Press, Princeton.

Terborgh, J., and Janson, C. H. (1986). The socioecology of primate groups. *Annual Review of Ecology and Systematics*, **17**, 111–35.

Thalmann, O., Hebler, J., Poinar, H. N., *et al.* (2004). Unreliable mtDNA data due to nuclear insertions: a cautionary tale from analysis of humans and other great apes. *Molecular Ecology*, **13**, 321–35.

The Mountain Gorilla Veterinary Project Employee Health Group. (2004). Risk of disease transmission between conservation personnel and the Mountain gorillas: Results from an employee health program in Rwanda. *EcoHealth*, **1**, 351–61.

Thomas, C. D. (2010). Climate, climate change and range boundaries. *Diversity and Distributions*, **16**, 488–95.

Thomas, W. (1958). Breeding in captivity of lowland gorillas. *Zoologica*, **43**, 95–104.

Thomas, L., Buckland, S. T., Rexstad, E. A., *et al.* (2010). Distance software: design and analysis of distance sampling surveys for estimating population size. *Journal of Applied Ecology*, **47**, 5–14.

Thomas, C. D., Cameron, A., Green, R. E., *et al.* (2004). Extinction risk from climate change. *Nature*, **427**, 145–8.

Thompson, M. E., and Wrangham, R. W. (2008). Diet and reproductive function in wild female chimpanzees (*Pan troglodytes schweinfurthii*) at Kibale National Park, Uganda. *American Journal of Physical Anthropology*, **135**, 171–81.

Thoren, S., Quietzsch, F., and Radespiel, U. (2010). Leaf nest use and construction in the golden-brown mouse lemur (*Microcebus ravelobensis*) in the Ankarafantsika National Park. *American Journal of Primatology*, **72**, 48–55.

Thorn, J. S., Nijman, V., Smith, D., and Nekaris, K. A. I. (2009). Ecological niche modeling as a technique for assessing threats and setting conservation priorities for Asian slow lorises (Primates: *Nycticebus*). *Diversity and Distributions, 15*, 289–98.

Tiddi, B., Aureli, F., di Sorrentinoa, E. P., *et al.* (2011). Grooming for tolerance? Two mechanisms of exchange in wild tufted capuchin monkeys. *Behavioral Ecology, 22*, 663–9.

Tieguhong, J. C., and Zwolinski, J. (2009). Supplies of bushmeat for livelihoods in logging towns in the Congo Basin. *Journal of Horticulture and Forestry, 1*, 65–80.

Tinbergen, N. (1963). On aims and methods of Ethology. *Zeitschrift für Tierpsychologie, 20*, 410–33.

Todd, A., Kuehl, H., Cipolett, C., and Walsh, P. (2008). Using dung to estimate gorilla density: modeling dung production rate. *International Journal of Primatology, 29*, 249–54.

Tolkamp, B. J., Emmans, G. C., Yearsley, J., and Kyriazakis, I. (2002). Optimization of short-term animal behaviour and the currency of time. *Animal Behavior, 64*, 945–95.

Tomasello, M., and Call, J. (2011). Methodological challenges in the study of primate cognition. *Science, 334*, 1227–8.

Tracy, L. N., and Jamieson, I. G. (2011). Historic DNA reveals contemporary population structure results from anthropogenic effects, not pre-framentation patterns. *Conservation Genetics, 12*, 517–26.

TRAFFIC (2008). *What's driving the wildlife trade? A Review of Expert Opinion on Economic and Social Drivers of the Wildlife Trade and Trade Control Efforts in Cambodia, Indonesia, Lao PDR, and Vietnam.* East Asia and Pacific Region Sustainable Development Discussion Papers. East Asia and Pacific Region Sustainable Development Department, World Bank, Washington, DC.

Travis, D. A. (2009). *Primate Necropsy Protocols Developed for Field Workers.* Lincoln Park Zoo, Chicago.

Travis, D. A., Hungerford, L., Engel, G. A., and Jones-Engel, L. (2006). Disease risk analysis: A tool for primate conservation planning and decision making. *American Journal of Primatology, 68*, 855–67.

Trishchenko, A. P., Chilar, J., and Li, Z. (2002). Effects of spectral response function on surface reflectance and NDVI measured with moderate resolution satellite sensors. *Remote Sensing of Environment, 81*, 1–18.

Trivers, R. L. (1972). Parental investment and sexual selection. In B. Campbell, ed. *Sexual Selection and the Descent of Man*, pp. 1871–971. Aldine, Chicago.

Trent, B. K., Tucker, M. E., and Lockard, J. S. (1977). Activity changes with illumination in slow loris *Nycticebus coucang*. *Applied Animal Ethology, 3*, 281–6.

Tucker, C. J. (1979). Red and photographic infrared linear combinations for monitoring vegetation. *Remote Sensing of Environment, 8*, 127–50.

Tung, J., Alberts, S. C., Wray G. A. (2010). Evolutionary genetics in wild primates: combining genetic approaches with field studies of natural populations. *Trends in Genetics, 26*, 353–62.

Ukizintambara, T. (2010). Forest Edge Effects on the Behavioral Ecology of l'Hoest's Monkey (*Cercopithecus lhoesti*) in Bwindi Impenetrable National Park, Uganda (Ph.D. Dissertation), Antioch University, Keene, NH.

Unwin, S., Ancrenaz, M., and Bailey, W. (2011). Handling, anesthesia, health evaluation and biological sampling. In J. M. Setchell, and D. J. Curtis, eds. *Field and Laboratory Methods in Primatology,* second edn, pp. 147–67. Cambridge University Press, Cambridge.

Valenta, K., and Fedigan, L. M. (2008). How much is a lot? Seed dispersal by white-faced capuchins and implications for disperser-based studies of seed dispersal systems. *Primates,* **49,** 169–75.

Valiere, N. (2002). GIMLET: a computer program for analysing genetic individual identification data. *Molecular Ecology Notes,* **2,** 377–9.

van der Merwe, N., and Medina, E. (1991). The canopy effect, carbon isotope ratios and foodwebs in Amazonia. *Journal of Archaeological Science,* **18,** 249–59.

Van Heuverswyn, F., Li, Y., Neel, C., *et al.* (2006). SIV infection in wild gorillas. *Nature,* **444,** 164.

van Holt, T., Townsend, W. R., and Cronkleton, P. (2010). Assessing local knowledge of game abundance and persistence of hunting livelihoods in the Bolivian Amazon using consensus analysis. *Human Ecology,* **38,** 791–801.

van Schaik, C. P. (1986). Phenological changes in a Sumatran rainforest. *Journal of Tropical Ecology,* **2,** 327–47.

van Schaik, C. P. (1989). The ecology of social relationships amongst female primates. In V. Standen, and R. A. Foley, eds. *Comparative Socioecology: The Behavioural Ecology of Humans and Other Mammals,* pp. 195–218. Blackwell Scientific Publications, Oxford.

van Schaik, C. P., Assink, P. R., and Salafsky, N. (1992). Territorial behavior in southeast Asian langurs: resource defense or mate defense? *American Journal of Primatology,* **26,** 233–42.

van Schaik, C. P., and Brockman, D. K. (2005). Seasonality in primate ecology, reproduction, and life history: an overview. In D. K. Brockman, and C. P. van Schaik, eds. *Seasonality in Primates: Studies of Living and Extinct Human and Non-Human Primates,* pp. 3–20. Cambridge University Press, Cambridge.

van Schaik, C. P., and Kappeler, P. M. (1997). Infanticide risk and the evolution of male-female association in primates. *Proceedings of the Royal Society London B,* **264,** 1687–94.

van Schaik, C. P., and Pfannes, K. R. (2005). Tropical climates and phenology: a primate perspective. In D. K. Brockman, and C. P. van Schaik, eds. *Seasonality in Primates: Studies of Living and Extinct Human and Non-Human Primates,* pp. 23–54. Cambridge University Press, Cambridge.

van Schaik, C. P., and van Hooff, J. (1983). On the ultimate causes of primate social systems *Behaviour,* **85,** 91–117.

van Schaik, C. P., Terborgh, J. W., and Wright, S. J. (1993). The phenology of tropical forests: adaptive significance and consequences for primary consumers. *Annual Review of Ecology, Evolution, and Systematics,* **24,** 353–77.

van Schaik, C. P., Wich, S. A., Utami, S. S., and Odom, K. (2005). A simple alternative to line transects of nests for estimating orangutan abundance. *Primates,* **46,** 249–54.

Van Soest, P. J. (1994). *Nutritional Ecology of the Ruminant.* Cornell University Press, Ithaca, New York.

Verbeek, P., and de Waal, F. B. M. (1997). Postconflict behavior of captive brown capuchins in the presence and absence of attractive food. *International Journal of Primatology,* **18,** 703–25.

Vie, J. C., Hilton-Taylor, C., and Stuart, S. N., eds. (2009). *Wildlife in a Changing World, An Analysis of the 2008 IUCN Red List of Threatened Species.* IUCN, Gland, Switzerland.

Vigilant, L., and Guschanski, K. (2009). Using genetics to understand the dynamics of wild primate populations. *Primates, 50,* 105–20.

Villers, L. M., Jang, S. S., Lent, C. L., *et al.* (2008). Survey and comparison of major intestinal flora in captive and wild ring-tailed lemur *(Lemur catta)* populations. *American Journal of Primatology, 70,* 175–84.

Virgin, C. E., and Sapolsky, R. M. (1997). Styles of male social behavior and their endocrine correlates among low-ranking baboons. *American Journal of Primatology, 42,* 25–39.

Visalberghi, E., Addessi, E., Truppa, V., *et al.* (2009). Selection of effective stone tools by wild bearded capuchin monkeys. *Current Biology, 19,* 213–17.

Visalberghi, E., and Limongelli, L. (1994). Lack of comprehension of cause-effect relations in tool-using capuchin monkeys *(Cebus apella). Journal of Comparative Psychology, 108,* 15–22.

Voelkl, B., Kasper, C., and Schwab, C. (2011). Network measures for dyadic interactions: stability and reliability. *American Journal of Primatology, 73,* 731–40.

Vogel, E. R. (2005). Rank differences in energy intake rates in white-faced capuchin monkeys, *Cebus capucinus*: the effects of contest competition. *Behavioral Ecology and Sociobiology, 58,* 333–44.

Vogel, E. R., Munch, S. B., and Janson, C. H. (2007). Understanding escalated aggression over food resources in white-faced capuchin monkeys. *Animal Behaviour, 74,* 71–80.

Vogel, E. R., van Woerden, J. T., Lucas, P. W., *et al.* (2008). Functional ecology and evolution of hominoid molar enamel thickness: *Pan troglodytes schweinfurthii* and *Pongo pygmaeus wurmbii. Journal of Human Evolution, 55,* 60–74.

Vogler, A. P., and DeSalle, R. (1994). Diagnosing units of conservation management. *Conservation Biology, 8,* 354–63.

Vogt, M., and Forster, B. (2010). The Primate Reintroduction Program in Phong Nha-Ke Bang National Park, Central Vietnam. In T. Nadler, B. M. Rawson, and Van Ngoc Thinh, eds. *Conservation of Primates in Indochina,* pp. 245–50. Frankfurt Zoological Society and Conservation International, Hanoi, Vietnam.

Waddington, C. H. (1942). Canalization of development and the inheritance of acquired characters. *Science, 3811,* 563–5.

Wallen, K., and Hassett, J. M. (2009). Neuroendocrine mechanisms underlying social relationships. In P. T. Ellison, and P. B. Gray, eds. *Endocrinology of Social Relationships,* pp. 32–53. Harvard University Press, Cambridge, MA.

Wallis, J. (1997). From ancient expeditions to modern exhibitions: the evolution of primate conservation in the zoo community. In J. Wall, ed. *Special Topics in Primatology Volume 1, Primate Conservation: The Role of Zoological Parks,* pp. 1–27. The American Society of Primatologists, USA.

Walsh, P. D., and White, L. J. T. (1999). What will it take to monitor forest elephant populations? *Conservation Biology, 13,* 1194–202.

Wandeler, P., Hoeck, P. E. A., and Keller, L. F. (2007). Back to the future: museum specimens in population genetics. *Trends in Ecology & Evolution, 22,* 634–42.

Wanyama, F., Muhabwe, R., Plumptre, A. J., *et al.* (2010). Censusing large mammals in Kibale National Park: evaluation of the intensity of sampling required to determine change. *African Journal of Ecology*, **48**, 953–61.

Wartmann, F. M., van Schaik, C. P., and Purves, R. S. (2010). Modelling ranging behaviour of female orang-utans: a case study in Tuanan, Central Kalimantan, Indonesia. *Primates*, **51**, 119–30.

Waser, P. M. (1976). *Cercocebus albigena*: site attachment, avoidance, and intergroup spacing. *American Naturalist*, **110**, 911–35.

Waser, P. M. (1981). Sociality or territorial defense? The influence of resource renewal. *Behavioral Ecology and Sociobiology*, **8**, 231–7.

Wasser, S. K., Risler, L., and Steiner, R. A. (1988). Excreted steroids in primate feces over the menstrual cycle and pregnancy. *Biology of Reproduction*, **39**, 862–72.

Wasser, S. K., Thomas, R., Nair, P. P., *et al.* (1993). Effects of dietary fiber on fecal steroid measurements in baboons (*Papio cynocephalus cynocephalus*). *Journal of Reproduction and Fertility*, **97**, 569–74.

Watts, D. P. (1990). Mountain gorilla life histories, reproductive competition, and socio-sexual behavior and some implications for captive husbandry. *Zoo Biology*, **9**, 185–200.

Watts, D. P., Mitani, J. C., and Sherrow, H. M. (2002). New cases of inter-community infanticide by male chimpanzees at Ngogo, Kibale National Park, Uganda. *Primates*, **43**, 263–70.

Watts, D. P., Potts, K. B., Lwanga, J. S., and Mitani, J. C. (2012). Diet of chimpanzees (*Pan troglodytes schweinfurthii*) at Ngogo, Kibale National Park, Uganda, 2. temporal variation and fallback foods. *American Journal of Primatology*, **74**, 130–44.

WAZA (World Association of Zoos and Aquariums). (2011). *Who we are*. WAZA, Gland, Switzerland. Available at: <http://www.waza.org/en/site/about-waza/who-we-are> (accessed september 2011).

Weatherall, D. (2006). *The Use of Non-Human Primates in Research*. Academy of Medical Sciences, Medical Research Council, The Royal Society and Wellcome Trust, London.

Weathers, W. W., Buttemer, W. A., Hayworth, A. M., and Nagy, K. A. (1984). An evaluation of time-budget estimates of daily energy expenditure in birds. *Auk*, **101**, 459–72.

Weaver, I. C. G., Cervoni, N., Champagne, F. A., *et al.* (2004). Epigenetic programming by maternal behavior. *Nature Neuroscience*, **7**, 847–54.

Weber, A. W., and Vedder, A. (1983). Population dynamics of the Virunga gorillas: 1959–1978. *Biological Conservation*, **26**, 341–66.

Weidt, A., Hagenah, N., Randrianambinina, B., *et al.* (2004). Social organization of the golden brown mouse lemur (*Microcebus ravelobensis*). *American Journal of Physical Anthropology*, **123**, 40–51.

Weimerskirch, H., Martin, J., Clerquin, Y., *et al.* (2001). Energy saving in flight formation. *Nature*, **413**, 697–8.

Wiese, R. J. and Hutchins, M. (1997). The role of North American Zoos in Primate conservation. In J. Wallis, ed. *Primate Conservation: The Role of Zoological Parks*. pp. 291. American Society of Primatologists, Norman, OK.

Weller, S. C. (2007). Cultural consensus theory: Applications and frequently asked questions. *Field Methods*, **19**, 339–68.

West, C., and Dickie, L. A. (2007). Introduction: Is there a conservation role for zoos in a natural world under fire? In A. Zimmerman, M. Hatchwell, L. Dickie, and C. West, eds. *Zoos in the 21st Century, Catalysts for Conservation?*, pp. 3–11. Cambridge University Press, Cambridge, UK.

Western, D. (1986). The role of captive populations in global conservation. In K. Benirschke, ed. *Primates the Road to Self-Sustaining Populations*, pp. 13–20. Springer-Verlag, New York.

Wharton, D. (2000). Gorilla Management for the 21st Century. *American Zoo and Aquarium Association Annual Conference Proceedings*, pp. 323–7. Disney's Animal Kingdom, Orlando.

Wharton, D. (2010). *North American Studbook for the Western Lowland Gorilla (Gorilla gorilla gorilla)*. AZA and the Chicago Zoological Society Brookfield Zoo, Silver Spring, MD and Chicago.

Wheatley, B. P. (1999). *The sacred monkeys of Bali*. Waveland Press, Prospect Heights.

White, G. C., and Burnham, K. P. (1999). Program MARK: survival estimation from populations of marked animals. *Bird Study*, 46, Supplement, 120–38.

White, L., and Edwards, A. (2000). *Conservation Research in the African Rain Forests: A Technical Handbook*. Wildlife Conservation Society, New York.

Whitehead, H. (2009). SOCPROG programs: analysing animal social structures. *Behavioral Ecology and Sociobiology*, 63, 765–78.

Whitesides, G. H., Oates, J. F., Green, S. M., and Kluberdanz, R. P. (1988). Estimating primate densities from transects in a West African rain forest: a comparison of techniques. *Journal of Animal Ecology*, 57, 345–67.

Whittaker, M. (2008). *North American Studbook for the Diana Monkey (Cercopithecus diana)*. AZA and the Dallas Zoo, Silver Spring, MD, and Dallas.

Whitten, P. L., Brockman, D. K., and Stavisky, R. C. (1998). Recent advances in noninvasive techniques to monitor hormone-behavior interactions. *Yearbook of Physical Anthropology*, 41, 1–23.

Wich, S. A., and Boyko, R.H. (2011). Which factors determine orangutan nests' detection probability along transects? *Tropical Conservation Science*, 4, 53–63.

Wich, S. A., Fredriksson, G., and Sterck, E. H. M. (2002). Measuring fruit patch size for three sympatric Indonesian primate species. *Primates*, 43, 19–27.

Wich, S. A., Steenbeek, R., Sterck, E. H. M., *et al.* (2007). Demography and Life History of Thomas Langurs (*Presbytis thomasi*). *American Journal of Primatology*, 69, 641–51.

Wich, S. A., and Sterck, E. H. M. (2007). Familiarity and threat of opponents determine variation in Thomas langur (*Presbytis thomasi*) male behaviour during between-group encounters. *Behaviour*, 144, 1583–98.

Wich, S. A, Utami-Atmoko, S. S., Mitra Setia, T., *et al.* (2004). Life history of wild Sumatran orangutans (*Pongo abelii*). *Journal of Human Evolution*, 47, 385–98.

Wich, S. A., and van Schaik, C. P. (2000). The impact of El Niño on mast fruiting in Sumatra and eslewhere in Malesia. *Journal of Tropical Ecology*, 16, 563–77.

Wich, S. A., Vogel, E. R., Larsen, M. D., *et al.* (2011). Forest fruit production is higher on sumatra than on borneo. *PLoS ONE*, 6, e21278.

Wiederholt, R., Fernandez-Duque, E., Diefenbachd, D. R., and Rudrane, R. (2010). Modeling the impacts of hunting on the population dynamics of red howler monkeys (*Alouatta seniculus*). *Ecological Modelling*, 221, 2482–90.

Wiederholt, R., and Post, E. (2010). Tropical warming and the dynamics of endangered primates. *Biology Letters*, **6**, 257–60.

Wiese, R. J., and Gray, J. (2010). The one curator—one species challenge. In *World Association of Zoos and Aquariums (WAZA) Proceedings of 65th Annual Conference*, pp. 24. WAZA, Gland, Switzerland.

Wilcox, A. S., and Nambu, D. M. (2007). Wildlife hunting practices and bushmeat dynamics of the Banyangi and Mbo people of Southwestern Cameroon. *Biological Conservation*, **134**, 251–61.

Wild, D. G. (2005). *The immunoassay handbook*, 3rd edition. Elsevier, New York.

Wilensky, U. (1999). *NetLogo*. <http://ccl.northwestern.edu/netlogo/> (accessed october 2012). Center for Connected Learning and Computer-Based Modeling, Northwestern University, Evanston, IL.

Wilkie, D. S., and Godoy, R. A. (2001). Income and price elasticities of bushmeat demand in lowland Amerindian societies. *Conservation Biology*, **15**, 751–69.

Willems, E. P., Barton, R. A., and Hill, R. A. (2009). Remotely sensed productivity, regional home range selection, and local range use by an omnivorous primate. *Behavioral Ecology*, **20**, 985–92.

Williams, B. K., Nichols, J. D., and Conroy, M. J. (2001). *Analysis and Management of Animal Populations*. Academic Press, New York.

Williams, S. E., Marsh, H., and Winter, J. (2002). Spatial scale, species diversity, and habitat structure: small mammals in Australian tropical rain forest. *Ecology*, **83**, 1317–29.

Williams, C. T., Sheriff, M. J., Schmutz, J. A., *et al.* (2011). Data logging of body temperatures provides precise information on phenology of reproduction events in a free-living arctic hibernator. *Journal of Comparative Biochemistry and Physiology B*, **181**, 1101–9.

Williamson, E. A., and Feistner, A. T. C. (2011). Habituating primates: processes, techniques, variables and ethics. In J. M. Setchell, and D. J. Curtis, eds. *Field and Laboratory Methods in Primatology: A Practical Guide*, pp. 33–49. Cambridge University Press, Cambridge.

Willoughby, D. P. (1978). *All About Gorillas*. A. S. Barnes and Co., Inc., Cranbury, NJ.

Wilson, G. A., and Rannala, B. (2003). Bayesian inference of recent migration rates using multilocus genotypes. *Genetics*, **163**, 1177–91.

Wilson, M. L., Hauser, M. D., and Wrangham, R. W. (2001). Does participation in intergroup conflict depend on numerical assessment, range location, or rank for wild chimpanzees? *Animal Behaviour*, **61**, 1203–16.

Wilson R. P. and McMahon C. R. (2006). Measuring devices on wild animals: what constitutes acceptable practice? *Ecology and the Environment*, **4**, 147–54.

Wingfield, J. C., Hegner, R. E., Dufty, A. M., and Ball, G. F. (1990). The challenge hypothesis—Theoretical implications for patterns of testosterone secretion, mating systems, and breeding strategies. *American Naturalist*, **136**, 829–46.

Wingfield, J. C., Lynn, S. E., and Soma, K. K. (2001). Avoiding the "costs" of testosterone: Ecological bases of hormone-behavior interactions. *Brain Behavior and Evolution*, **57**, 239–51.

Winn, R. M. (1989). The aye-ayes, *Daubentonia madagascariensis*, at the Paris Zoological Gardens: Maintenance and preliminary behavioural observations. *Folia Primatologica*, **52**, 109–23.

Wolf, T. J., Ellington, C. P., Davis, S., and Feltham, M. J. (1966). Validation of doubly labelled water technique for bumblebees *Bombus terrestris* (L.). *Journal of Experimental Biology*, 199, 959–72.

Woodroffe, R., Thirgood, S., and Rabinowitz, A., eds. (2005). *People and Wildlife: Conflict or Coexistence?* Cambridge University Press, Cambridge.

Woolhouse, M., and Gaunt, E. (2007). Ecological origins of novel human pathogens. *Critical Reviews in Microbiology*, 33, 1–12.

Worobey, M., Gemmel, M., Teuwen, D. E., *et al.* (2008). Direct evidence of extensive diversity of HIV-1 in Kinshasa by 1960. *Nature*, 455, 661–4.

Worton, B. J. (1989). Kernel methods for estimating the utilization distribution in home-range studies. *Ecology*, 70, 164–8.

Worton, B. J. (1995). Using Monte-Carlo simulation to evaluate kernel-based home-range estimators. *Journal of Wildlife Management*, 59, 794–800.

Wrangham, R. W. (1979). On the evolution of Ape Social Systems. *Social Science Information*, 18, 336–68.

Wrangham, R. W. (1980). An ecological model of female bonded primate groups. *Behaviour*, 75, 262–300.

Wrangham, R. W., Chapman, C. A., Clark-Arcadi, A. P., and Isabirye-Basuta, G. (1996). Social ecology of Kanyawara chimpanzees: implications for understanding the costs of great ape groups. In W. C. McGrew, L. F. Marchant, and T. Nishida, eds. *Great Ape Societies*, pp. 45–57. Cambridge University Press, Cambridge.

Wrangham, R. W., Conklin-Brittain, N. L., and Hunt, K. D. (1998). Dietary response of chimpanzees and cercopithecines to seasonal variation in fruit abundance. I. Antifeedants. *International Journal of Primatology*, 19, 949–70.

Wright, B. W. (2005). Craniodental biomechanics and dietary toughness in the genus Cebus. *Journal of Human Evolution*, 48, 473–92.

Wright, T. (2011). International Studbook for the aye-aye (*Daubentonia madagascariensis*), Number 3. Durrell Wildlife Conservation Trust, Jersey, Channel Islands.

Wright, P. C., Andriamihaja, B. R., and Raharimiandra, S. A. (2005). Tanala synecological relations with lemurs in southeastern Madagascar. In J. Paterson, and J. Wallis, eds. *Commensalism and conflict: The Primate-Human Interface*, pp. 119–45. American Society of Primatologists, Norman, OK.

Wright, S. J., and Calderón, O. (2006). Seasonal, El Nino and longer term changes in flower and seed production in a moist tropical forest. *Ecology Letters*, 9, 35–44.

Wurster, E. C., Murrish, D. E., Sulzman, F. M. (1985). Circadian rhythms in body temperature of the pigtailed macaque (*Macaca nemestrina*) exposed to different ambient temperatures. *American Journal of Primatology*, 9, 1–13.

WWF/TRAFFIC. (2002). *Switching Channels: Wildlife trade routes into Europe and the UK*. University of Wolverhampton, Wolverhampton, UK.

Wyler, L. S., and Sheikh, P. A. (2008). *International Illegal Trade in Wildlife: Threats and U.S. Policy*. CRS Report for Congress, Washington, DC.

Yeager, C. P., Silver, S. C., and Dierenfeld, E. S. (1997). Mineral and phytochemical influences on foliage selection by the proboscis monkey (*Nasalis larvatus*). *American Journal of Primatology*, 41, 117–28.

Yildirim, S., Yeoman, C. J., Sipos, M., *et al.* (2010). Characterization of the fecal microbiome from non-human wild primates reveals species specific microbial communities. *PLoS ONE*, 5, e13963.

YoG (Year of the Gorilla). (2009). *Year of the Gorilla.* UNEP Convention on Migratory Species, UNEP/UNESCO Great Apes Survival Partnership, and WAZA, Gland, Switzerland. Available at: <http://www.yog2009.org/> (accessed october 2012).

Young, L. J., and Wang, Z. X. (2004). The neurobiology of pair bonding. *Nature Neuroscience*, 7, 1048–54.

Yu, Y., Harris, A. J., and He, X. (2010). S-DIVA (Statistical dispersal-vicariance analysis): A tool for inferring biogeographic histories. *Molecular Phylogenetics and Evolution*, 56, 848–50.

Zhao, Q. (2005). Tibetan macaques, visitors, and local people at Mt. Emei: Problems and countermeasures In A. Fuentes, and L. D. Wolfe, eds. *Primates face to face: The Conservation Implications of Human-Nonhuman Primate Interconnections*, pp. 377–99. Cambridge University Press, Cambridge.

Zhao, Q., and Tan, C. L. (2010). Inter-unit contests within a provisioned troop of Sichuan snub-nosed monkeys (*Rhinopithecus roxellana*) in the Qinling Mountains, China. *American Journal of Primatology*, 71, 1–8.

Ziegler, T., Hodges, J. K., Winkler, P., and Heistermann, M. (2000). Hormonal correlates of reproductive seasonality in wild female Hanuman langurs (*Presbytis entellus*). *American Journal of Primatology*, 51, 119–34.

Ziegler, T. E., and Wittwer, D. J. (2005). Fecal steroid research in the field and laboratory: Improved methods for storage, transport, processing, and analysis. *American Journal of Primatology*, 67, 159–74.

Zihlman, A., Bolter, D., and Boesch, C. (2004). Wild chimpanzee dentition and its implications for assessing life history in immature hominin fossils. *Proceedings of the National Academy of Sciences USA*, 101, 10541–3.

Zinner, D., Arnold, M. L., and Roos, C. (2011). The strange blood: Natural hybridization in primates. *Evolutionary Anthropology*, 20, 96–103.

Zinner, D., Hindahl, J., and Schwibbe, M. (1997). Effects of temporal sampling patterns of all-occurrence recording in behavioural studies: many short sampling periods are better than a few long ones. *Ethology*, 103, 236–46.

Zinner, D., Pelaez, F., and Torkler, F. (2002). Distribution and habitat of grivet monkeys (*Cercopithecus aethiops aethiops*) in eastern and central Eritrea. *African Journal of Ecology*, 40, 151–8.

Zinner, D. P., van Schaik, C. P., Nunn, C. L., and Kappeler, P. M. (2004). Sexual selection and exaggerated sexual swellings of female primates. In P. M. Kappeler, and C. P. van Schaik, eds. *Sexual Selection in Primates: New and Comparative Perspectives*, pp. 71–89. Cambridge University Press, Cambridge, UK.

Index